Complete Automotive Welding: Metals and Plastics

Complete Automotive Welding: Metals and Plastics

Robert Scharff
Dave Caruso

WE ENCOURAGE PROFESSIONALISM

THROUGH TECHNICIAN CERTIFICATION

Delmar Publishers Inc.®

NOTICE TO THE READER

Publisher does not warrant or guarantee any of the products described herein or perform any independent analysis in connection with any of the product information contained herein. Publisher does not assume, and expressly disclaims, any obligation to obtain and include information other than that provided to it by the manufacturer.

The reader is expressly warned to consider and adopt all safety precautions that might be indicated by the activities described herein and to avoid all potential hazards. By following the instructions contained herein, the reader willingly assumes all risks in connection with such instructions.

The publisher makes no representations or warranties of any kind, including but not limited to, the warranties of fitness for particular purpose or merchantability, nor are any such representations implied with respect to the material set forth herein, and the publisher takes no responsibility with respect to such material. The publisher should not be liable for any special, consequential or exemplary damages resulting, in whole or in part, from the readers' use of, or reliance upon, this material.

Scharff Staff
Production Manager: Marilyn Strouse-Hauptly
Editor: Lois Breiner
Layout Design: Karen Weaver
Cover Design: Eric Schreader
Logo Design: Ed Foulk

Delmar Staff
Editor-in-Chief: Mark W. Huth
Administrative Editor: Joan Gill
Project Editor: Carol Micheli

For information, address Delmar Publishers Inc.
3 Columbia Circle, PO Box 15015
Albany, New York 12212-5015

Printed in the United States of America
Published simultaneously in Canada
by Nelson Canada,
a division of The Thomson Corporation

10 9 8 7 6 5 4 3

Library of Congress Cataloging-in-Publication Data

Scharff, Robert.
Complete automotive welding: metals and plastics / Robert Scharff, Dave Caruso.
p. cm.
ISBN 0-8273-3622-5.—ISBN 0-8273-3623-3 (instructor's guide).
1. Automobiles—Welding. I. Caruso, Dave. II. Title.
TL 154.S333 1990
629.28'72—dc20 89-28980
CIP

CONTENTS

PREFACE

"They don't make 'em like they used to."

Unfortunately, this has become the catchphrase for today's car owner who's left shaking his head and staring at his bumper (or hood, or door, or trunk lid) after it has been mangled practically beyond recognition as the result of a seemingly minor collision. As American cars have become sleeker, lighter, and more fuel efficient, they have had to sacrifice a certain degree of crash worthiness. Gone are the days when everything that rolled off the Detroit assembly lines was an indestructible tank, and no one has felt this fundamental change more than the auto body technician.

Virtually all cars manufactured today are unibody vehicles. This change from traditional frame construction has had a major impact on the duties, knowledge, procedures, and responsibilities of today's body technician. Before the introduction of unibody cars, the technician's primary concern was to straighten the frame and replace or fix damaged panels or sections. Nowadays the responsibilities are much greater.

With the introduction of unibody vehicles, the body technician has had to learn about the use of new welding equipment and techniques. The conventional body-over-frame vehicle was made primarily of low carbon or mild steel. These materials were usually welded or cut with the standard oxyacetylene gas torch. However, high-strength steel is used in unibody construction, and the vehicle manufacturers mandate that MIG welding must be used on all structural parts repair. In addition to MIG, this textbook also covers TIG, resistance spot, and oxyacetylene welding; plasma arc cutting; sectioning; and brazing. Each of these repair methods has its own special relevance to maintaining the structural integrity of today's automobiles.

A substantial portion of this textbook is devoted to the subject of automotive plastics. The use of plastics—in both unibody and traditional frame construction— has literally exploded onto the scene in recent years. It is imperative that the auto body technician of the nineties adds plastic repair techniques (including fiberglass) to his or her repertoire. Without this vital skill, it will simply not be possible to meet the needs of the body shop customer in the years to come.

ACKNOWLEDGMENTS

Grateful acknowledgment is made to the following companies for reference material and information used in the preparation of this book:

American Technology, Inc.
Body Shop by Jim Schlier
Century Mfg. Co.
Henrob Corp.
Inter-Industry Conference on Auto Collision Repair
Laramy Products Co.
Lenco, Inc.
Lincoln Electric Co.
Lors Machinery, Inc.
MAC Tools, Inc.
Miller Electric Mfg. Co.
Nissan Motor Corp.
Seelye, Inc.
Thermadyne Industries, Inc.
T. J. Snow Co., Inc.
Triple-A Specialty Co.
Toyota Motor Company
Unican Corp.
Urethane Supply Company, Inc.

Thanks is extended to the following people for reviewing the manuscript and for their helpful comments:

Mr. John Deyarmin
NEC-Vale Tech Campus
135 West Market Street
Blairsville, Pennsylvania 15717

Mr. Rodger Randall
Cleveland Technical College Auto Body Shop
137 South Post Road
Shelby, North Carolina 28150

Mr. Tom Rydalch
Compton Community College
1111 East Artesia Boulevard
Compton, California 90221

Mr. James M. Walker
AMTECH Institute
4011 East 31st Street, South
Wichita, Kansas 67210

Special thanks to J. G. Doneski of Lors Machinery, Inc., whose ideas and expertise helped broaden the scope of this textbook.

CHAPTER ONE

WELDING SAFETY

Objectives

After reading this chapter, you should be able to

- Describe what types of protective clothing and devices are available for the welder.
- Explain the various methods of ventilation.
- Describe how to maintain a safe work area.
- Explain how to use hand tools and electrical equipment safely.
- Describe the proper method for handling and storing cylinders.
- Explain how to prevent and put out fires in the shop.

Specific potential safety problems are associated with welding. If injuries to personnel and damages to property are to be avoided during welding, it is essential that the strictest safety measures and precautions are maintained.

There are federal and state specifications, codes, standards, and rules (largely interchangeable terms) related to the welding industry that must be followed. These rules released by a governmental agency, at any level, have the full force and effect of law. For example, the Occupational Safety and Health Administration (OSHA) of the U.S. Department of Labor administers federal safety laws. Keep in mind that in most instances, OSHA regulations are not just safety recommendations, they are laws.

Many trade and professional associations offer recommendations for the various welding processes. The American Welding Society (AWS), American National Standards Institute (ANSI), American Society for Testing and Materials (ASTM), National Fire Protection Association (NFPA), Compressed Gas Association (CGA), and National Safety Council (NSC) have established many voluntary safety standards for the welding trade. Although these standards are not laws, they are held in high esteem by many in the industry. Actually, many OSHA regulations are based on AWS recommendations.

Although there are serious threats to safety in the welding shop, welding can be a safe occupation if the proper precautionary measures are instituted. Under these situations, welding is no more hazardous or injurious to health than any other metalworking occupation.

BURNS

Burns are one of the most common injuries in the welding shop. They are classified according to their degree of severity into one of three categories—first-degree, second-degree, and third-degree. To reduce the danger of infection, every burn must be given proper medical treatment.

When the surface of the skin is reddish in color, tender, and painful, a first-degree burn has occurred. First aid consists of applying a burn cream, ointment, or spray to the affected area. Professional help will be needed if the burn is large or becomes infected.

A second-degree burn has occurred when the skin blisters and possibly even breaks. The priority in first aid is to reduce the pain by preventing air from getting to the affected area. If a small area is affected, apply a burn cream or ointment. If a large area is affected, wrap it with a clean, wet, lint-free cloth. Then seek medical help immediately.

When the skin (and possibly the tissue below it) is burned black, a third-degree burn has occurred.

First aid consists of preventing air from getting to the affected area. Immerse the injured area in cool (not cold) water or wrap the area with a clean, wet, lint-free cloth. Then seek medical help immediately.

Burns can occur from some types of light as well as from contact with hot welding materials. Ultraviolet light and infrared light, both invisible to the unaided human eye, can cause burns. Arc welding produces both of these as well as visible light. Gas welding produces only visible and infrared light.

Ultraviolet light is the most dangerous. A welder cannot see or feel ultraviolet light while being exposed to it. It can cause first- and second-degree burns to a welder's eyes or exposed skin. The higher the current and the closer the welder is to the arc, the quicker a burn can occur. During some welding processes, the ultraviolet light is so intense that a welder's eye can receive a flash burn within seconds, and the skin can be burned within minutes. Therefore, the welder must stay protected when in the area of any of the arc welding processes. Ultraviolet light can penetrate damaged or poorly maintained arc welding helmets, thin clothing, light-colored clothing, and loosely woven clothing.

On occasion welders and others will have their eyes exposed to ultraviolet light for a short period of time. This will result in what is known as *arc burn.* It is very similar to a sunburn of the eyes. It is sometimes called an *arc flash,* and for a period of approximately 24 hours, the welder will have the painful sensation of sand in the eyes. The condition is normally temporary and should not last more than 48 hours. The welder who receives an arc flash might not be aware of it at the time. The first indication of an arc burn can come in the middle of the night. Temporary relief can be obtained by using eye drops and eye washes. Apply a few drops of sterile oil, (such as olive oil), and take an aspirin for the pain. Bandages or slightly tinted glasses (shade 3) can be worn for additional relief. Complete rest of the eyes is important. If the painful sensation lasts beyond one day, a doctor should be consulted for treatment.

Infrared light is felt as heat. Consequently, burns can easily be avoided, since a person will immediately feel the heat of this light.

PERSONAL PROTECTION

Proper equipment and clothing should be used by welders and any other persons near the welding station to protect them from burns, spatter, and the radiant or light energy of the welding processes.

GENERAL WORK CLOTHING

Choosing general work clothing that will minimize the possibility of getting burned is important. High-temperature sparks, metal, and slag are produced during welding, and it is not always possible to wear special protective clothing.

Although rare, 100 percent wool clothing is the first choice because it is the least easily ignited material. A good second choice is 100 percent cotton clothing. Avoid synthetic materials such as nylon, rayon, and polyester. They are easily burned and produce a hot, sticky ash that can increase the severity of any burn. Some also produce poisonous gases. The clothing chosen should be dark, thick, and tightly woven so that ultraviolet light cannot pass through it.

Care should be taken to cover as much of the skin as possible. Shirts should be long sleeved and have a high buttoned collar so that the arms and neck are protected (Figure 1-1). They should also be long enough to tuck into the pants to protect the waist. Shirt pockets should have flaps to keep out sparks. Otherwise they should be removed. Pant legs must be long enough to cover boot tops and be free of cuffs that could catch sparks. A cap that is thick enough to prevent sparks from burning the top of the head should be worn. Low-cut shoes should not be worn unless the ankles are covered with protective spats. In fact, boots with steel safety toes (Figure 1-2), covered with smooth, nonseamed leather, are the best footwear for welding.

All tears, holes, or frayed edges of clothing should be removed or repaired. Remember that folds, creases, and large seams in clothing create spark-catching areas, leading to flammability. Therefore, clothing should be relatively tight-fitting. It is most important that clothing is free from oil and grease. This lessens the probability of igniting clothing and causing painful burns. Clothing should always be kept dry to avoid electrical shock.

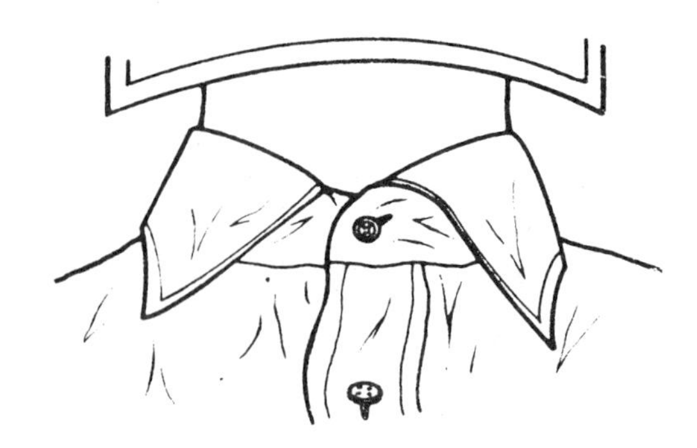

FIGURE 1-1 The high collar should be buttoned.

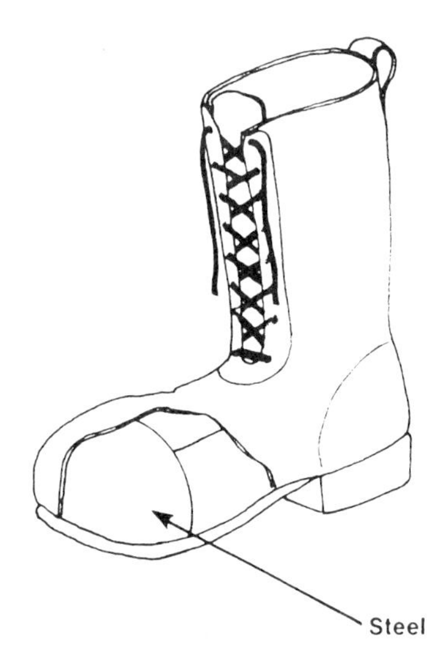

FIGURE 1-2 Steel toe boots

Butane lighters and matches must always be removed from the welder's pockets and placed a safe distance away before starting to weld. Otherwise, they might catch fire or explode if subjected to welding heat and sparks. Also, avoid wearing rings, necklaces, and any other types of jewelry.

SPECIAL PROTECTIVE CLOTHING

In addition to the general work clothing already mentioned, extra protection is needed for any person who is in direct contact with hot materials. Leather is lightweight, burn-resistant, flexible, and readily available. This makes it the best material to be used in most instances. Capes, jackets, aprons, sleeves, gloves, caps, pants, knee pads, and spats are among the leather protection items available. Synthetic insulating materials are also available.

FIGURE 1-3 Full leather jacket *(Courtesy of Mac Tools, Inc.)*

Body Protection

Either a full leather jacket or a leather cape and apron should be worn for any out-of-position work in order to protect the welder's shoulders, arms, and chest (Figure 1-3). The jacket offers protection for the welder's back and chest. Although the cape offers less protection than the jacket, it is open and much cooler to wear. Wearing a bib apron with the cape provides additional protection as well as ventilation for the back.

Bib aprons or full aprons protect both a welder's waist and lap. This protection is especially needed when the welder is squatting, sitting, bending over, or leaning against an object.

Arm and Hand Protection

In some vertical welding situations, a full or half sleeve can protect a welder's arm (Figure 1-4). These sleeves can be used when the work level is not above the welder's chest. When the work level is above chest level, a jacket or cape is required to keep sparks off the welder's shoulders.

All welders must wear protective gloves (Figure 1-5). If possible, they should be the gauntlet type

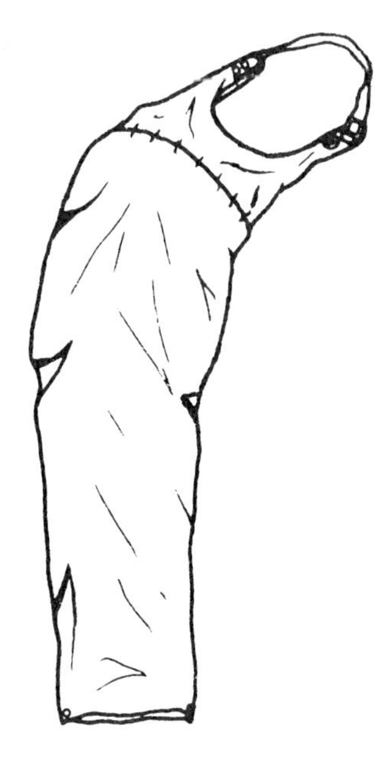

FIGURE 1-4 Full leather sleeve

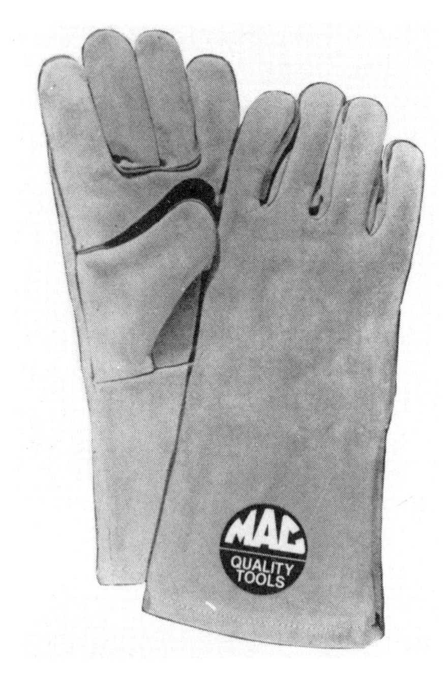

FIGURE 1-5 All-leather gauntlet-type welding gloves *(Courtesy of MAC Tools, Inc.)*

that provide protection for the wrists. For hot work, leather gauntlet gloves with a cloth liner for insulation are best. When more flexibility is needed, noninsulated gloves can be used. Some leather gloves are available with a canvas gauntlet top. These should not be used for anything except light work.

For those tasks in brazing, soldering, and other delicate processes where manual dexterity is critical, deerskin gloves may be used (Figure 1-6). All-cotton gloves should be used only for very light welding.

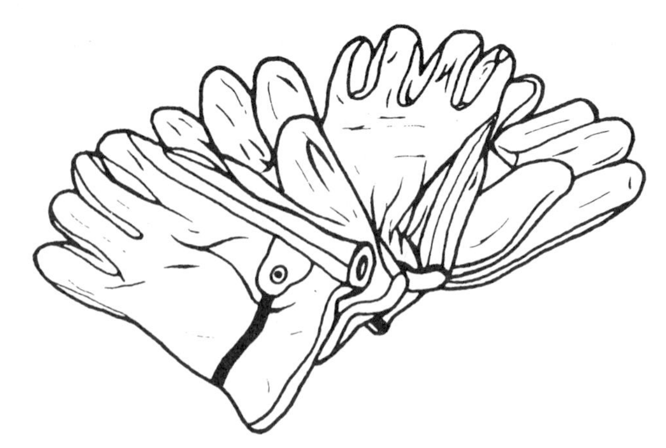

FIGURE 1-6 Deerskin gloves for jobs requiring manual dexterity

Leg and Foot Protection

When cutting or welding is being done and numerous sparks are falling, the welder's legs and feet should be protected with leather pants and spats. Leather aprons with leggings can be used instead when the weather is hot and full leather pants would be uncomfortable. Leggings leave the back open; they are strapped to the legs. Spats protect the front of lace-up boots from sparks that would burn through them (Figure 1-7).

Face and Eye Protection

Some type of eye protection must be worn in the shop at all times if injuries from flying debris and arc radiation are to be avoided.

The welding arc or flame should never be observed at close quarters with unprotected eyes. The nearest safe distance for viewing an arc with the unprotected eye is 40 feet. Failure to observe this rule can result in various degrees of eye burn. Excessive exposure to arc radiation is not noticed in its early stages; damage to the eye can occur without warning of impending danger. For this reason, care must be taken in choosing the appropriate fillers or goggles for the welding process being used.

The eye can be burned in two ways by ultraviolet light (Figure 1-8). The retina at the back of the eye can be injured. When this injury occurs, it is not painful, but it might cause some loss of eyesight. The whites of the eyes can also be burned. Since this

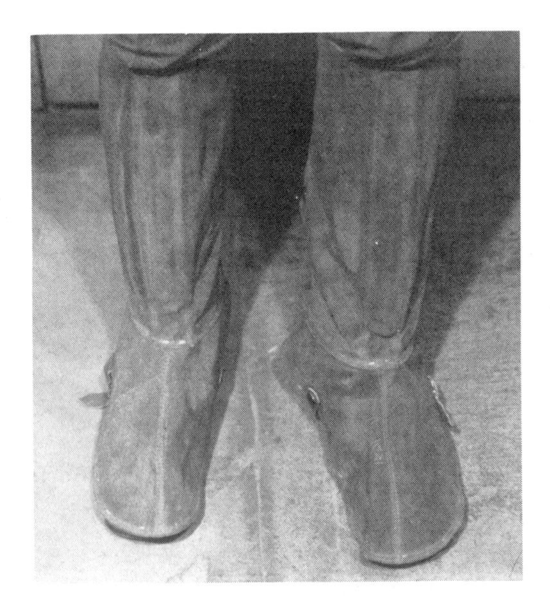

FIGURE 1-7 Typical spats worn by welders

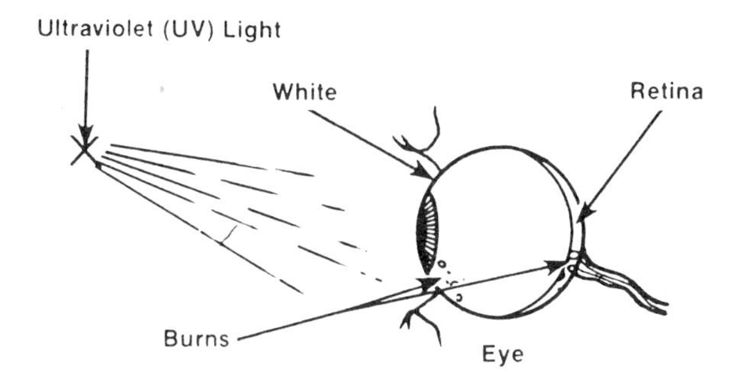

FIGURE 1-8 UV light can burn the white of the eye and the retina.

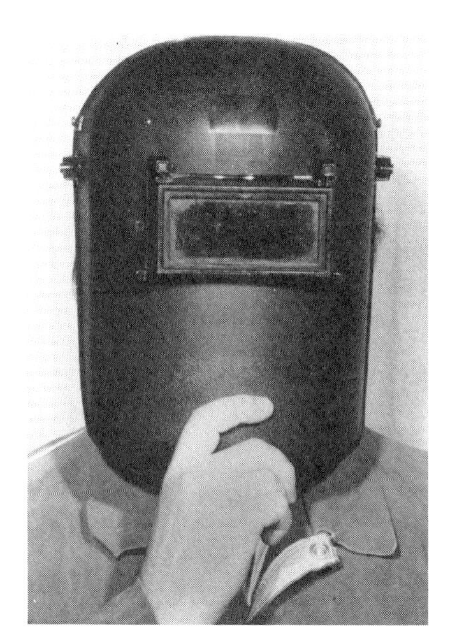

A

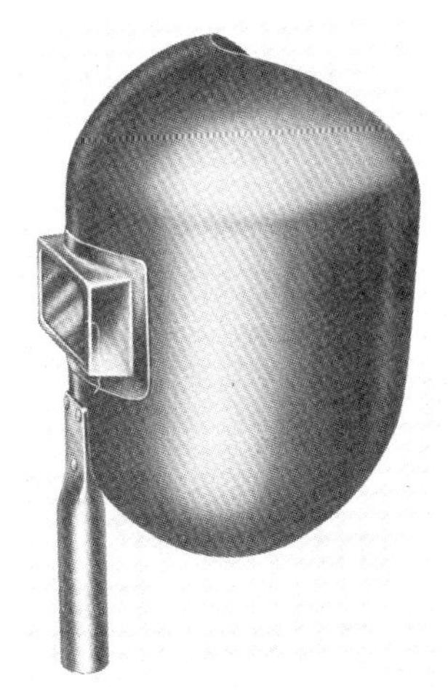

B

FIGURE 1-9 (A) Helmet-type head shield required for protecting the welder's eyes and face. (B) Hand-held face shield is convenient for use by forepeople, inspectors, and other onlookers.

part of the eye is very sensitive, the injury is extremely painful. When the eye is burned, it is extremely susceptible to infection. So, again, be sure to take the proper safety precautions to protect the eyes.

A helmet-type head shield, as shown in Figure 1-9A, is standard equipment for protecting the welder's face and eyes from the direct rays of the arc. A hand-held face shield, as shown in Figure 1-9B, is convenient for the use of onlookers. (Sunglasses or gas-welding goggles are not adequate protection.) These shields are generally made from a nonflammable insulating material and are black or gray in color to minimize reflection. They are shaped to protect the face, neck, and ears from direct radiant energy from the arc. In recent years, the curved front welding helmet has gained preference over the straight front because it reduces the amount of welding fumes in the welder's breathing zone. Figure 1-10 shows both types of welding helmets. Welding helmets must be attached to safety hard hats for construction and maintenance work (Figure 1-11).

In selecting the helmet and hand shield bodies, it is important to select only those made of materials that are temperature and electricity insulated . The helmets and shields should be noncombustible (or self-extinguishing) and opaque to visible ultraviolet and infrared radiation. Helmets, shields, and goggles must be capable of withstanding disinfection. It is important that the helmets and shields, like all protective devices and clothing, be made of a material that is nontoxic and will neither irritate nor discolor the skin.

A piece of ordinary colored glass might look like welding lens, but it would not have the necessary light screening characteristics needed for eye protection. Special equipment is required to measure

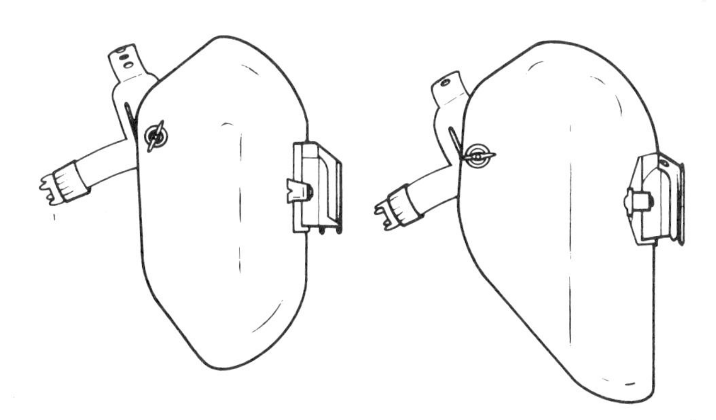

FIGURE 1-10 Helmets: curved and straight front types

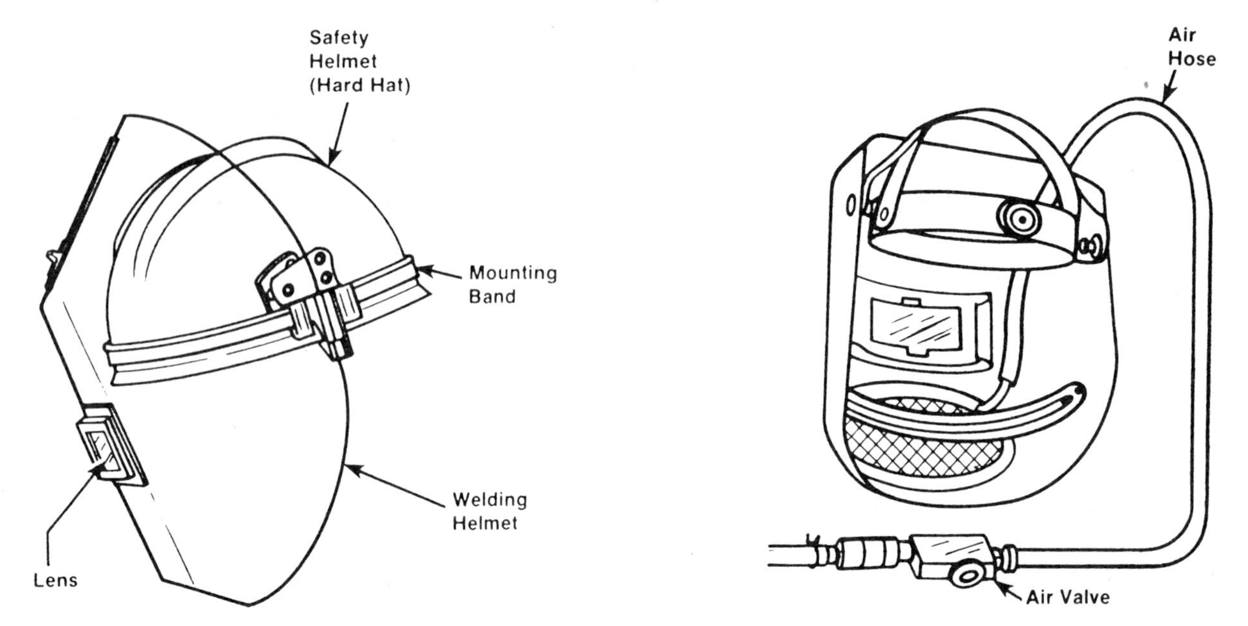

FIGURE 1-11 One method of fastening the hard hat and helmet

TABLE 1-1: EYE PROTECTION FILTER SHADE SELECTOR FOR WELDING OR CUTTING (GOGGLES OR HELMET)

Welding or Cutting Operation	Electrode Size Metal Thickness or Welding Current	Filter Shade Number
Torch soldering		2
Torch brazing		3 or 4
Oxyfuel cutting		
Light	Under 1"	3 or 4
Medium	1 to 6"	4 or 5
Heavy	Over 6"	5 or 6
Oxyfuel welding		
Light	Under 1/8"	4 or 5
Medium	1/8" to 1/2"	5 or 6
Heavy	Over 1/2"	6 or 8
Shielded metal arc welding	Electrodes under 5/32"	10
	Electrodes 5/32" to 1/4"	12
	Electrodes over 1/4"	14
Gas metal arc welding (MIG)		
Nonferrous base metal	All	11
Ferrous base metal	All	12
Gas tungsten arc welding (TIG)	All	12
Atomic hydrogen welding	All	12
Carbon arc welding	All	12
Plasma arc welding	All	12
Carbon arc air gouging		
Light		12
Heavy		14
Plasma arc cutting		
Light	Under 300 Amps	9
Medium	300 to 400 Amps	12
Heavy	Over 400 Amps	14

the amount of infrared and ultraviolet rays a lens will absorb. Welding lenses should be purchased only from suppliers who can be depended upon to furnish quality products.

Lenses are available in a number of shades for various types of work. Recommended shade numbers for common welding and cutting operations are listed in Table 1-1. Welding helmets have lens holders for inserting the cover glass and filter glass or plate (Figure 1-12A). The standard size filter plate is 2 by 4-1/4 inches. In some helmets, the lens holders will open or flip upward. Helmets that accommodate larger-size filter lenses are also available and are used for light-duty work. The larger filter glasses are 4-1/2 by 5-1/4 inches and are more expensive. Filter glasses or plates come in various optical densities to filter out more or less of the light.

Generally, a cover plate is placed on the outside of the filter glass to protect it from weld spatter (Figure 1-12B). Plastic or glass plates are used. Some welders also use magnifier lenses behind the filter plate to provide clearer vision. The filter glass must be tempered so that it will not break if hit by flying spatter, and the like. (Remember, slag has a tendency to pop off when it cools.) Filter glasses must also be marked showing the manufacturer, the shade number, and the letter "H," indicating that it has been treated for impact resistance.

Use tape to temporarily repair worn or cracked spots on a helmet until it can be permanently repaired or replaced. Ultraviolet light can make its way through small undetectable leaks in an arc welding helmet and cause a welder's eyes to itch or feel sore after welding all day. To prevent these leaks, make sure the lens gasket is installed properly (Figure 1-13). Check the lens for cracks by twisting it between your fingers (Figure 1-14).

In addition to a helmet or shield, goggles with side shields should also be worn during arc welding or cutting operations. Goggles provide protection from spatter or rays, particularly at times when the

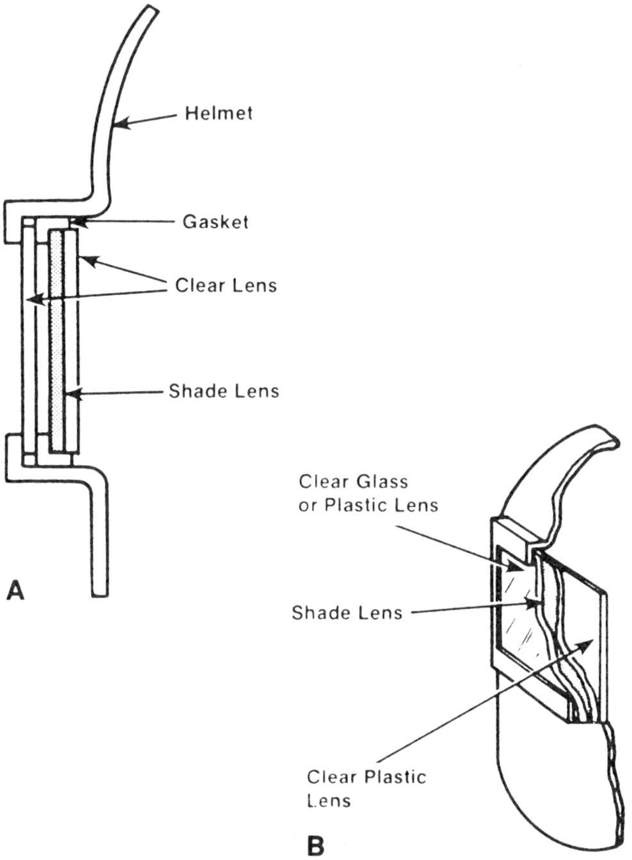

FIGURE 1-12 (A) The gasket around the shade lens will stop ultraviolet light from bouncing around the lens assembly. (B) It is important for the third lens to be plastic.

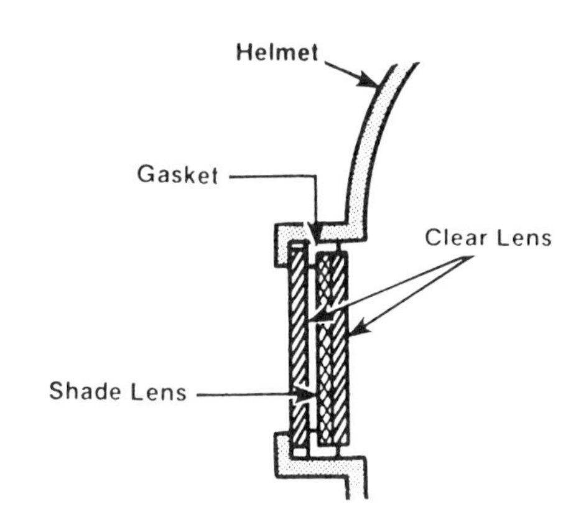

FIGURE 1-13 Correct placement of the gasket around the shade lens is important.

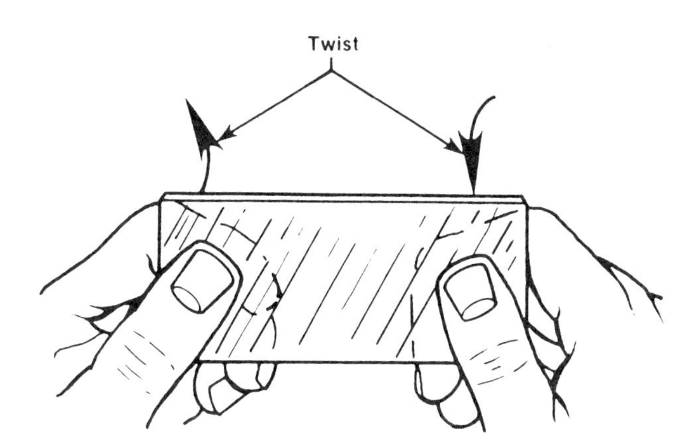

FIGURE 1-14 Twist the lens to check for possible cracks.

shield is removed, as is necessary when replacing electrodes, removing slag, or inspecting the weld (Figure 1-15). Goggles should be worn by welders' helpers and others working near the arc to protect their eyes from occasional flashes. Goggles should be lightweight, ventilated, and sterilizable, with the frames made of a heat-insulating material. Clear, spatter-resistant cover glasses and tinted lenses should be used in the goggles.

Contact lenses should not be worn while welding or when working around welders. Safety glasses may be required to be worn underneath the welding helmet. These are required since the helmet is usually lifted when slag is chipped or welds are ground. OSHA requirements do not include safety glasses. However, for the welder's own protection it is prudent to use them when chipping or grinding. Tinted safety glasses with side shields are recommended. People working around welders should also wear tinted safety glasses with side shields.

Ear Protection

On occasion it will be necessary to protect the ears against excess noise and/or hot sparks. Sometimes the noise level in the welding environment is high enough to cause pain and some loss of hearing if the welder's ears are unprotected. The unprotected ear is also susceptible to severe burns from hot sparks.

Two forms of ear protection are available. There are plugs that fit into the ear canal (Figure 1-16). These will protect one's hearing. The second form of protection is earmuffs that cover the outer ear completely (Figure 1-17). Earmuffs protect a person's hearing as well as the outer ears from burns.

Respiratory Protection

Undesirable by-products such as fumes and gases are produced by all arc welding processes. They are unavoidable since the intense heat developed by the arc heats the metals and fluxes above the temperatures at which they boil or decompose. Although most of the by-products are recondensed in the weld, some manage to escape into the atmosphere. It is these by-products that produce the haze that can be seen in poorly ventilated welding shops. Some fluxes produce fumes that irritate the welder's

A

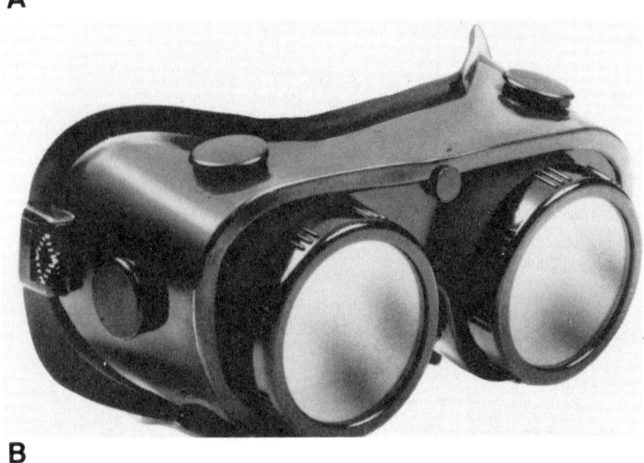

B

FIGURE 1-15 Goggles (A) for workers not wearing RX glasses, and (B) for workers wearing RX glasses *(Courtesy of MAC Tools, Inc.)*

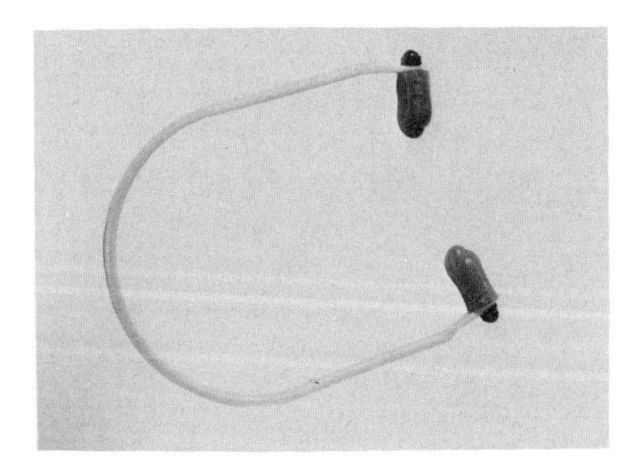

FIGURE 1-16 Earplugs used as protection from noise only *(Courtesy of Willson Safety Products Inc.)*

FIGURE 1-17 Earmuffs provide complete ear protection.

respiratory passages. If the zinc fumes from galvanized sheet metal are inhaled over an extended period of time, they can be harmful.

Welders must recognize that fumes of any type, regardless of their source, should not be inhaled. The best solution to the problem of fumes is to provide adequate ventilation. If this is not possible, then personal respiratory protection devices should be used. The welder should wear either an air line respirator, a hose mask, or a respirator with an appropriate canister/filter. Make sure the air is supplied from an oilless compressor. Examples of typical protection devices used in poorly ventilated or confined quarters are shown in Figures 1-18, 1-19, and 1-20. When working on lead, beryllium, cadmium, or their alloys, ordinary filter-type masks are inadequate and nothing less than an air supply respirator is proper or safe.

Fumes from chlorinated solvents, such as those used for degreasing metals, must be prevented from drifting into a welding shop. When these fumes decompose in the arc, they produce a potentially dangerous compound called *phosgene.* Fumes drifting into the welding shop from a nearby degreasing operation have been known to cause illnesses. Also, never cut into a tank or other vessel that has had solvent in it.

Potentially dangerous gases are also present in a welding shop. Ozone can be present in excessive amounts when making gas tungsten arc welds in confined spaces. Under these conditions, adequate ventilation or respirators should be provided because this gas will irritate respiratory passages. Care also must be taken to avoid the leakage of such gases as argon or carbon dioxide when welding in confined spaces. These gases are odorless, colorless, and denser than air. They can displace the air and cause asphyxiation. Here, again, proper ventilation is essential. In fact, no matter what welding process is being used, proper ventilation or respirators are necessary.

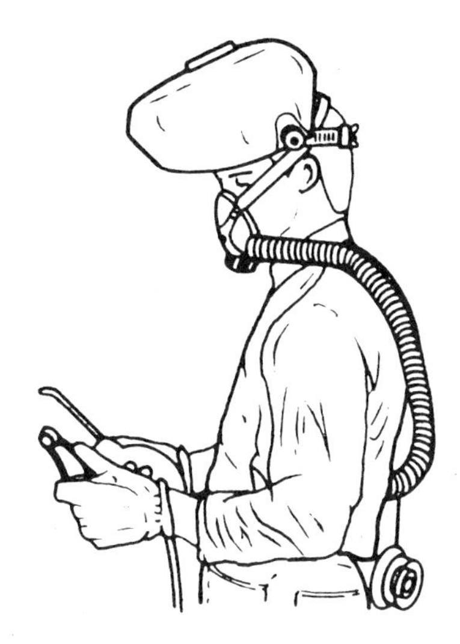

FIGURE 1-19 Typical belt-mounted respirator for use in contaminated environments

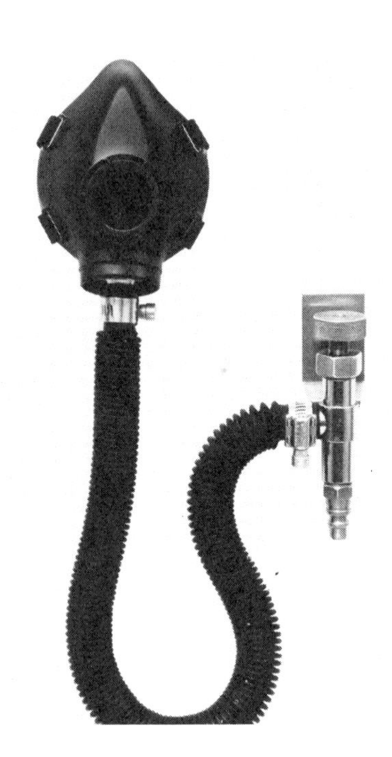

FIGURE 1-20 Typical air line respirator attached to an external air supply

FIGURE 1-18 Typical respirator for contaminated environment

VENTILATION

Ventilation for welding can be accomplished in three different ways: natural ventilation, mechanical ventilation, or local exhaust ventilation of specific welding operations. In general, if the room is sufficiently large or if the welding takes place outdoors, natural ventilation will remove the contaminants from the welder's breathing zone. If the natural ventilation is not sufficient because the area is too small or because there are too many welders in the area, or if there is the possibility of welding on contaminating materials, then mechanical ventilation must be provided. It is necessary if the space is less than 10,000 cubic feet per welder, if the ceiling height of the room is less than 16 feet, or if the welding is done in a confining state that contains partitions, balconies, or other structural barriers. Mechanical ventilation should be at the rate of 200 cubic feet per minute (cfm) unless local ventilation is supplied.

If mechanical ventilation is not suitable, then local ventilation might be required. Local exhaust ventilation can be obtained by one of the following methods:

- Movable hoods placed as near to the welding operation as possible
- Portable blower or blowers to provide a proper airflow
- Low-volume, high-velocity fume exhaust device attached to the welding gun (Figure 1-21)

The latter device is based on collecting the fumes as close as possible to the point of generation or at the arc. This method of fume exhaust has become quite popular for the semiautomatic welding process. Smoke exhaust systems incorporated in semiautomatic guns provide the most economical exhaust system, because they exhaust much less air and eliminate the need for massive air makeup units to provide heated or cooled air to replace what is lost (Figure 1-22). Local ventilation should have a rate of airflow sufficient to maintain a velocity away from the welder of not less than 100 feet per minute. Air velocity is easy to measure using a velometer or airflow meter, but these two systems can be extremely difficult to use when welding on other than small weldments. In all cases when local ventilation is used, the exhaust air should be filtered.

Welding in small and restricted spaces requires particular care to prevent the accumulation of toxic materials and to ensure that personnel have adequate air for breathing. All air that is exhausted must be replaced with clean air.

A

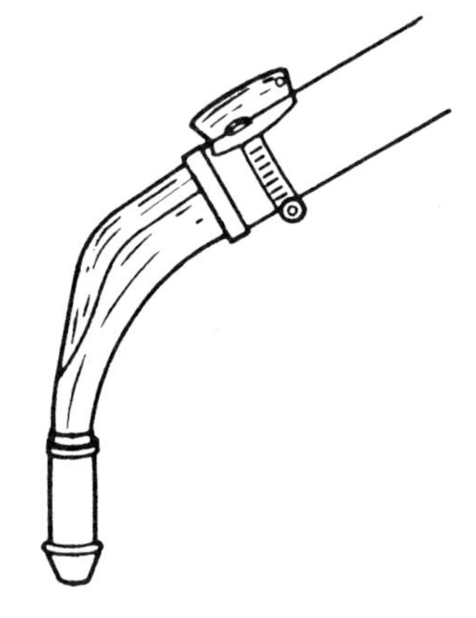

B

FIGURE 1-21 (A) Smoke exhaust nozzles are very useful on welding guns. (B) This unit enables welding fumes to be removed at a point very close to their source—the welding arc.

WORK AREA

Keep the work area picked up and swept clean. Although it is difficult to work around collections of steel, welding electrode stubs, wire, hoses, and cables, it is all too easy to trip in such a cluttered area. Hoses and cables can be hung on hooks and scrap steel should be thrown into scrap bins. Electrodes

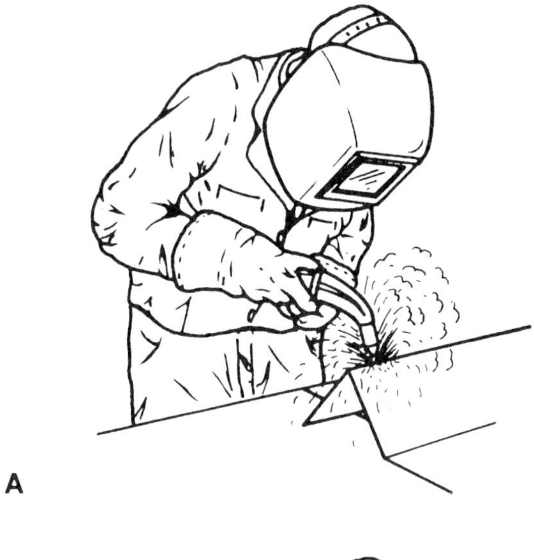

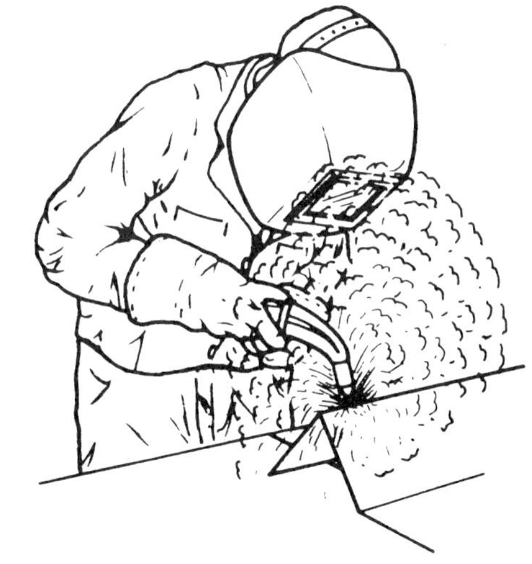

FIGURE 1-22 Smoke exhaust gun (A) on and (B) off

and stubs should be placed in an electrode caddy can (Figure 1-23).

Arc welding areas should be painted flat black so that as much ultraviolet light as possible is absorbed. Whenever arc welding is to be done outside of a welding booth, portable screens should be used.

If a piece of metal or any tool is hot, it should not be left unattended without first marking it with a heat crayon.

EQUIPMENT MAINTENANCE

Leaking coolant, loose wires, poor grounds, frayed insulation, and split hoses are all small problems that can frequently be found in a welding shop. If they are not fixed promptly, however, they can lead to a major equipment breakdown or injury. Valuable time will be saved by repairing small problems when they are first discovered. Potential problems can usually be detected if a routine schedule of equipment maintenance is kept.

Refer any maintenance that goes beyond routine external maintenance to a trained service technician. In most states, it is against the law for anyone but a factory-trained repair technician to work on regulators and anyone but a licensed electrician to work on arc welders. Exploding regulators and electrical shock are a potential safety hazard.

ELECTRICAL EQUIPMENT

Electric shock can cause injuries and even death. The shock hazard should be guarded against whenever any electrical equipment is in use. In addition to arc welding power supplies, there are many electrical devices used in the welding shop, such as grinders, cutting machines, drills, and extension lights. Most of this equipment is powered by alternating current sources having input voltages ranging from 115 to 460 volts. The usual cause of electrical shock is not contact with a welding torch, but accidental contact with bare or poorly insulated conductors operating at these voltages.

To prevent body shocks, keep hands, feet, and clothing dry. Never stand or lie in puddles of water, damp ground, or against grounded metal when welding without suitable insulation against shock.

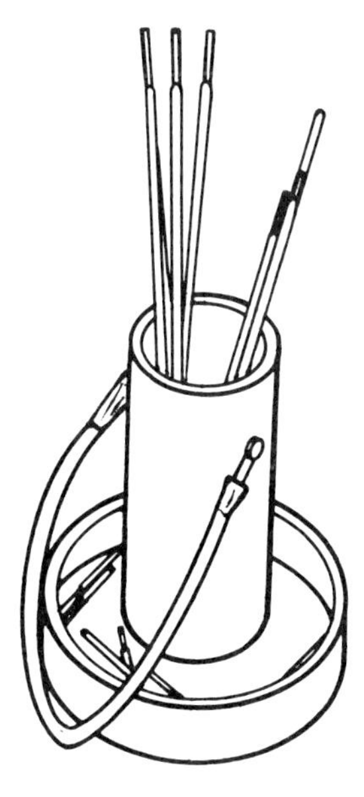

FIGURE 1-23 An easy-to-build electrode caddy can

Always find a dry board or rubber mat to stand on when water, moisture, or perspiration cannot be avoided. Dampness between the body and an energized or grounded metallic part lowers the resistance to the passage of current to the body, which may produce a harmful or fatal shock. Salt in perspiration or sea water also dangerously lowers contact resistances.

WELDING EQUIPMENT

Most industrial electrical welding machines meet the standard set by the National Electrical Manufacturers Association (NEMA). This is usually contained in the literature of the manufacturer and appears on the nameplate of the individual welding machine. In certain states of the United States, the approval of the Underwriters Laboratory (UL) is required for welding power sources of the transformer type. The U.S. government, through OSHA, has caused substantial changes in the manufacture of welding equipment to improve its safety. Output terminals are now covered by insulation devices. Ventilation holes are now so small that they make it impossible for the individual welder to come in contact with the high voltage within the case. OSHA has also caused a change in the welding machine casing so that now "tools" are required to open the case where high voltage is exposed. Figure 1-24 shows a typical warning label attached to welding equipment.

WARNING

READ AND UNDERSTAND THIS WARNING — PROTECT YOURSELF AND OTHERS

ELECTRIC SHOCK can kill.
- Do not permit electrically live parts or electrodes to contact skin . . . or your clothing or gloves if they are wet
- Insulate yourself from work and ground

FUMES AND GASES can be dangerous to your health.
- Keep fumes and gases from your breathing zone and general area
- Keep your head out of fumes
- Use enough ventilation or exhaust at the arc or both

ARC RAYS can injure eyes and burn skin.
- Wear correct eye, ear, and body protection

ENGINE EXHAUST GASES can kill.
- Use in open, well ventilated areas or vent the engine exhaust to the outside.

HIGH VOLTAGE can kill
- Do not touch electrically live parts.
- Stop engine before servicing.

MOVING PARTS can cause serious injury.
- Keep clear of moving parts.
- Do not operate with protective covers, panels, or guards removed.

- Only qualified personnel should install, use, or service this equipment.
- Read operating manual for details.

READ AND UNDERSTAND THE MANUFACTURER'S INSTRUCTIONS AND YOUR EMPLOYER'S SAFETY PRACTICES.

See, American National Standard Z49.1, "Safety in Welding and Cutting" published by the American Welding Society, 550 Le Jeune Rd., Miami, Florida 33126. OSHA Safety and Health Standards, 29 CFR 1910, available from U.S Government Printing Office, Washington, D.C. 20402.

DO NOT REMOVE THIS WARNING — E202

FIGURE 1-24 Read and obey the warning label attached to welding equipment.

Installation

All electric arc welding machines must be installed in accordance with the National Electrical Code and all local codes. Installation instructions are included in the manufacturer's manual that accompanies the welding machine. The manual also gives the size of power cable that should be used to connect the machine to the main line (Table 1-2). Motor generator welding machines feature complete separation of the primary power and the welding circuit, since the generator is mechanically connected to the electric motor. However, the metal frames and cases of motor generators must be grounded (Figure 1-25), since the high voltage from the main lines does enter the case. In transformer and transformer rectifier-type machines, the primary and secondary transformer windings are electrically isolated from each other by insulation. This insulation may become defective in time if proper maintenance practices are not observed. The metal frame and cases of transformers and transformer rectifier machines must be grounded to earth. The work terminal of the welding machine should not be grounded to earth.

Disconnect switches should be employed with all power sources so that they can be disconnected from the main lines for maintenance or if an emergency occurs. The disconnect switches should be placed in the *open* position when work is completed.

Three-phase arc welding machines are available (Figure 1-26) and require a three-phase power line. The hookup of such a line as well as any other electrical wiring should be made by a certified or qualified electrician. Any corrections to the electrical system must be made before any welding begins.

The welding electrode holders must be connected to machines with flexible cables designed for

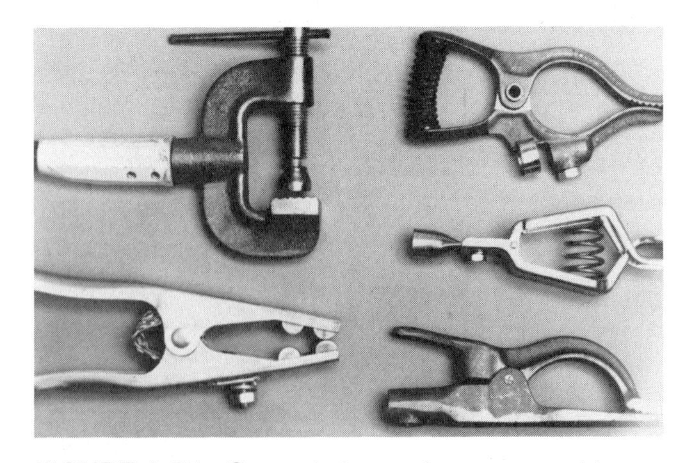

FIGURE 1-25 Ground clamps for an arc welder

TABLE 1-2: CABLE SIZE SELECTOR GUIDE BASED ON 4-VOLT DROP

	American Wire Gauge (AWG) Sizes											
	Distance in Feet from Welding Machine to Work											
Amperes	**50**	**75**	**100**	**125**	**150**	**175**	**200**	**225**	**250**	**300**	**350**	**400**
100	2	2	2	2	1	1/0	1/0	2/0	2/0	3/0	4/0	4/0
150	2	2	1	1/0	2/0	3/0	3/0	4/0				
200	2	1	1/0	2/0	3/0	4/0	4/0					
250	2	1/0	2/0	3/0	4/0							
300	1	2/0	3/0	4/0								
350	1/0	2/0	4/0									
400	1/0	3/0	4/0									
450	2/0	3/0										
500	2/0	4/0										
550	3/0											
600	3/0											

Increasing Cable Size (→)

Increasing Cable Size (↓)

Note: Recommended copper cable sizes will result in a voltage drop of no more than 4 volts between the welding machine and the work.

welding applications, and the electrical connections must be tight. Power cables and terminals for welding leads must be shielded from accidental contact by personnel or by metal objects. Do not use cables for applications beyond their current-carrying and duty cycle capacities; overheating and deterioration of the insulation would occur rapidly. There should be no splices in the electrode cable within 10 feet of the electrode holder. Splices, if used in work or electrode leads, must be insulated. Check the cables periodically to see if they have become frayed. If they have, they should be replaced immediately.

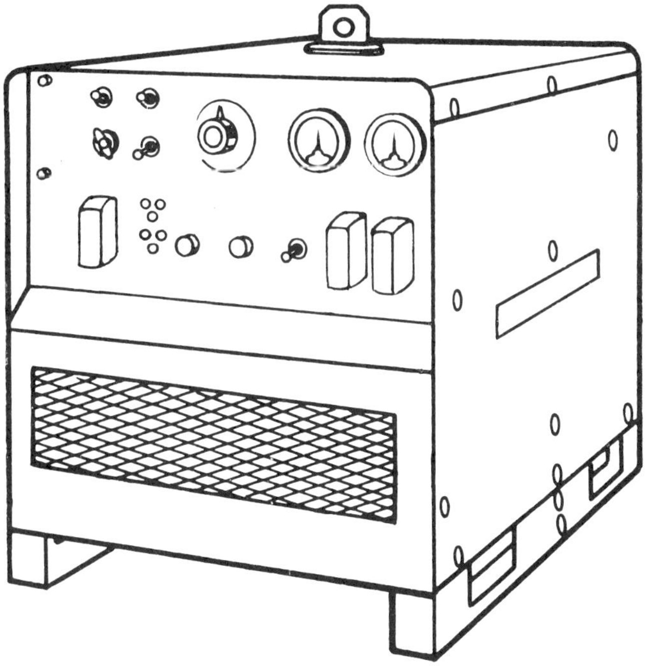

FIGURE 1-26 Typical three-phase arc welding machine

Use

Do not allow the metal parts of electrodes or electrode holders to touch your skin or wet clothing. Always wear rubber-soled shoes and dry gloves that are in good condition. Every precaution should be taken when working in cramped kneeling, sitting, or lying positions. Use insulated mats or dry, wooden boards for protection from the earth. Make certain the insulation is large enough to cover your full area of physical contact with the work and ground.

When changing or adjusting torches or guns, turn off the welding circuits. They should also be turned off when the workstation is left unattended. Wear dry gloves when changing coated electrodes because the electrode holder is energized.

Never simultaneously touch electrically "hot" parts of electrode holders connected to two welders. The voltage between the two can be the total of the open circuit voltages of both welders. Never dip the electrode in water for cooling.

GRINDERS

Grinders are required to do many welding jobs. Often it is necessary to remove rust or to smooth or remove a weld. Grinding stones have a maximum RPM listed on the paper blotter (Figure 1-27). Never use a stone on a machine that has a higher rated RPM. Keep in mind that a grinding stone can explode if it is turned too fast.

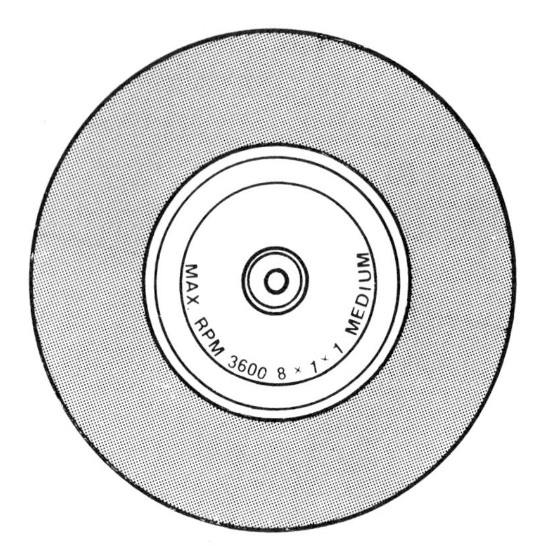

FIGURE 1-27 Maximum revolutions per minute are listed on the paper blotter.

Test a grinding stone for cracks before putting it on the machine. Tap the stone in four places and listen for a sharp ring, which indicates it is good (Figure 1-28). A grinding stone that makes a dull sound is cracked and should not be used.

After a stone has been installed and used, it may need to be trued and balanced. Figure 1-29 shows a special tool designed for that purpose. Truing keeps the stone face flat and sharp for better results.

Grinding stones are made for both ferrous and nonferrous metals. If a ferrous stone is used to grind nonferrous metals, such as aluminum, copper, and brass, the stone will become glazed as the surface clogs with metal, and it might explode due to the frictional heat building up on the surface. If a nonferrous stone is used to grind ferrous metal, such as iron, steel, or stainless steel, the stone will be quickly worn away.

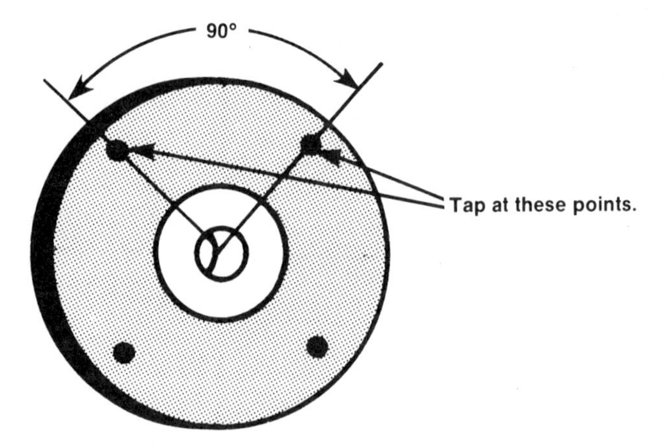

FIGURE 1-28 Tap the stone to check for cracks.

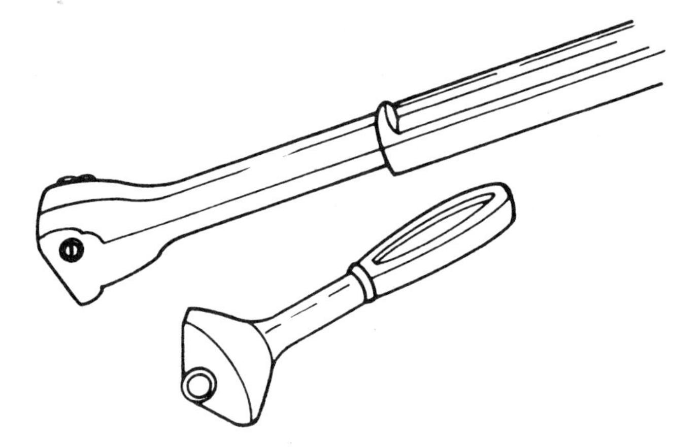

FIGURE 1-29 A redressing tool used to keep the stone in balance

Never wear gloves when grinding. The glove could get caught in the stone and pull in your whole hand. Keep the sparks from grinding down and away from all personnel and equipment.

DRILLS

Certain general procedures should be kept in mind when drilling. For starters, mark the position distinctly with a center punch to provide a seat for the drill point and to keep it from "walking" away from the mark when you apply pressure.

Unless the workpiece is stationary or large, fasten it in a vise or clamp. Holding a small item in your hand can cause injury if it is suddenly seized by the bit and whirled from your grip. This is most likely to happen just before the bit breaks through the hole at the underside of the work.

Carefully center the drill bit in the jaws as the chuck is securely tightened. Avoid inserting the bit off-center because it will wobble and probably break when it spins. After centering, place the drill bit tip on the exact point at which the hole is to be drilled, then start the motor by pulling the trigger switch. (Never apply a spinning drill bit to the work.) With a variable-speed drill, run it at a very slow speed until a pilot cut has been made. Then, gradually increase to the optimum drill speed.

Except when it is desirable to drill a hole at an angle, hold the drill perpendicular to the face of the work. Align the drill bit and the axis of the drill in the direction the hole is to go and apply pressure only along this line, with no sidewise or bending pressure. Changing the direction of this pressure will distort the dimensions of the hole, which could snap a small drill.

Use just enough steady and even pressure to keep the drill cutting. Guide the drill—do not force it. Too much pressure can cause the bit to break or the tool to overheat. Too little pressure will keep the bit from cutting and dull its edges due to the friction created by sliding over the surface.

If the drill becomes jammed in the hole, release the trigger immediately, remove the drill bit from the work, and determine the cause of the stalling or jamming. Do not squeeze the trigger on or off in an attempt to free a stalled or jammed drill; this will damage the motor. When using a reversing-type model, the direction of the rotation may be reversed to help free a jammed bit. Be sure the direction of the rotation is reset before attempting to continue the drilling.

Reduce the pressure on the drill just before the bit cuts through the work to avoid stalling in metal. When the bit has completely penetrated the work and is spinning freely, withdraw it from the work while the motor is still running, then turn off the drill.

SAFETY RULES FOR PORTABLE ELECTRIC TOOLS

In all tool operation, safety is simply the removal of any element of chance. A few safety precautions that should be observed are listed below. These are general rules that apply to all power tools. They should be strictly obeyed to avoid injury to the operator and damage to the power tool.

- *Know the tool.* Learn the tool's applications and limitations as well as its specific potential hazards.
- *Ground the portable power tool unless it is double-insulated.* If the tool is equipped with a three-prong plug, it should be plugged into a three-hole electrical receptacle. If an adapter is used to accommodate a two-pronged receptacle, the adapter wire must be attached to a known ground. Never remove the third prong.
- *Keep the work area clean.* Cluttered work areas invite accidents.
- *Avoid dangerous or explosive environments.* Do not expose the power tool to rain. Do not use a power tool in wet locations. Keep the work area well lighted. Avoid chemical or corrosive environments. Also, remember that it is characteristic of the motor usually used in portable electric tools to spark faintly at the places where the brushes contact the armature commutator. This sparking is quite normal and will not harm the tool. Because of this sparking, however, portable electric tools should never be started or run where there is any possibility of a fire or explosion due to the presence of manufactured or natural gas, gasoline, naphtha, or acetylene.
- *Do not force a tool.* It will do the job better and more safely if operated at the rate for which it was designed.
- *Use the right tool for the job.* Do not use a small tool or an attachment to do the job of a heavy-duty tool. Also, never use a tool for any purpose except that for which it was designed.
- *Wear proper apparel.* Do not wear loose clothing or jewelry that can get caught in moving parts.
- *Wear eye protectors.* Safety glasses or goggles will protect the eyes while operating power tools. Also wear a face or dust mask if the operation creates dust. All persons in the area where power tools are being operated should also wear eye protectors and a dust mask or respirator.
- *Do not abuse the power cord.* Never carry a tool by its cord or yank it to disconnect it from the receptacle. Keep the cord from heat, oil, and sharp edges. Replace a damaged or worn power cord and strain-reliever immediately.
- *Secure your work.* Use clamps to hold the work; it is safer than using your hands, and it frees both hands to operate the tool.
- *Do not overreach when operating a power tool.* Keep proper footing and balance at all times.
- *Maintain power tools with care.* Keep tools sharp and clean for the best and safest operation. Follow the manufacturer's instructions for lubricating and changing accessories. Replace all worn, broken, or lost parts immediately. Full details on the care and maintenance of portable power tools are given later in this chapter.
- *Disconnect the tools from the power source.* When not in use, when changing the accessories, or before servicing, tools should always be disconnected. Store idle tools in a dry location.
- *Remove adjusting keys and wrenches before operation.* Form the habit of checking to see that any keys or wrenches are removed from the tool before turning it on.

- *Avoid accidental starting.* Do not carry a plugged-in tool with your finger on the switch. Be sure the switch is off when plugging in the tool.
- *Be sure accessories and cutting tools are attached securely to the tool.* Extra care must be taken when using the tools at elevated locations (such as on a ladder, scaffold, or the like) to prevent injury to someone on a lower level in the event that the tool and/or accessory should be dropped.
- *Do not use tools with cracked or warped housing, handles, and so on, especially those made of plastic.* Dirt or moisture embedded in the cracks can provide a current path to the exterior of the tool. This could produce a shock in the event a short should develop within the tool. Do not use a tool if its switch does not turn the unit on and off properly.
- *When operating a portable power tool, give it your full and undivided attention.* After using the tool, turn the power off, remove it from the power source, and wait for all rotation of the tool to stop. Clean the tool before storing it.

ELECTRICAL SAFETY SYSTEM

For protection from electric shock, the standard portable tool is built with either of two equally safe systems: external grounding or double-insulation.

A tool with external grounding has a wire that runs from the housing, through the power cord, to a third prong on the power plug. When this third prong is connected to a grounded, three-hole, electrical outlet, the grounding wire will carry any current that leaks past the electrical insulation of the tool away from the user and into the ground. In most modern electrical systems, the three-prong plug fits into a three-prong, grounded receptacle. If the tool is operated at less than 150 volts, it has a plug like that shown in Figure 1-30A. If it is for use on 150 to 250 volts, it has a plug like that shown in Figure 1-30B. In either type, the green (or green and yellow) conductor in the tool cord is the grounding wire. Never connect the grounding wire to a live terminal.

An adapter (Figure 1-31) is available for connecting three-prong grounding-type plugs to two-prong receptacles. The rigid ear lug extending from the adapter is the grounding means and must be connected to a permanent ground, for example, to a properly grounded outlet box. If there is uncertainty about whether or not the receptacle in question is

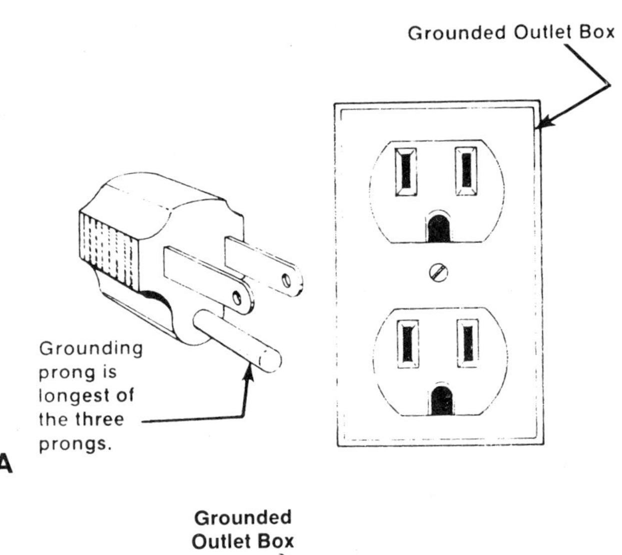

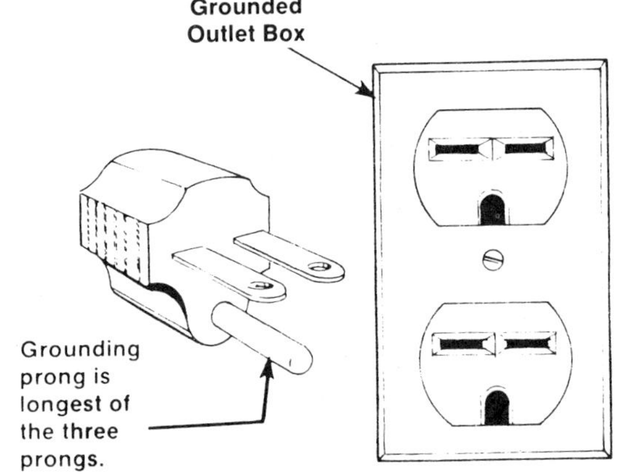

FIGURE 1-30 (A) A three-prong grounding plug for use with up to 150 volt tools and (B) a grounding plug for use with 150 to 250 volt tools

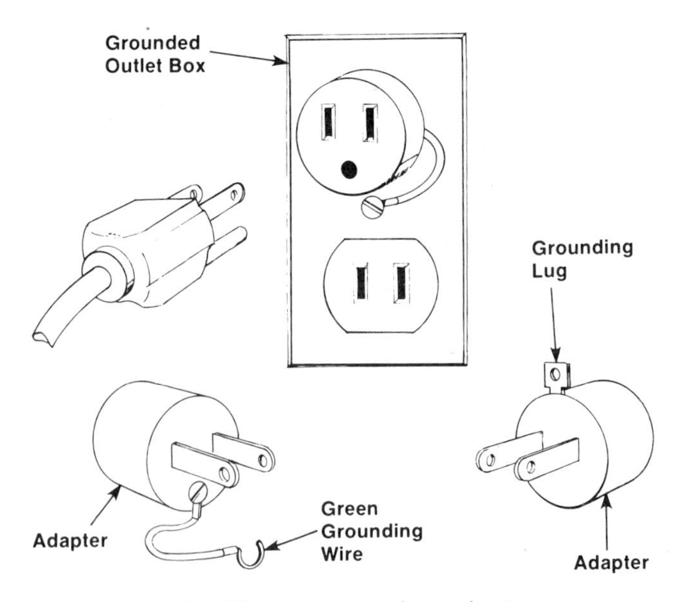

FIGURE 1-31 Three-prong plug adapters

properly grounded, have it checked by a certified electrician. No adapter is available for a plug of the type shown in Figure 1-30B.

A double-insulated tool has an extra layer of electrical insulation that eliminates the need for a three-prong plug and grounded outlet. Double-insulated tools do not require grounding and thus have a two-prong plug. In addition, double-insulated tools are always so indicated on their nameplate or case (Figure 1-32).

VOLTAGE WARNINGS

Before connecting a tool to a power source (a receptacle, outlet, or the like), be sure the voltage supplied is the same as that specified on the nameplate of the tool. A power source with a voltage greater than that specified for the tool can lead to serious injury to the user, as well as damage to the tool. If in doubt, do not plug in the tool. Using a power source with a voltage lower than the nameplate rating is harmful to the motor.

Tool nameplates also bear a figure with the abbreviation *amps* (for amperes, a measure of electric current). This refers to the input of electrical energy. On the surface, it would seem that the higher the input current, the more powerful the motor should be. But since this figure increases with the load (and is greatest on stall, when the motor is so burdened it cannot turn), the ampere rating on a nameplate is not of itself a true indication of a tool's capacity.

EXTENSION CORDS

If there is considerable distance from the power source to the work area, or if the portable tool is equipped with a so-called stub power cord, an extension cord must be used. When using extension cords on portable power tools, the size of the conductors must be large enough to prevent an excessive drop in voltage, which would cause loss of power, overheating, and possible motor damage. Table 1-3 shows the correct size to use, based on cord length and nameplate amperage rating. If in doubt, use the next heavier size.

FIGURE 1-32 Typical portable power tool nameplate

SHOP TALK

The smaller the gauge number of an extension cord, the heavier the cord.

Two-wire extension cords with two-prong plugs are acceptable for double-insulated tools. However, tools with three-prong, grounded plugs must only be used with three-wire, grounded extension cords connected to properly grounded three-wire receptacles (Figure 1-33). Current National Electrical Code and OSHA specifications call for outdoor receptacles to be protected with ground fault detector devices.

When using extension cords, keep in mind the following safety tips:

- Always connect the cord of a portable electric power tool into the extension cord before

TABLE 1-3: RECOMMENDED EXTENSION CORD SIZES FOR USE WITH PORTABLE ELECTRIC TOOLS

	Nameplate Ampere Rating															
Cord Length	0 to 5	6	8	8	9	10	11	12	13	14	15	17	17	18	19	20
25′	18	18	18	18	18	18	16	16	16	14	14	14	14	14	12	12
50′	18	18	18	18	18	18	16	16	16	14	14	14	14	14	12	12
75′	18	18	18	18	18	18	16	16	16	14	14	14	14	14	12	12
100′	18	18	18	16	16	16	16	16	14	14	14	14	14	14	12	12
125′	18	18	16	16	16	14	14	14	14	14	14	12	12	12	12	12
150′	18	16	16	16	14	14	14	14	14	12	12	12	12	12	12	12

Note: Wire sizes shown are AWG (American Wire Gauge) based on a line voltage of 120.

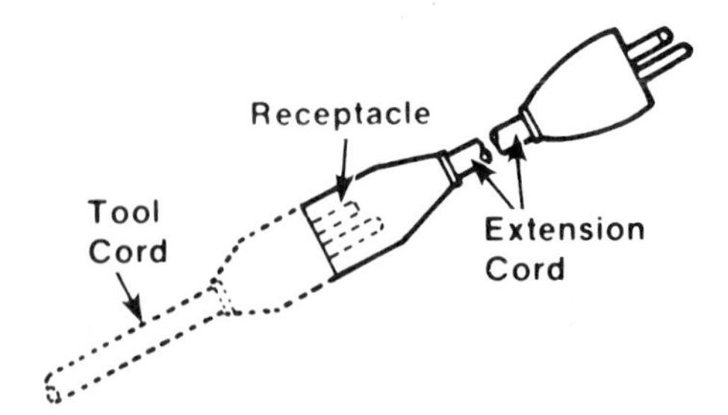

FIGURE 1-33 Typical three-wire extension cord

the extension cord is inserted into an outlet. Always unplug the extension cord from the receptacle before the cord of the portable power tool is unplugged from the extension cord.

- Extension cords should be long enough to make connections without being pulled taut, creating unnecessary strain and wear.
- Be sure that the extension cord does not come in contact with sharp objects or hot surfaces. The cords should not be allowed to kink, nor should they be dipped in or splattered with oil, grease, or chemicals.
- Before using a cord, inspect it for loose or exposed wires and damaged insulation. If a cord is damaged, it must be replaced. This also applies to the tool's power cord.
- Extension cords should be checked frequently while in use to detect unusual heating. Any cable that feels more than comfortably warm to the bare hand placed outside the insulation should be checked immediately for overloading.
- See that the extension cord is in a position to prevent tripping or stumbling.
- To prevent the accidental separation of a tool cord from an extension cord during operation, make a knot as shown in Figure 1-34A or use a cord connector as shown in Figure 1-34B.

HAND TOOL SAFETY

Several hand tools are used in the welding shop. These tools should be treated properly and not abused. Many accidents can be avoided by using the right tool for the job. For instance, a tool that is the correct weight for the work should be used instead of a tool that is too heavy or too light.

Keep hand tools clean to protect them against the damage caused by corrosion. Wipe off any accumulated dirt and grease. Dip the tools occasionally in cleaning fluids or solvents and wipe them clean. Lubricate adjustable and other moving parts to prevent wear and misalignment.

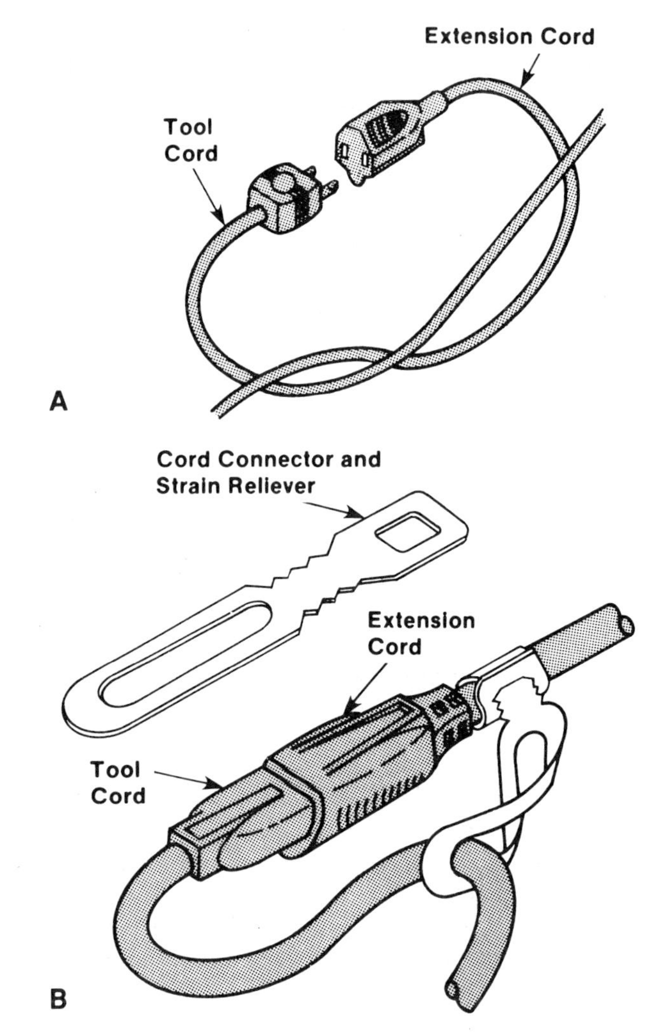

FIGURE 1-34 (A) A knot will prevent the extension cord from accidentally pulling apart from the tool cord driving operation. (B) A cord connector will serve the same purpose.

Make sure that hand tool cutting edges are sharp. Sharp tools make work easier, improve the accuracy of the work, save time, and are safer than dull tools. When sharpening, redressing, or repairing tools, shape, grind, hone, file, fit, and set them properly, using other tools suited to each purpose. For sharpening tools, either an oilstone or a grindstone is preferable. If grinding on an abrasive wheel is required, grind only a small amount at a time, with the tool rest set not more than 1/8 inch from the wheel. Hold the tool lightly against the wheel to prevent overheating, and frequently dip the part being ground in water to keep it cool. This will protect the hardness of the metal and help to retain the sharpness of the cutting edge. Tools struck by hammers, such as chisels or punches, should have their head ground periodically to prevent mushrooming (Figure 1-35). Be sure to wear safety goggles when sharpening or redressing tools.

Keep handles secure and safe. Do not rely on friction tape to secure split handles or to prevent handles from splitting. Check wedges and handles frequently. Be sure heads are wedged tightly on handles. Keep handles smooth and free of rough or jagged surfaces. Protect their tips before driving them into tools or use a proper mallet to avoid splitting or mushrooming them. Replace handles that are split, chipped, or that cannot be refitted securely.

When swinging any tool, be absolutely certain that no one is within range or can come within range of the swing or be struck by flying material. Always allow plenty of room for arm and body movements and for handling the work. When carrying tools, protect the cutting edges and carry the tools in such a way that you will not endanger yourself or others. Carry pointed or sharp-edged tools in pouches or holsters. Use only nonsparking tools in the presence of flammable materials or explosive dusts and vapors.

STRIKING TOOL SAFETY

The following safety precautions generally apply to all striking tools (Figure 1-36):

- Check to see that the handle is tight before using any striking tool. Never use a striking or struck tool with a loose or damaged handle.
- Always use a striking tool of suitable size and weight for the job.
- Discard any striking or struck tool if the face shows excessive wear, dents, chips, mushrooming, or improper redressing.
- Rest the face of the hammer on the work before striking to get the feel or aim; then, grasp the handle firmly with the hand near the extreme end of the handle. Get the fingers out of the way before striking with force.
- A hammer blow should always be struck squarely, with the hammer face parallel to the surface being struck. Always avoid glancing blows and over-and-under strikes.
- For striking another tool (cold chisel, punch, wedge, and so on), the face of the hammer should be proportionately larger than the head of the tool. For example, a 1/2-inch cold chisel requires at least a 1-inch hammer face.
- Never use one hammer to strike another hammer.
- Do not use the end of the handle of any striking tool for tamping or prying; it might split.

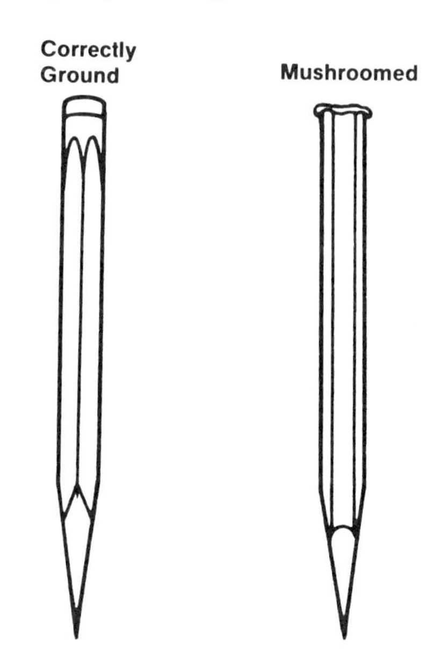

FIGURE 1-35 Ground the head periodically.

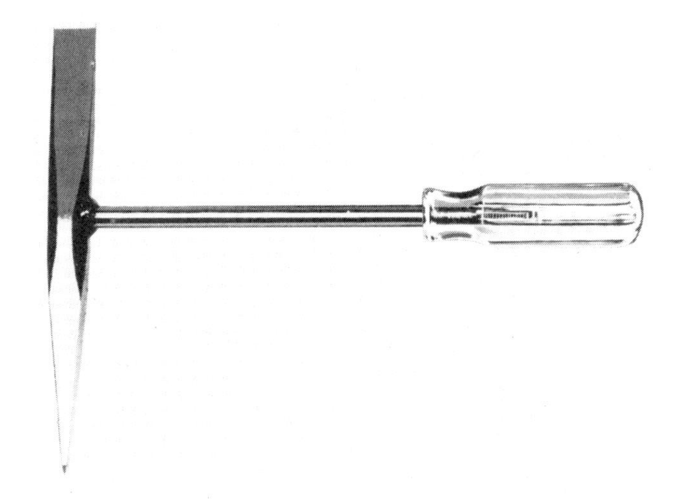

FIGURE 1-36 Welder's chipping hammer *(Courtesy of MAC Tools, Inc.)*

HOSES

Do not use hoses for anything except the gas or liquid for which they were designed. Red hoses are to be used exclusively for acetylene or other fuel gases. Green hoses are to be used only for oxygen. Avoid using unnecessarily long lengths of hoses.

Keep hoses out of the direct line of sparks. Oil, grease, lead, and other pipe-fitting compounds should never be used for any joints in the hoses. Repair or replace any hoses with leaking or bad joints (Figure 1-37).

FIGURE 1-37 Welding hose must be in good condition. *(Courtesy of MAC Tools, Inc.)*

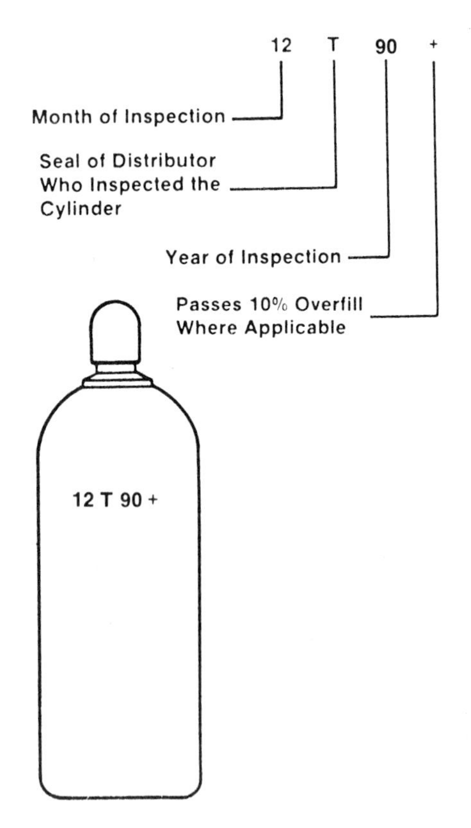

FIGURE 1-38 Department of Transportation markings

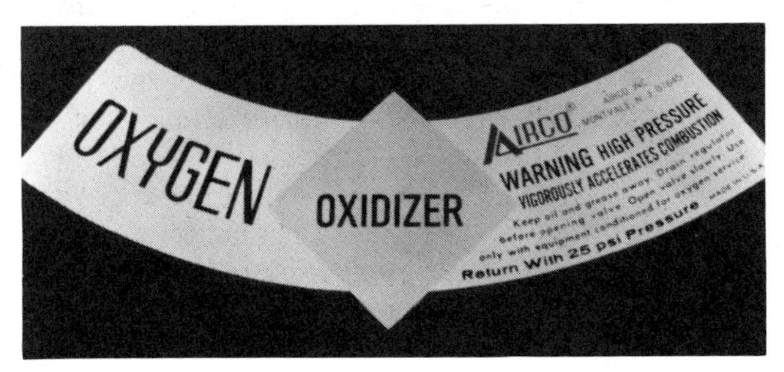

FIGURE 1-39 Carefully read the marking or decal label on the cylinder (usually on its shoulder or neck) for the true identity of the gas contents.

HANDLING AND STORAGE OF CYLINDERS

Certain precautions are necessary when handling oxygen and fuel gases. For instance, be sure that all cylinders have Interstate Commerce Commission (ICC) markings indicating the dates of bottle pressure tests (Figure 1-38). The date of inspection should be within five years. Special care must also be taken in the identification and selection of cylinders in order that the proper type of gas is used. For example, compressed oxygen should never be used in place of compressed air to operate pneumatic tools. Identification should be made from the tag (label) attached to the bottle or from the stenciled information on the bottle instead of depending on the bottle color code (Figure 1-39).

Cylinders must be handled carefully. They should not be dropped or jarred. Proper lifting procedures and equipment must be used when lowering or raising cylinders from one working level to another. Make sure the cylinder valves are completely closed before moving any cylinder. Cylinders should be lowered, raised, or stored in the upright (vertical) position with their protective caps on. Never use these caps for lifting the cylinders.

Cylinders are designed to store gases and should never be used as rollers, props, and the like. Never move a cylinder by dragging, sliding, or rolling it on its side (Figure 1-40). Avoid striking it

FIGURE 1-40 The rolling motion not only is less dangerous, it is also an easy way to handle a heavy cylinder.

against any object that might create a spark. Even if there is only a very small gas leak, an explosion could result.

Grease or oil should never come in contact with the cylinder valves. Keep in mind that while oxygen itself is nonflammable, it will quickly aid in the combustion of flammable materials. Cylinders must be kept in a well-ventilated area or outdoors. Oxygen cylinders are pressurized to between 2200 and 2700 psi, and thus should be handled carefully.

When storing cylinders outdoors, the loading and storage platform (which safe practice recommends) must be used so that cylinders can be transferred between the delivery truck and platform without being jarred or dropped. It is recommended that the platform be provided with a gradual ramp to ground level and a suitable cover from ice, snow, and direct rays of the sun. The full and empty cylinders of each type of gas must be stored separately.

Cylinders must be stored so that they will not be knocked over or damaged by falling objects, passing vehicles, or persons. Secure them with a chain or some other device. Keep them away from halls, stairwells, and exits so that in case of emergency they will not block an escape route. They must not be stored near radiators, stoves, or any other source of heat. Cylinders must not be permitted to come into contact with electrical wires. The storage area should be marked with a sign that says **Danger—No smoking, matches, or open flames** (Figure 1-41).

Cylinders containing oxygen must not be stored close to acetylene cylinders inside of a building. These storage areas must either be separated by 20 feet or by a 5-foot-high wall with at least a half hour burn rating (Figure 1-42). Inert gas cylinders can be stored separately or with either fuel cylinders or

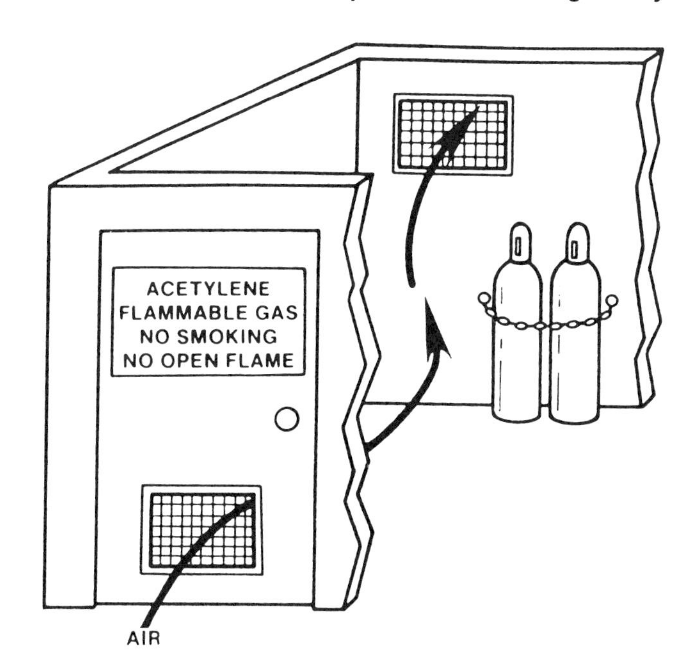

FIGURE 1-41 Storage areas must be well-ventilated and marked with a warning sign.

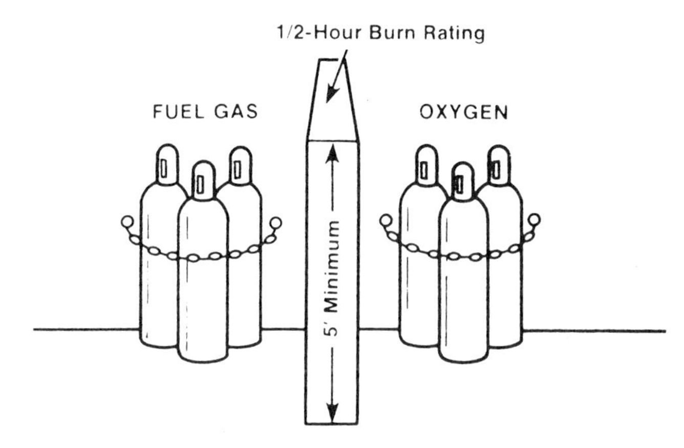

FIGURE 1-42 If fuel gas and oxygen are stored within 20 feet of each other, they must be separated by a 5-foot-high wall with at least a half hour burn rating.

oxygen cylinders. All storage rooms must be well-ventilated. Never expose a cylinder to excessive heat, such as placing it near a heat source or the sparks of a cutting torch.

When work is finished, cylinder valves must be closed and valve protection caps (Figure 1-43) should be in place before cylinders are moved or placed in storage. Special portable cylinder carts must be used for moving cylinders (Figure 1-44). All cylinders must be used, transported, and stored in an upright position. The cylinder valves must be closed and the regulators and the hose lines released of pressure when stopping work or when

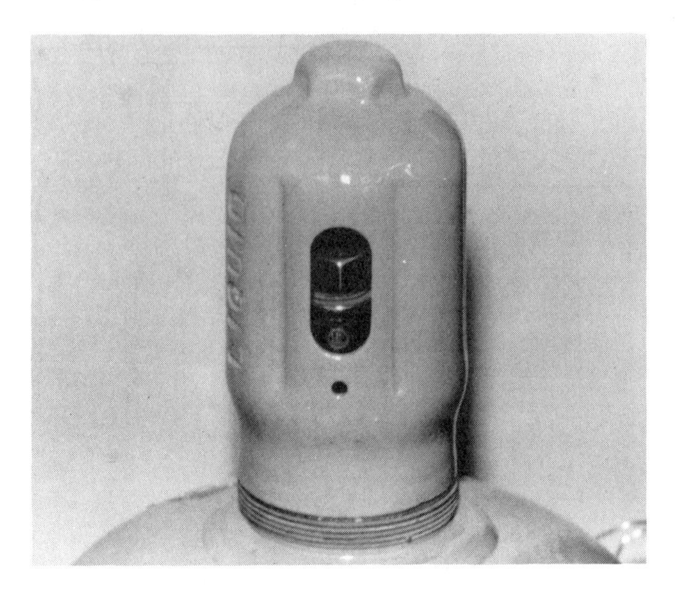

FIGURE 1-43 Typical oxygen protective cap

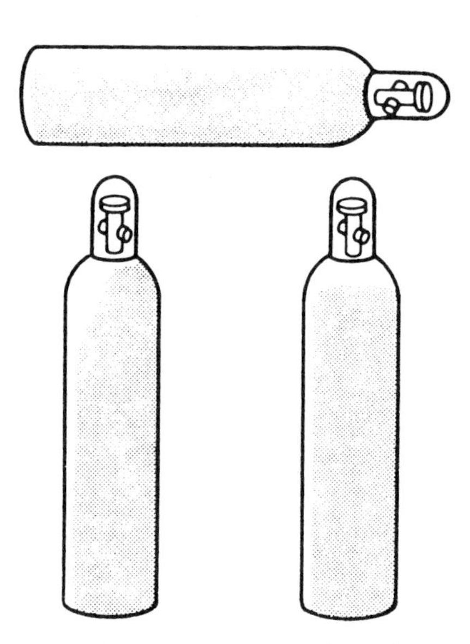

FIGURE 1-45 Acetone must settle before an acetylene cylinder can be used safely.

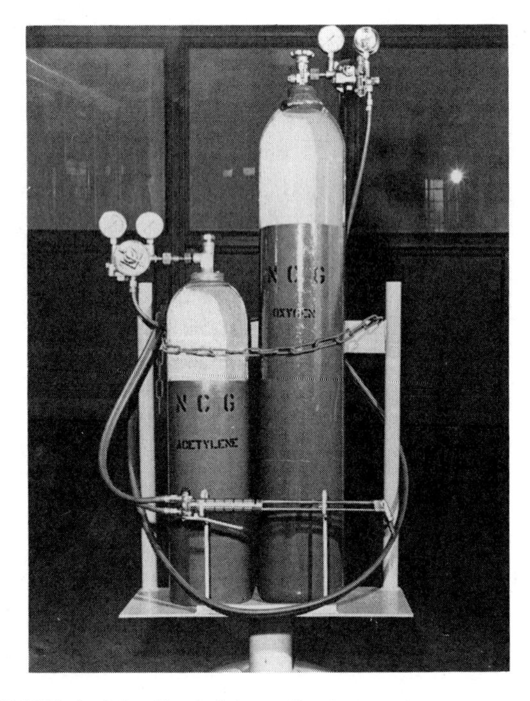

FIGURE 1-44 Portable cylinder cart

moving welding equipment in portable carts. Cylinders should be chained or wired to the cart. If an acetylene cylinder has been lying on its side, it must stand upright for 15 minutes or more before it is used. The acetylene is absorbed by acetone, which is absorbed in a filler. The filler prevents the liquid from settling back away from the valve very quickly (Figure 1-45). If the acetone is not given enough time to settle, it can be drawn out of the cylinder. Besides lowering the flame temperature, acetone can damage the regulator or torch valve settings.

In a fire, there are special precautions that should be taken, especially with acetylene cylinders. All acetylene cylinders are equipped with one or more safety relief devices filled with a low melting point metal. This fusible metal melts at about the boiling point of water (212 degrees Fahrenheit or 100 degrees Celsius). If fire occurs on or near an acetylene cylinder, the fuse plug will melt. The escaping acetylene may be ignited and will burn with a roaring sound. Immediately evacuate all people from the area. It is difficult to put out such a fire. The best action is to spray water on the cylinder to keep it cool and to keep all other acetylene cylinders in the area cool. Attempt to remove the burning cylinder from close proximity to other acetylene cylinders, from flammable or hazardous materials, or from combustible buildings. It is best to allow the gas to burn rather than to allow acetylene to escape, mix with air, and possibly explode.

If the fire on a cylinder is a small flame around the hose connection, the valve stem, or the fuse plug, try to put it out as quickly as possible. A wet glove, wet heavy cloth, or mud slapped on the flame will frequently extinguish it. Thoroughly wetting the gloves and clothing will help protect the person approaching the cylinder. Avoid getting in line with the fuse plug (Figure 1-46), which might melt at any time.

If a cylinder is damaged and is leaking, it should be removed to a vacant lot or open air, away from all

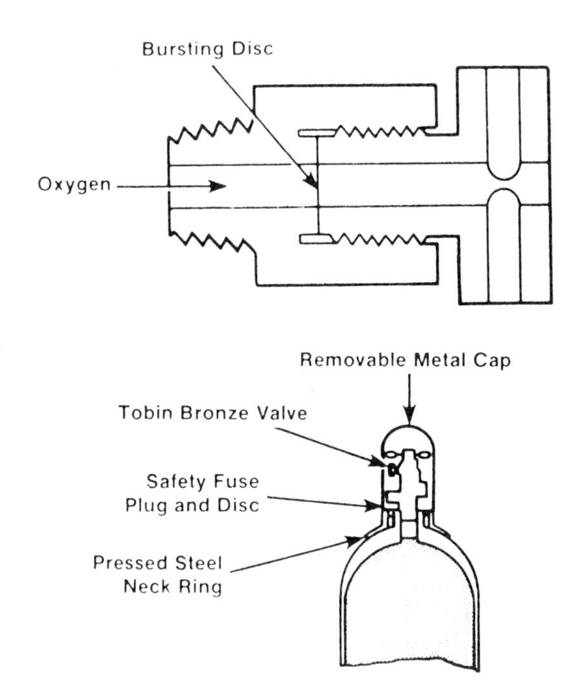

FIGURE 1-46 This cross-sectional sketch shows parts of the oxygen cylinder safety plug and where it is located in a cylinder.

FIGURE 1-47 Move a leaking fuel gas cylinder out of the building or any work area and post a warning.

possible sources of ignition (Figure 1-47). Post a warning on or near the cylinder. Also, the manufacturer should be notified immediately, giving the cylinder number stamped on the side. Under no circumstances should any attempt be made to repair a leaking cylinder by welding or by any other means. Also, under no circumstances should you tamper with or attempt to repair faulty cylinder valves.

Gas cylinders are generally owned by the companies that furnish the gases contained in them. A small rental fee is charged for the use of the cylinders after a reasonable rent-free period. When returning cylinders to the manufacturer, the valves should be tightly closed and the cap, where provided, screwed on the valve collar. The cylinders should be plainly marked "MT," and a tag should be attached giving the customer's name and the manufacturer's name and address. Also, any cylinder that leaks or has a bad valve or damaged threads should be identified and reported to the supplier. While awaiting shipment, empties should be stored apart from full cylinders to avoid dangerous confusion. Cylinders should be used in the order of receipt and returned promptly within the free loan period to avoid late charges.

FIRE PROTECTION

Fire is a constant danger in the welding shop. Although the possibilities of fire cannot always be removed, they can be minimized. Keep highly combustible materials 35 feet or more away from any welding. When combustible materials cannot be removed from the welding area, they should be wet down with water or covered with sand or noncombustible insulating blankets. A fire extinguisher must be close at hand. It should be the type required to put out a fire on the type of combustible materials near the welding.

FIRE EXTINGUISHERS

A fire can be extinguished by depriving it of its essential ingredients, which are heat, fuel, and oxygen. Most extinguishers work by cooling the fire and removing the oxygen. If the extinguisher is going to be used effectively, it must be aimed at the base of the flame where the fuel is located (Figure 1-48).

There are four basic types of fire extinguishers: type A, type B, type C, and type D. Although each type is designed to put out fires on particular types of materials, some fire extinguishers might have the ability to put out more than one type of fire. Nevertheless, using the wrong type of fire extinguisher can cause the spread of the fire, electric shock, or explosion. Table 1-4 gives a breakdown of the different classes of fires and the type of extinguisher that should be used.

Type A Extinguishers

The symbol for a type A extinguisher is a letter A in the center of a green triangle. It is used for combustible solids such as wood, cloth, and paper.

TABLE 1-4: GUIDE TO EXTINGUISHER SELECTION

	Class of Fire	Typical Fuel Involved	Type of Extinguisher
Class A Fires (green)	**For Ordinary Combustibles** Put out a class A fire by lowering its temperature or by coating the burning combustibles.	Wood Paper Cloth Rubber Plastics Rubbish Upholstery	Water*[1] Foam* Multipurpose dry chemical[4]
Class B Fires (red)	**For Flammable Liquids** Put out a class B fire by smothering it. Use an extinguisher that gives a blanketing, flame-interrupting effect; cover whole flaming liquid surface.	Gasoline Oil Grease Paint Lighter fluid	Foam* Carbon dixoide[5] Halogenated agent[6] Standard dry chemical[2] Purple K dry chemical[3] Multipurpose dry chemical[4]
Class C Fires (blue)	**For Electrical Equipment** Put out a class C fire by shutting off power as quickly as possible and by always using a nonconducting extinguishing agent to prevent electric shock.	Motors Appliances Wiring Fuse boxes Switchboards	Carbon dioxide[5] Halogenated agent[6] Standard dry chemical[2] Purple K dry chemical[3] Multipurpose dry chemical[4]
Class D Fires (yellow)	**For Combustible Metals** Put out a class D fire of metal chips, turnings, or shavings by smothering or coating with a specially designed extinguishing agent.	Aluminum Magnesium Potassium Sodium Titanium Zirconium	Dry power extinguishers and agents only

*Cartridge-operated water, foam, and soda-acid types of extinguishers are no longer manufactured. These extinguishers should be removed from service when they become due for their next hydrostatic pressure test.

Notes:

(1) Freeze in low temperatures unless treated with antifreeze solution, usually weighs over 20 pounds, and is heavier than any other extinguisher mentioned.

(2) Also called ordinary or regular dry chemical. (sodium bicarbonate)

(3) Has the greatest initial fire-stopping power of the extinguishers mentioned for class B fires. Be sure to clean residue immediately after using the extinguisher so sprayed surfaces will not be damaged. (potassium bicarbonate)

(4) The only extinguishers that fight A, B, and C classes of fires. However, they should not be used on fires in liquefied fat or oil of appreciable depth. Be sure to clean residue immediately after using the extinguisher so sprayed surfaces will not be damaged. (ammonium phosphates)

(5) Use with caution in unventilated, confined spaces.

(6) May cause injury to the operator if the extinguishing agent (a gas) or the gases produced when the agent is applied to a fire is inhaled.

Type B Extinguishers

The symbol for a type B extinguisher is a letter B in the center of a red square. Type B extinguishers are used for combustible liquids, such as paint thinner, gas, and oil.

Type C Extinguishers

The symbol for a type C extinguisher is a letter C in the center of a blue circle. Type C extinguishers are used for electrical fires, such as those involving motors, fuse boxes, and welding machines.

Type D Extinguishers

The symbol for a type D extinguisher is a letter D in the center of a yellow star. Type D extinguishers are used on fires involving combustible metals, such as zinc, magnesium, and titanium.

Location of Fire Extinguishers

The type of fire extinguisher located in a particular area must be determined by the types of combustible materials that are there. Place the extinguishers where they can be easily removed without reaching

FIGURE 1-48 Point the extinguisher at the fuel source, not at the flames.

over combustible material. Also, place them at a level low enough to be easily lifted off the mounting (Figure 1-49). Mark the location of fire extinguishers with red paint and signs. These markings must be high enough to be seen from a distance over people and equipment. The location of the extinguishers should also be marked near the floor (Figure 1-50). These low markings will make the extinguishers easier to be found if the room should fill with smoke.

VEHICLE HANDLING IN THE SHOP

When handling a vehicle in the shop, keep the following safety precautions in mind:

- Set the parking brake when working on the vehicle. If the car has an automatic transmission, set it in *park* unless instructed otherwise for a specific service operation. If the vehicle has a manual transmission, it should be in *reverse* (engine off) or *neutral* (engine on) unless instructed otherwise for a specific service operation.

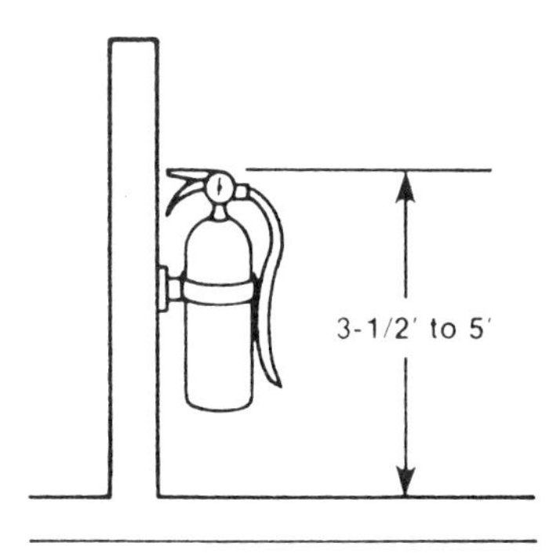

FIGURE 1-49 Mount the fire extinguisher low enough for easy access.

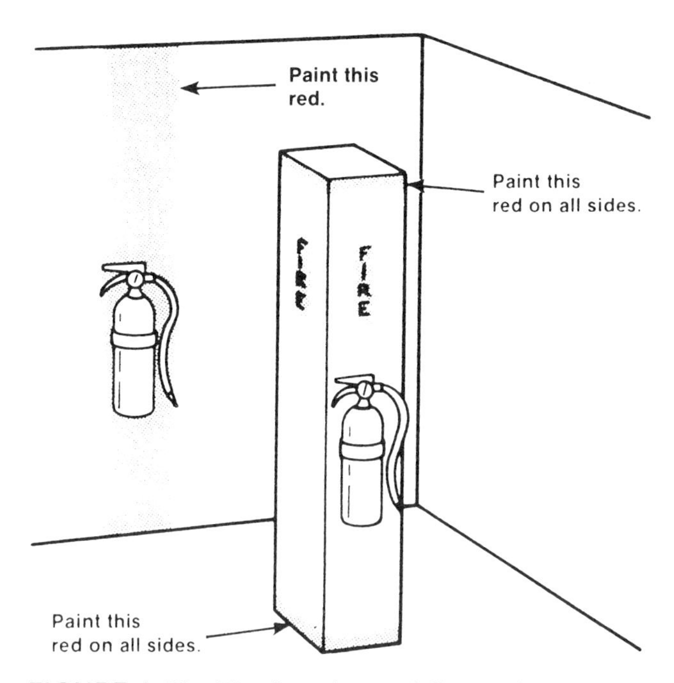

FIGURE 1-50 The locations of fire extinguishers should be marked.

- Use safety stands (Figure 1-51) whenever a procedure requires work under the vehicle.
- To prevent serious burns, avoid contact with hot metal parts such as the radiator, exhaust manifold, tail pipe, catalytic converter, and muffler.
- Keep clothing and yourself clear from moving parts when the engine is running, especially the radiator fan blades (Figure 1-52) and belts. Electric cooling fans can start to operate at any time by an increase in temperature under the hood, even though the igni-

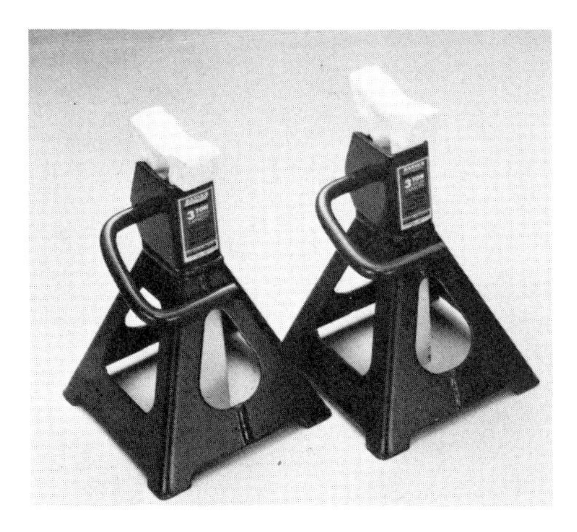

FIGURE 1-51 Use safety stands whenever working under a vehicle.

FIGURE 1-52 Keep fingers and other body parts away from moving parts. Do not turn on ignition switch (only required by the checking procedure).

tion is in the *off* position. Therefore, care should be taken to insure that the electric cooling fan is completely disconnected when working under the hood.

- Be sure that the ignition switch is always in the *off* position, unless otherwise required by the procedure.
- When moving a vehicle around the shop, be sure to look in all directions and make certain that nothing is in the way.
- Do not smoke while working on vehicles.

HORSEPLAY

Horseplay, running, and otherwise fooling around should not be part of any shop. One thing can lead to another and eventually cause an injury. Horseplay is also distracting and wastes time.

REVIEW QUESTIONS

1. Which type of burn blisters and breaks the skin?
   a. first-degree
   b. second-degree
   c. third-degree
   d. all of the above

2. Which type of light is the most dangerous to the welder?
   a. visible
   b. ultraviolet
   c. infrared
   d. none of the above

3. Which of the following types of general work clothing offers the welder the most protection?
   a. rayon
   b. cotton
   c. polyester
   d. wool

4. Which part of the eye, when it is burned by ultraviolet light, does not feel pain but does suffer some loss of vision?
   a. pupil
   b. white of the eye
   c. retina
   d. none of the above

5. When chlorinated solvents, such as those used for degreasing metals, decompose in the arc, which of the following is produced?
   a. ozone
   b. argon
   c. phosgene
   d. none of the above

6. Ventilation for welding can be accomplished ______________.
   a. naturally
   b. mechanically
   c. locally
   d. all of the above
   e. none of the above

7. Welder A says the usual cause of electric shock in a welding shop is contact with a welding torch. Welder B says the usual cause of shock is contact with a bare or poorly insulated conductor. Who is right?
   a. Welder A
   b. Welder B
   c. Both A and B
   d. Neither A nor B

8. Welder A does not wear gloves when grinding. Welder B does. Who is right?
   a. Welder A
   b. Welder B

c. Both A and B
d. Neither A nor B

9. Which of the following safety systems is used on portable electric tools?
   a. external grounding
   b. double-insulation
   c. both a and b
   d. none of the above

10. Which of the following is true?
    a. The smaller the gauge, the heavier the extension cord needed.
    b. Two-prong plugs are suitable for double-insulated tools.
    c. The extension cord is connected to the tool before the extension cord is inserted in the outlet.
    d. Both a and b
    e. All of the above

11. When should the handle on a hand tool be replaced?
    a. when it is cracked
    b. when it is chipped
    c. when it cannot be refitted securely
    d. all of the above

12. A striking tool should be discarded when it is ______________.
    a. improperly redressed
    b. mushrooming
    c. chipped
    d. all of the above

13. Green hoses are to be used only with ______________.
    a. oxygen
    b. acetylene
    c. argon
    d. none of the above

14. What should the welder do to repair a leaking gas cylinder?
    a. find the leak and weld it shut
    b. apply a pipe-fitting compound to the leak
    c. apply duct tape
    d. none of the above

15. A fire is extinguished by depriving it of ______________.
    a. heat
    b. fuel
    c. oxygen
    d. all of the above

16. Which of the following regulations are, in most cases, laws, not just recommendations?
    a. AWS
    b. OSHA
    c. ASTM
    d. ANSI

17. In the event of a third-degree burn, which of the following should not be done?
    a. Wrap the affected area in a clean, wet, lint-free cloth.
    b. Seek medical help.
    c. Prevent air from getting to the affected area.
    d. Immerse the affected area in cold water.

18. Ultraviolet light can penetrate what type of clothing?
    a. loosely woven
    b. light colored
    c. thin
    d. all of the above

19. Welder A wears a leather cape and apron for out-of-position work; Welder B wears a full leather jacket. Who is right?
    a. Welder A
    b. Welder B
    c. Both A and B
    d. Neither A nor B

20. What is the recommended eye protection filter shade number for torch brazing?
    a. 3 or 4
    b. 6 or 8
    c. 10
    d. 14

21. What color should arc welding areas be painted to absorb as much ultraviolet light as possible?
    a. flat white
    b. flat black
    c. glossy white
    d. glossy black

22. A grinding stone being tested for cracks makes a dull sound when tapped. Welder A says this means the stone is cracked and

therefore should not be used. Welder B says it means the stone is in good condition. Who is right?
   a. Welder A
   b. Welder B
   c. Both A and B
   d. Neither A nor B

23. Which of the following statements concerning drilling is incorrect?
   a. Never apply a spinning drill bit to the work.
   b. Reduce the pressure on the drill just before the bit cuts through the work to avoid stalling.
   c. When the drill becomes jammed, squeeze the trigger on and off until it is freed.
   d. Hold the drill perpendicular to the surface of the work.

24. To prevent the accidental separation of a tool cord from its extension cord during use, Welder A uses a cord connector. Welder B makes a knot using the two cords. Who is right?
   a. Welder A
   b. Welder B
   c. Both A and B
   d. Neither A nor B

CHAPTER TWO

MIG WELDING THEORY

Objectives

After reading this chapter, you should be able to

- Name the three main welding categories.
- List five characteristics of welding.
- Explain why MIG welding is recommended for all automotive structural collision repairs.
- List and describe the various shielding gases used with MIG welding.
- Explain the process of short circuiting metal transfer.
- Name the basic components found in any MIG welding setup.
- Explain the slope, current rating, and duty cycle of the MIG power supply.
- List the beneficial characteristics of the flux-cored arc welding process.
- Describe the setup of MIG equipment.
- Explain how variables such as welding current, arc voltage, and tip-to-base metal distance affect a welding job.

The two basic methods of joining metal in the automobile assembly are:

- Mechanical, including nuts and bolts and riveting (Figure 2-1)
- Welding (Figure 2-2)

As defined by the American Welding Society, a weld is "a localized coalescence (the growing together of the grain structure of the materials being welded) of metals or nonmetals produced by heating the materials to suitable temperatures, with or without the application of pressure, and with or without the use of filler materials." To elaborate on this definition, a weld is formed when separate pieces of material are fused together through the application of heat. The heat applied is high enough in temperature to cause the softening or melting of

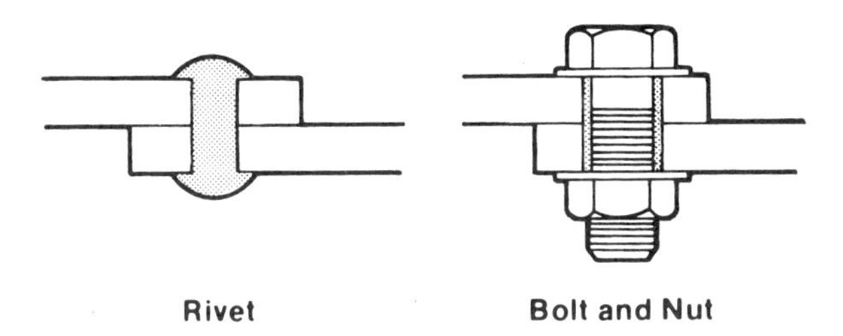

FIGURE 2-1 Mechanical joining methods

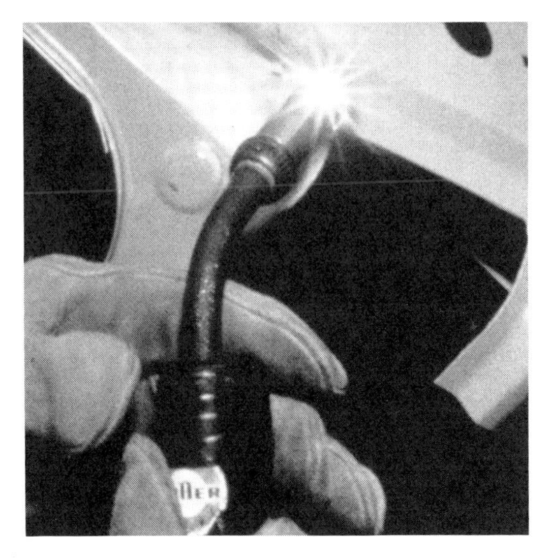

FIGURE 2-2 One metal welding method used in automobile work

the pieces to be joined. Once the pieces are heated, pressure may or may not be used to force them together. In some cases, filler material is added to the weld joint when needed. For practical purposes, welding can be divided into three main categories.

1. *Pressure Welding.* The pieces of metal are heated to a softened state by electrodes,

pressure is applied, and the metal is joined. Of the various types of pressure welding, electric resistance welding (spot welding) is used most often in automobile manufacturing, and to a lesser degree in repair operations.

2. *Fusion Welding.* The pieces of metal are heated to the melting point, joined together (usually with a filler rod), and allowed to cool.
3. *Braze Welding.* Metal with a melting point lower than the pieces of base metal to be joined is melted over the joint of the pieces—without melting the base metal. Braze welding is classified as either soft or hard, depending on the temperature at which the brazing material melts. Soft brazing is done with material that melts at temperatures below 850 degrees Fahrenheit; hard brazing is done above 850 degrees Fahrenheit.

As shown in Table 2-1, there are distinct methods used within each respective category. Many of these methods can be implemented in automotive applications.

WELDING CHARACTERISTICS

The characteristics of welding can be summarized as follows:

- Since the shape of welding joints is virtually limitless, it is the perfect method for joining a

TABLE 2-1: WELDING METHODS

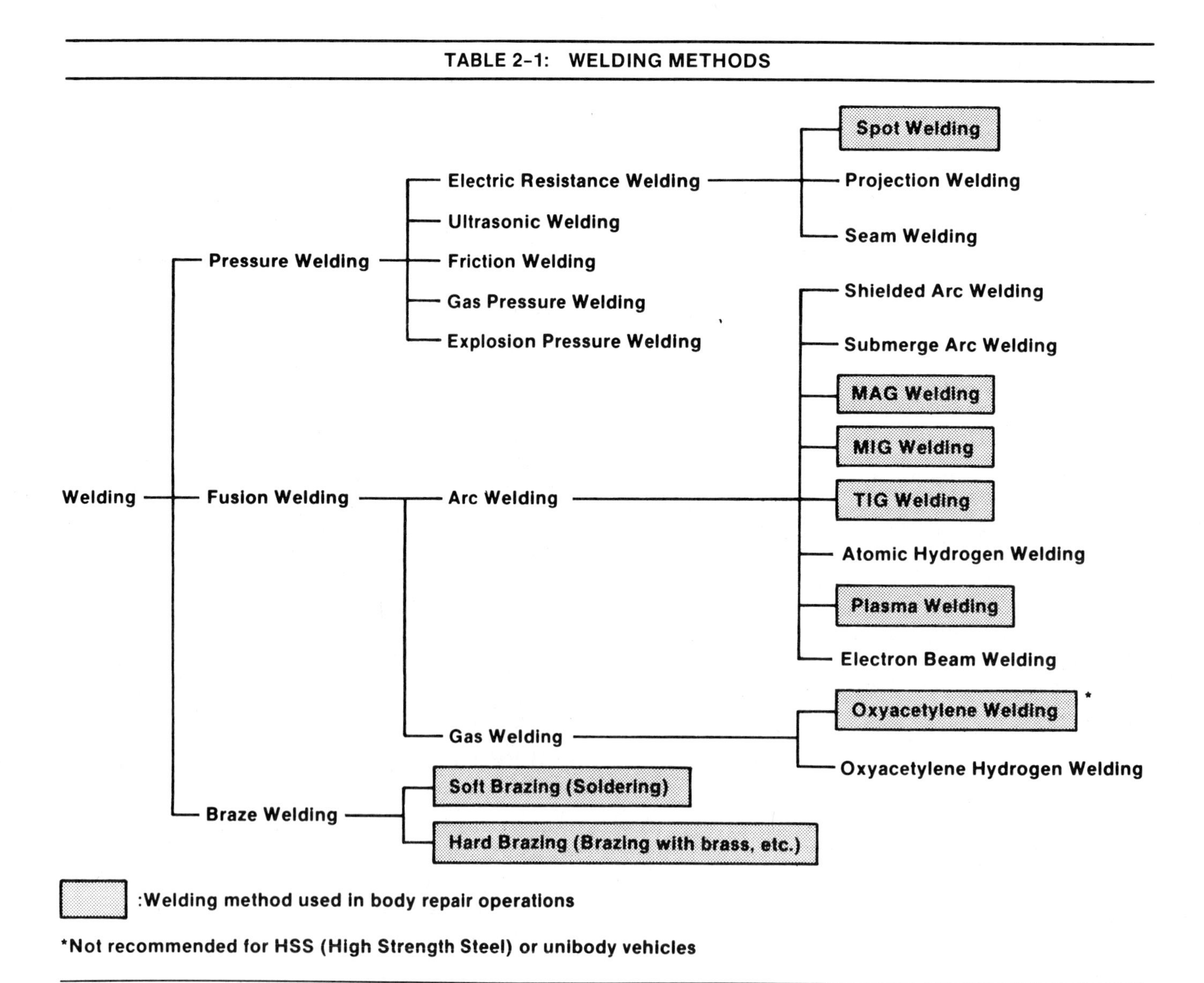

unibody structure and still maintain body integrity.
- Weight can be reduced because no fasteners are necessary.
- Air and water tightness are excellent.
- Production efficiency is very high.
- The strength of a welded joint is greatly influenced by the skill level of the operator. For instance, the surrounding panels will warp if too much heat is used.

AUTOMOTIVE WELDING

New welding techniques and equipment have emerged in recent years, replacing the once popular arc and oxyacetylene processes (Figure 2-3). The reason is simply that the new steel alloys used in today's cars cannot be welded properly with these two processes. It demands a technique that leaves a narrow heat-affected zone. This is especially true with coated steels; oxyacetylene welding would burn off too much of the galvanized zinc coating in the weld area. Also, unibody car designs require a 100 percent fusion weld for most repairs, normally impossible with oxyacetylene.

At the present time, metal inert gas (MIG) and resistance spot welding offer more advantages than other methods for welding high-strength steels (HSS) and the high-strength, low-alloy (HSLA) steel component parts used in unibody cars. Most of the applications of HSS and HSLA steels are confined to body structures, reinforcement gussets, brackets, and supports rather than large panels or outer skin panels.

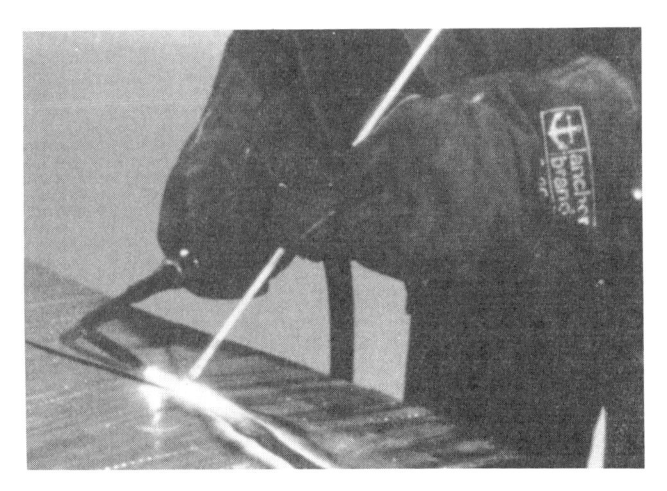

FIGURE 2-3 The once popular oxyacetylene welding process is no longer recommended for automobile work.

The advantages of MIG welding (Figure 2-4) over conventional stick electrode arc welding (Figure 2-5) are so numerous that car manufacturers now recommend that it be used in all of their dealerships, not only for HSS and unibody repair, but for all structural collision repair. The advantages of MIG welding are as follows:

- MIG welding is easy to learn. The typical welder can learn to use MIG equipment and actually reach a peak level of proficiency with just a few hours of instruction and practice. Moreover, experience shows that even

FIGURE 2-4 MIG is now the number one welding method in auto repair.

FIGURE 2-5 Conventional stick electrode or shielded arc welding

an average MIG welder can produce higher quality welds faster and more consistently than a highly skilled welder using conventional stick electrode welds.

- MIG produces 100 percent fusion in the base metals. This means MIG welds can be ground down flush with the surface (for cosmetic reasons) without loss of strength.
- Low current can be used to MIG weld thin metals. This prevents heat damage to adjacent areas that can cause strength loss and warping.
- The arc is smooth and the weld puddle small, so it is easily controlled (Figure 2-6). This ensures maximum metal deposit with minimum splatter.
- MIG welding is tolerant of gaps and misfits. Several gaps can be spot welded by making several spots on top of each other—immediately, with no slag to remove. Therefore, the area can be easily refinished.
- Almost all steels can be MIG welded with one common type of weld wire. What is in the machine is generally right for any job.
- Metals of different thicknesses can be MIG welded with the same diameter of wire. Again, what is in the machine is right for any job.
- The MIG welder can control the temperature of the weld and the time the weld takes.
- With MIG welding, the small area to be welded is heated for a short period of time, therefore reducing metal fatigue, warpage, and distortion of the panel. Vertical and/or overhead welding is possible because the metal is molten for a very short time.
- MIG welding is safer to use when making repairs under the vehicle next to brake lines, fuel lines, and fuel tanks, because the wire is not energized until the gun trigger is depressed.
- With MIG welding, there is minimum waste of welding consumables.

The other type of welding that is being recommended in automotive applications is resistance spot welding (Figure 2-7). This method is used to form spot weld attachments such as production welds. To use spot welding equipment, the operator must install the proper extensions and electrodes on the welder to provide access to the area being welded. The clamping force on the so-called squeeze-type resistance spot welder for the panels being welded must be properly adjusted. On some equipment, the amperage current flow and timing are all made with one adjustment. After the adjustment is made, the spot welder is positioned for the panels being joined, making sure the electrodes are directly opposite each other. The trigger is squeezed and the spot weld is made.

The resistance spot welder provides very fast, high-quality welds while maintaining the best control of temperature buildup in adjacent panels. It also requires the least skill to operate. When reference is made in the body shop to resistance-type spot welders, it generally means the type of welding that requires the weld to take place on both sides of the panels at the same time, not the type that welds panels together from the same side at the same time. Opposite side spot welding is a structural weld.

Be sure to consult the car manufacturer's recommendations in the vehicle's service manual before welding. When replacing body panels, the new welds should be similar in size to the original factory welds. Except when spot welding, the number of

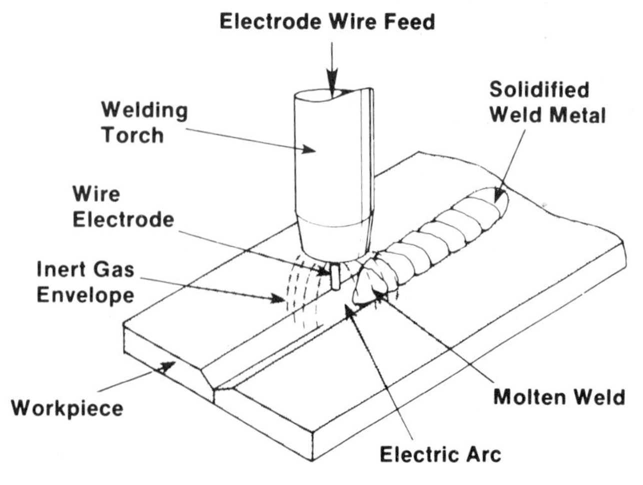

FIGURE 2-6 Basic MIG welding process

FIGURE 2-7 Using a portable squeeze-type resistance spot welding gun to weld a left rear quarter panel to the lower back body panel

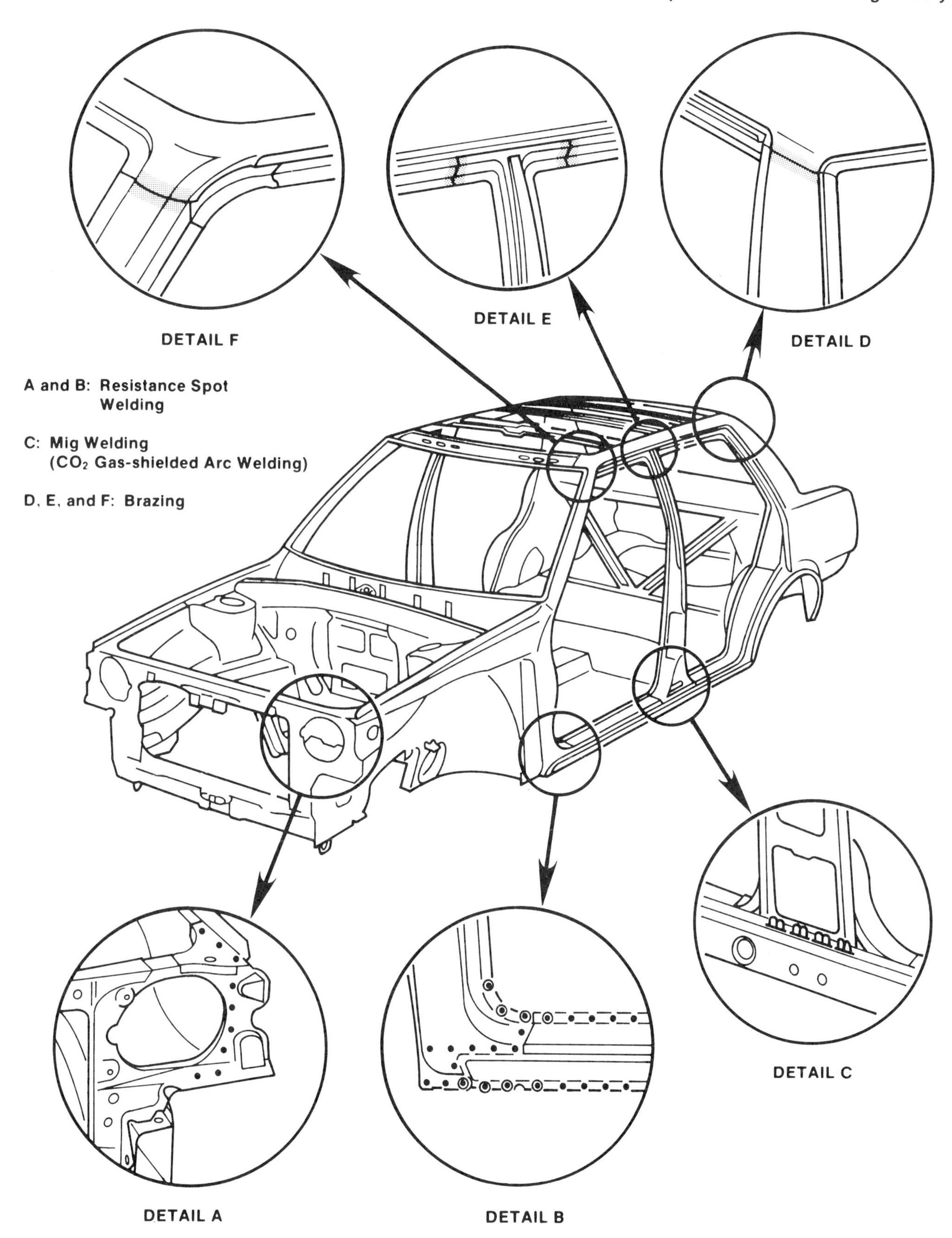

FIGURE 2-8 Body welding on a production vehicle *(Courtesy of Nissan Motor Corp.)*

replacement welds should be the same as the original number of production welds. (See Chapter 6 for further details on spot welding.)

Strength and durability requirements differ depending on the location of the part that is to be welded to the body. The factory decides what is the best assembly welding method (Figure 2-8) by determining the intended use, the physical characteristics, and the location of the part as it is assembled onto the auto body.

It is essential that the welding methods used do not reduce the original strength and durability of the car body. This will be accomplished if the following basic points are observed:

- Try to use either spot welding or MIG/MAG (metal inert gas/metal active gas) welding.
- Do not braze any body components other than those brazed at the factory.
- Do not use an oxyacetylene torch for welding late-model auto bodies.

SHOP TALK

Regardless of the type of welding that is being done, the surface must be properly cleaned before starting the welding procedure. Remove all surface materials back to the bare metal (Figure 2-9). When dirt, rust, sealers, or paint are left in the area of the weld, these materials will burn during the application of heat and the ash or oxidized material can become a part of the weld. Dirt and foreign material cause the weld to be weakened and, in some cases, prevent a proper weld from being made.

MIG WELDING

As already mentioned, MIG welding became popular when auto manufacturers began using high-strength, low-alloy (HSLA), thin-gauge steels. Car makers insisted that the only correct way to weld HSLA and other thin-gauge steel was with MIG (or a similar gas metal arc welding system). Once the MIG welder was in place, it was easy to see that it provided clean, fast welds for all applications. Welding a rear quarter panel with an oxyacetylene welder takes about 4 hours; a MIG welder can do the same job in barely more than a half hour.

MIG welding is not limited to body repairs alone. It is also ideal for exhaust work, repairing mechanical supports, installing trailer hitches and bumpers,

FIGURE 2-9 Make sure the surfaces to be welded are completely free of rust and scale. Remove dirt and foreign materials by grinding, sanding, or sandblasting.

and any other welds that would be done with either an arc or gas welder. In addition, it is possible to weld aluminum castings such as cracked transmission cases, cylinder heads, and intake manifolds.

MIG PRINCIPLES AND CHARACTERISTICS

MIG welding uses a welding wire that is fed automatically at a constant speed as an electrode. A short arc is generated between the base metal and the wire, and the resulting heat from the arc melts the welding wire and joins the base metals together. Since the wire is fed automatically at a constant rate, this method is often mistakenly called semiautomatic arc welding. During the welding process (Figure 2-10), either an inert or active gas shields the weld from the atmosphere and prevents oxidation of the base metal. The type of inert or active gas used depends on the base material to be welded.

SHIELDING GASES AND METALS

Several types of shielding gases and shielding gas mixtures are used with the gas metal arc (GMAW) process. Some of the gases have a broad range of application; others are restricted in their use. This section deals with the various shielding gases and their uses with weldable metals.

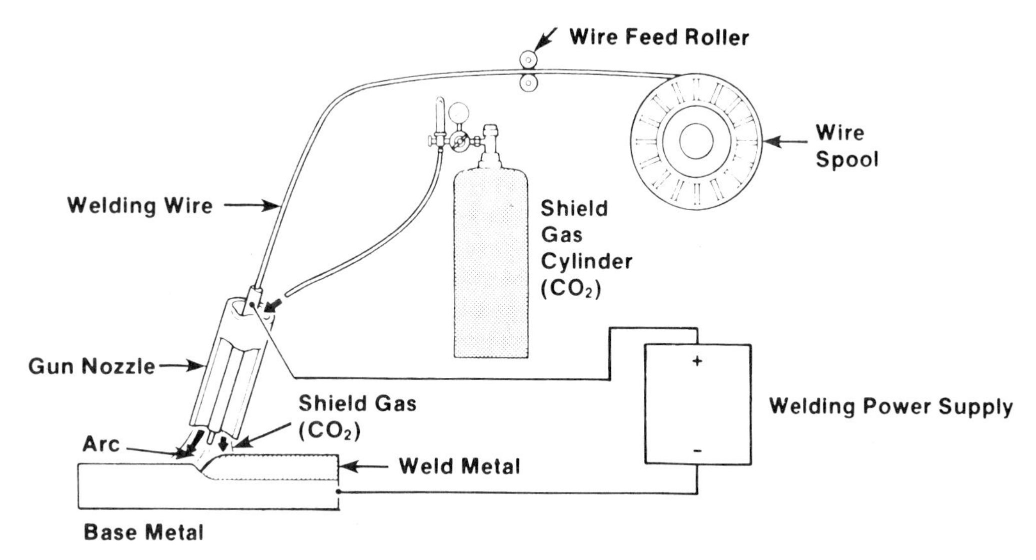

FIGURE 2-10 The principle of MIG welding *(Courtesy of Toyota Motor Corp.)*

Argon

Argon is a chemically inert gas that will not combine with the products of the weld zone. It has an ionization potential of 15.7 volts. Ionization potential is the energy required to remove an electron from the gas atom, thereby making it an ion, or charged atom. The charged atom then becomes a better path for the welding current.

Argon has low thermal conductivity. The arc plasma is constricted, with the result that high arc densities are present. The high arc density permits more of the available energy to go into the work as heat. The result is a relatively narrow bead width with deep penetration at the weld center. Argon causes a more concentrated arc than any of the other commonly used gases employed with the gas metal arc process. It is for this reason that argon has a reputation for cleaning the work area. Actually, it is the concentration of the arc plasma, and therefore heat energy, that causes the refractory oxides to be loosened. This phenomenon is particularly noticeable when welding aluminum.

Argon is in abundant supply since it comprises 8/10 of 1 percent of the Earth's atmosphere. The gas is obtained as a by-product in the manufacture of oxygen. Argon is used as the shielding gas when welding many types of metals. Its primary use is in the welding of nonferrous metals and alloys such as aluminum, magnesium, alloys of the two, and copper. In some metal applications, argon does not provide the penetration characteristics required for heavier weldments. In these cases, argon-helium mixtures are sometimes used. This gas mixture will be discussed under a separate heading.

Helium

Helium is also an inert gas and can be compared to argon in that respect. There the similarity ends. Helium has an ionization potential of 24.5 volts. It is lighter than air and has high thermal conductivity. The helium arc plasma will expand under heat (thermal ionization), reducing the arc density. With helium, there is a simultaneous change in arc voltage where the voltage gradient of the arc length is increased by the discharge of heat from the arc stream or core. This means that more arc energy is lost in the arc itself and is not transmitted to the work. The result is that, with helium, there will be a broader weld bead with relatively shallower penetration than with argon (Figure 2-11). (For TIG welding, the opposite is true.) This also accounts for the higher arc voltage, for the same arc length, that is obtained with helium.

Helium is derived from natural gas. The process by which it is obtained is similar to that of argon. First the natural gas is compressed and cooled. The hydrocarbons are drawn off, then nitrogen, and finally the helium. This is a process of liquefying the

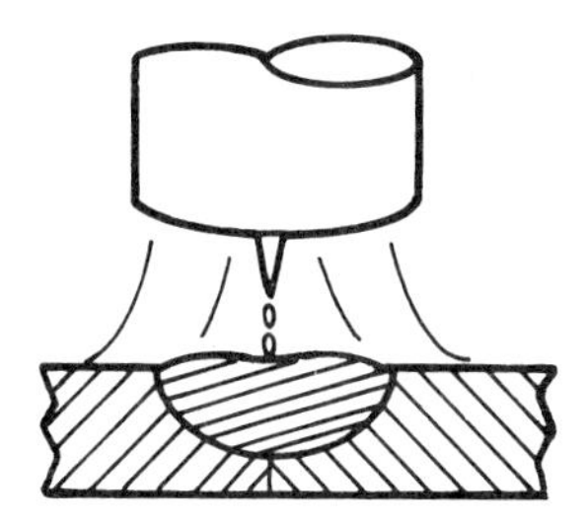

FIGURE 2-11 Helium shielded weld

various gases until, at 452 degrees Fahrenheit, helium is produced.

Helium has sometimes been found in short supply due to governmental restrictions and, therefore, has not been used as much as it might have been for welding purposes. It is difficult to initiate an arc in a helium atmosphere with the tungsten arc process. The problem is less acute with the gas metal arc process.

Helium is used primarily for nonferrous metals such as aluminum, magnesium, and copper. It is also used in combination with other shielding gases.

Argon/Oxygen

Argon is an excellent shielding gas for the gas metal arc process because it allows the use of spray-type metal transfer. When depositing flat or horizontal fillet welds, however, the typical deep central penetration does not allow the weld metal to "wet out" at the toes of the weld (Figure 2-12). This is particularly noticeable when welding steel or stainless steel. This phenomenon will invariably cause undercut at the edges of the weld bead.

The tendency to undercut can be prevented by the addition of 1 to 5 percent oxygen to the argon. The oxygen permits a controlled oxidation to take place and increases the temperature of the molten metal transferred across the arc. The additional time at liquidus (the lowest temperature at which a metal or an alloy is completely liquid) allows the hotter molten metal to "wet out" at the toes of the weld. This action produces a featheredge at the junction of the weld and the parent metal.

FIGURE 2-12 The addition of oxygen to argon allows the weld metal to "wet out" at the toes of the weld.

Argon/oxygen mixtures are very common for welding stainless steels. They can be used for mild and low-alloy steels, but the cost is usually prohibitive. Argon/oxygen shielding gases are usually purchased as premixed gases.

Argon/Helium

Argon/helium mixtures are usually used to achieve a favorable combination of the welding characteristics of both gases. The addition of helium to argon can be in percentages of 20 to 90 percent or more and is usually done by the user to suit his or her requirements. Purchase a cylinder of each type of gas and use flowmeters and a Y-connection to control the percentage of each gas in the mixture. This permits the user to get exactly the welding characteristics that are needed from the shielding gas.

Carbon Dioxide (CO_2)

Carbon dioxide is a compound gas. The primary elements are carbon and oxygen. Carbon dioxide is not an inert gas such as helium or argon; it has a feature that neither of these gases has—the ability to dissociate and recombine. It is this factor that permits more heat energy to be absorbed in the gas. It also uses the free oxygen in the arc area to superheat the weld metal transferring from the electrode to the work. Carbon dioxide has a wider arc plasma than argon but less than helium. Depending on the type of metal transfer used, the weld deposit cross section will show a medium-narrow, deeply penetrating weld. Care must be taken when using open arc CO_2 welding techniques so that plastic cold shuts, gas entrapment, and intrabead cracking do not occur. Such faults in the weld could be caused by the penetration characteristics and/or the globular transfer of metal across the arc. The use of welding grade CO_2 shielding gas and electrode wires with the proper deoxidizing elements will decrease the chances of such flows occurring in the weld.

Carbon dioxide has a tendency to cause a rather spattery, unstable arc when used for open arc transfer of metals. The spatter can be contained by maintaining an extremely close arc—by burying the arc in the work. The "buried arc" technique functions when the gun nozzle is very near the surface of the base metal. The arc is actually below the surface of the metal, hence the name "buried arc." The arc digs a hole within which almost all of the spatter is contained and used as filler metal. This technique is usually used when high-speed welding of heavy steel materials is required. Carbon dioxide also works well with the short-circuiting process of metal transfer to be discussed later.

SHOP TALK

Carbon dioxide arc welding is sometimes called MIG welding. Actually, since carbon dioxide is not a completely inert gas, it is more accurately called MAG welding (metal active gas). Although most auto body welding is done with carbon dioxide gas as the shield gas, the term MIG is used to describe all gas metal arc welding processes. In fact, many welders on the market can use carbon dioxide (a semi-active gas) or argon (an inert gas) by simply changing the gas cylinder and the regulator.

Argon/CO_2

For some applications of steel welding, welding grade CO_2 does not provide the arc characteristics needed for the job. This will usually manifest itself, where surface appearance is a factor, in the form of intolerable spatter in the weld area. In such cases a mixture of argon/CO_2 has usually eliminated the problem (Figure 2-13). Some welding authorities believe that the mixture should not exceed 25 percent CO_2. Others think that mixtures with up to 80 percent CO_2 are practical.

The reason for wanting to use as much CO_2 as possible in the mixtures is primarily cost. By using a cylinder of each type of gas, argon and CO_2, the mixture percentages may be varied by the use of flowmeters. This method precludes the possibility of gas separation such as might occur in premixed cylinders. Because premixed argon/CO_2 gas is sold at the price of pure argon, the cost factor also makes it more practical to mix the gases on the premises. The price of CO_2 is approximately 15 percent that of argon in most areas of the country.

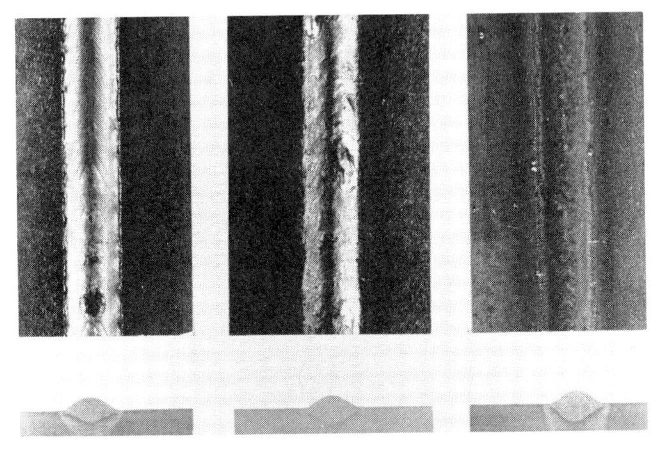

FIGURE 2-13 Weld penetration with various gas mixtures

Argon/CO_2 shielding gas mixtures are employed for welding high-strength steel, mild steel, low-alloy steel, and, in some cases, stainless steel.

TYPES OF METAL TRANSFER

There are a number of methods in MIG welding for transferring the filler, or welding wire, to the base metals.

Spray Type

Spray-type metal transfer is accomplished by the movement of a stream of tiny droplets across the arc from the electrode to the base metals (Figure 2-14). The arc has a characteristic sizzling sound or, in some cases, a buzzing sound. Spray-type transfer usually employs relatively high voltages and current.

Shielding gases that are used with this type of metal transfer include argon, helium, argon/helium mixtures, argon/oxygen mixtures, and helium/oxygen mixtures. The shielding gases noted are those in common use at this time.

Globular Type

As the name implies, globular-type metal transfer across the arc occurs in large, irregularly shaped drops or "globs" (Figure 2-15). This type of metal transfer is indicative of CO_2 shielded welding of steel. It also can exist because of low arc voltage, or low amperage, when using other shielding gases.

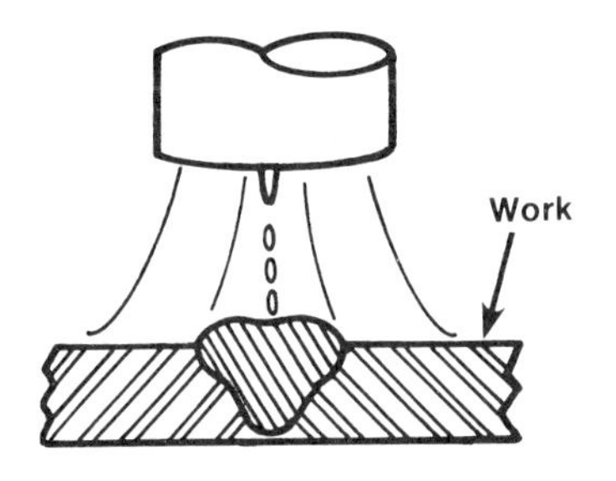

FIGURE 2-14 Spray-type metal transfer

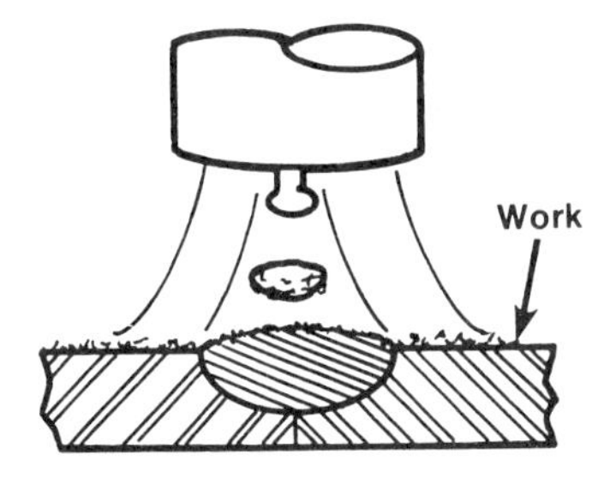

FIGURE 2-15 Globular-type metal transfer

Short-Circuiting Type

The short-circuiting method is unique in that no metal transfers across the arc. Instead, the metal transfer occurs when the electrode wire makes contact with the base metals. Welding of thin automobile sheet metal can cause strain, blow holes, and warped panels. To prevent these problems, it is necessary to limit the amount of heat near the weld. The short-circuit arc method uses very thin welding rods, a low current, and low voltage. With this technique, the amount of heat introduced into the panels is kept to a minimum and penetration of the base metal is quite shallow.

As shown in Figure 2-16, the end of the wire is melted by the heat of the arc and forms into a drop, which then contacts the base metal and creates a short circuit. When this happens, a large current flows through the metal and the shorted portion is torn away by the pinch force or burnback, which reestablishes the arc. That is, the bare wire electrode is fed continuously into the weld puddle at a controlled, constant rate, where it short circuits, and the arc goes out. When the arc is out, the puddle flattens and cools; but the wire continues to feed, shorting to the workpiece again. This heating and cooling occurs on an average of 100 times a second. The metal is transferred to the workpiece with each of these short circuits. Generally, if current is flowing through a cylindrically shaped fluid (in this case molten metal) or through an arc, the current is pulled toward the weld. This works as a constricting force in the direction of the center of the cylinder. This action is known as the pinch effect, and the size of the force is called the pinch force (Figure 2-17). In summary, the process works like this:

- At the weld site the wire undergoes a split-second sequence of short circuiting, burnback, and arcing (Figure 2-18).
- Each sequence produces a short arc transfer of a small drop of electrode metal from the tip of the wire to the weld puddle.
- A gas curtain or shield surrounds the wire electrode. This gas shield prevents contamination from the atmosphere and helps to stabilize the arc.
- The continuously fed electrode wire contacts the metal and sets up a short circuit, and resistance heats the wire and the weld site.
- As the heating continues, the wire begins to melt and thin out.
- Increasing resistance in the molten neck accelerates the heating in this area.
- The molten neck burns through, depositing a puddle on the workpiece and starting the arc.
- The arc tends to flatten the puddle and burn back the electrode.
- With the arc gap at its widest, it cools; this allows the wire feed to move the electrode closer to the work.
- The short end starts to heat up again, enough to further flatten the puddle but not enough

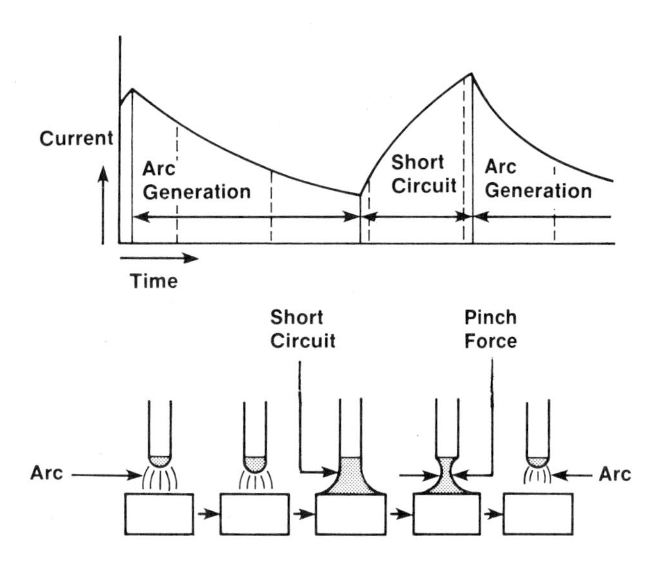

FIGURE 2-16 Short-circuiting arc method of metal transfer

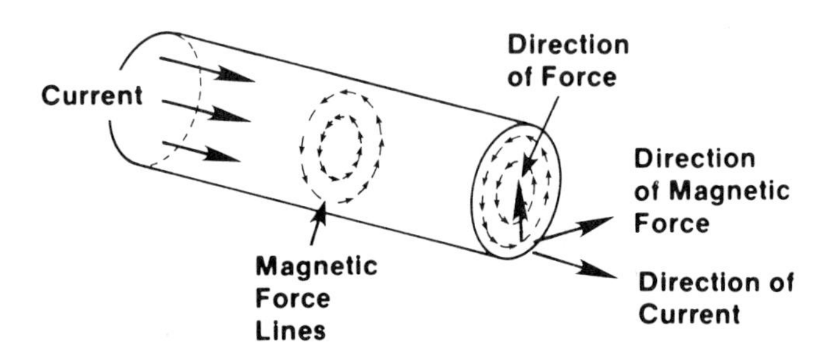

FIGURE 2-17 Typical pinch force and how it is formed

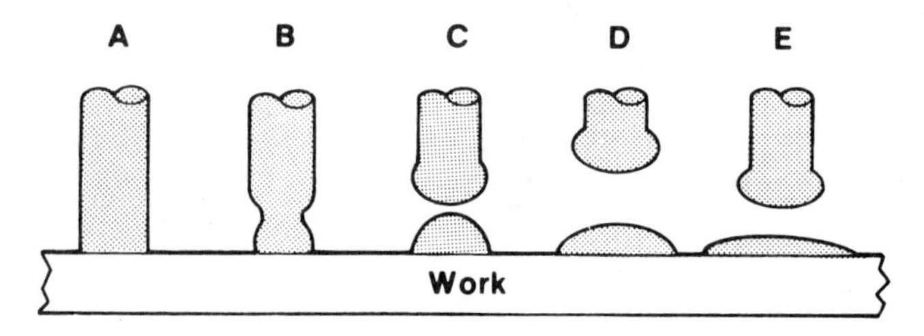

FIGURE 2-18 Typical action of the welding wire as it burns back from the work during the MIG welding process

to keep the electrode from recontacting the workpiece. This extinguishes the arc, reestablishes the short circuit, and restarts the process. The complete cycle occurs automatically at a frequency ranging from 50 to 200 cycles a second.

As already mentioned, carbon dioxide works well with this short-circuiting process. It produces the forceful arc needed during the arcing interval to displace the weld pool. Helium can also be used, but pure argon tends to be too sluggish to be effective. However, a mixture of 75 percent argon and 25 percent carbon dioxide produces an arc that is less harsh and a flatter, more pleasing weld profile. This gas mixture is preferred even though it is more costly.

The power supply, critical in the short-circuiting process, must have a constant potential output and sufficient inductance to slow the time rate of current increase during the short-circuit interval. The inductance can be in the form of an external choke. Also, the power supply must sustain an arc long enough to premelt the wire tip in anticipation of the transfer at recontact with the weld pool. Too much inductance causes the system to become sluggish. The short-circuiting rate decreases enough to make the process difficult to use. Too little inductance causes spatter due to high current surges.

MIG WELDING EQUIPMENT

MIG welding processes can be performed as semiautomatic (SA), machine (ME), or automatic (AU) welding (Table 2-2). However, most MIG welding that is used in and designed for automobile repair work is considered semiautomatic. This means that the machine's operation is automatic, but the gun is hand controlled. Before starting to weld, the operator sets the following:

- Voltage for the arc
- Wire speed
- Shielding gas flow rate

Once the power button is pressed, the operator has complete freedom to concentrate entirely on the weld site, the molten puddle, and the welding technique.

Regardless of the brand of MIG equipment used, it will comprise the following basic components (Figure 2-19).

- Power supply
- Electrode (wire) feed unit
- Electrode (wire) supply of a specified type and diameter
- Supply of shielding gas with a flow regulator to protect the molten weld pool from contamination
- Control circuit
- Work cable and clamp assembly
- Welding gun and cable assembly that is held by the welder to direct the wire to the weld area

POWER SUPPLIES FOR SHORT-CIRCUITING TRANSFER

Although this GMAW power supply is said to have a constant potential, it is not perfectly constant. There is a slight decrease in voltage as the amperage increases. This rate of decrease, known as *slope,* is expressed as the voltage decrease per 100-ampere increase; for example, 10V/100A. Some power supplies for short-circuiting welding are equipped to allow changes in the slope by steps or continuous adjustment.

For auto body applications, these power sources are usually self-contained, meaning the power source, wire feed equipment, and wire are all in one unit (Figure 2-20). The controls found on this type of power source are a voltage adjustment and wire feed speed adjustment. Some optional controls available (Figure 2-21) are a burnback control and probe control.

Voltage adjustment and wire feed speed must be set according to the diameter of the wire being used. It should be noted that when setting these

TABLE 2-2: METHODS OF PERFORMING GMAW WELDING PROCESSES

Function	Semiautomatic (SA)	Machine (ME)	Automatic (AU)
Maintaining the arc	Machine	Machine	Machine
Feeding the filler metal	Machine	Machine	Machine
Provide the joint travel	Welder	Machine	Machine
Provide the joint guidance	Welder	Welder	Machine

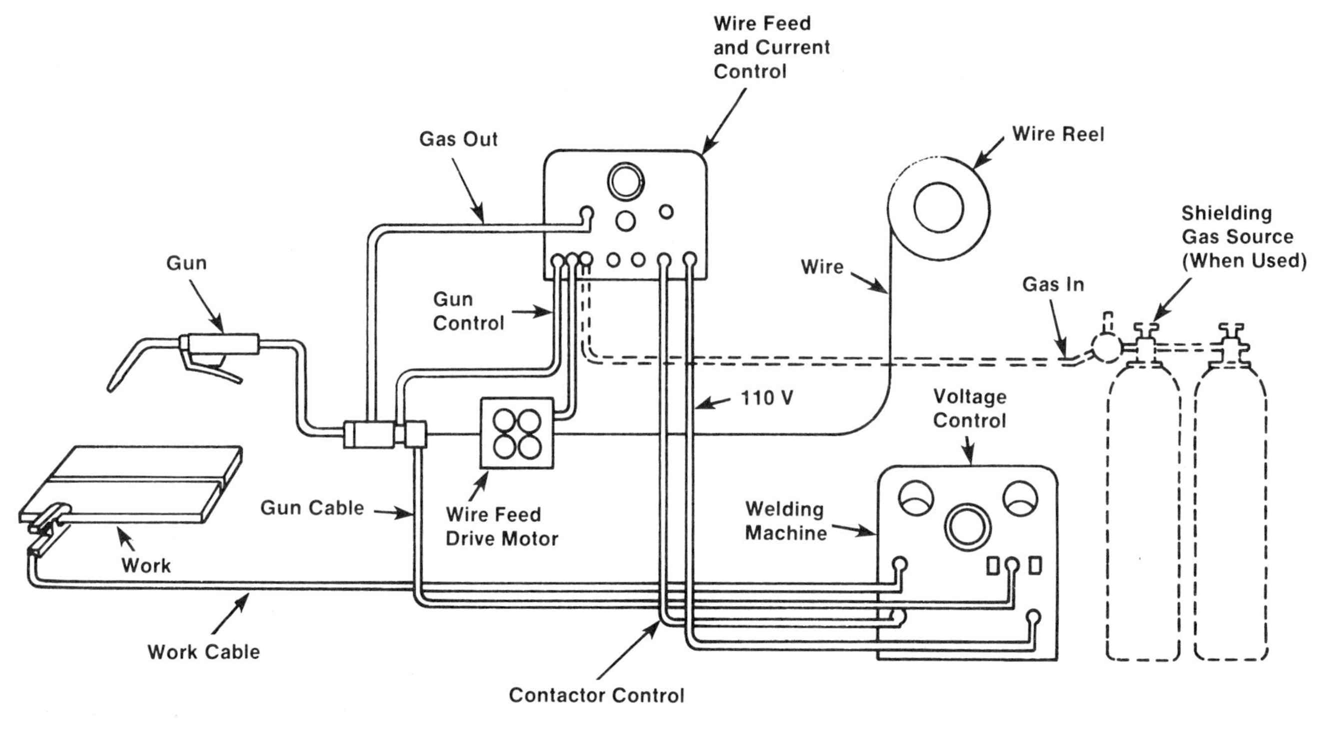

FIGURE 2-19 Equipment setup for GMAW

FIGURE 2-20 A self-contained MIG unit

FIGURE 2-21 Typical MIG welder control panel

parameters, manufacturers' recommendations should be followed. When rough parameters are selected, change only one variable at a time until the machine is fine tuned for an optimum welding condition. MIG welders can be tuned using both visual and audio signals.

The pulse control allows continuous seam welding with less chance of burn-through or distortion. This is accomplished by starting and stopping the wire for preset times without releasing the trigger. The weld "on" and "off" times can be set for the operator's preference and metal thickness. The burnback control (found on most MIG welders) provides an adjustable burnback of the electrode to prevent it from sticking in the puddle at the end of a weld.

In MIG welding, the polarity of the power source is important in determining the amount of penetration to the workpiece. DC power sources used for MIG welding typically use DC reverse polarity. This means the wire is positive and the workpiece is negative. Weld penetration is greatest using this connection.

In straight polarity, the wire is negative and the workpiece positive. This puts more heat in the wire, providing less penetration. The disadvantage of using straight polarity would be a high rope-like bead, requiring more grinding. However, it can be used effectively with flux-cored wire because of this wire's greater penetration. If a machine is used for both solid and flux-cored wires, it must have simple polarity switching capabilities. Figure 2-22 shows the positive and negative terminals inside one machine. In this case, the side cover can be lifted and the positive and negative connections can be changed easily to give straight or reverse polarity.

SLOPE

Although the slope is sometimes called the volt-ampere curve, it is really a straight line. Since the slope is a straight line, it can be determined by finding two points (Figure 2-23). The first point is referred to as the *open circuit voltage.* It is the set voltage as read from the voltmeter when the gun switch is activated but no welding is being done. The second point is the voltage and current as read during a weld. While the voltage is not adjusted during the test, the amperage can be changed. Of course, to make calculations easier, the wire feed should be adjusted so that the current is an even 100 increment, such as 100 amperes, 200 amperes, 300 amperes, and so on. After determining the voltage difference between the first and second readings, determine the slope by simply dividing that difference by 1 for 100 amperes, 2 for 200 amperes, and so forth.

Circuit resistance (resulting from a number of factors, including poor connections, long leads, or a dirty contact tube) affects the machine slope. A

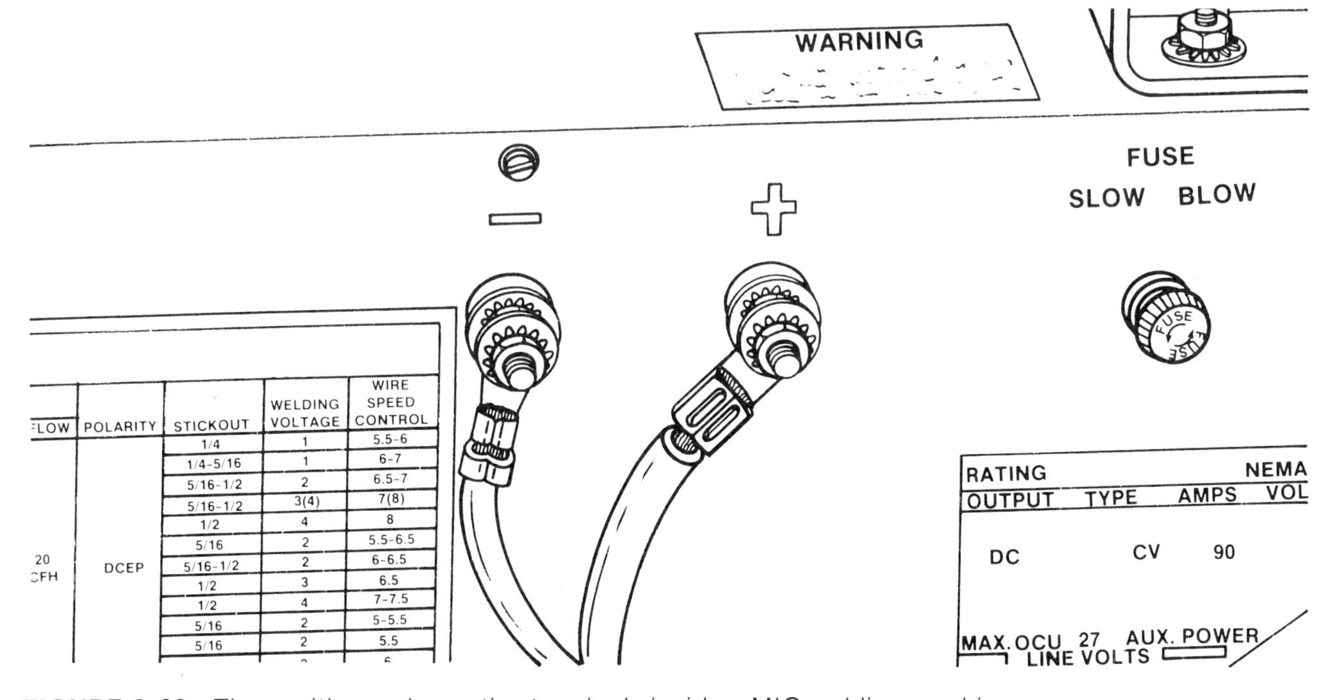

FIGURE 2-22 The positive and negative terminals inside a MIG welding machine

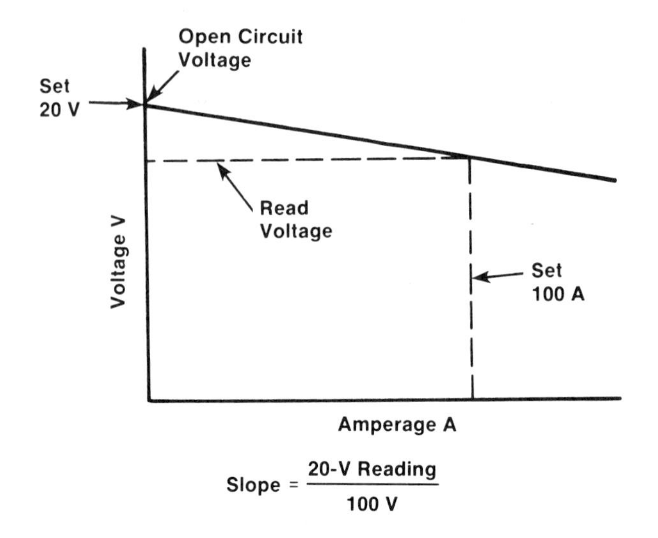

FIGURE 2-23 Calculating the slope

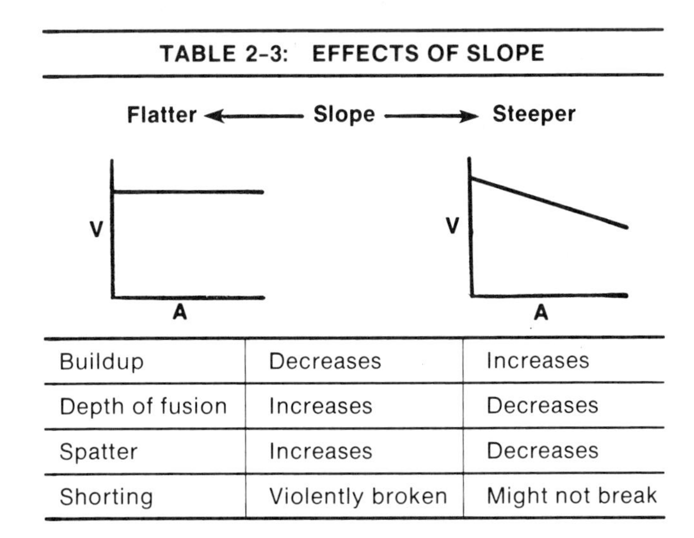

TABLE 2-3: EFFECTS OF SLOPE

	Flatter ← Slope	→ Steeper
Buildup	Decreases	Increases
Depth of fusion	Increases	Decreases
Spatter	Increases	Decreases
Shorting	Violently broken	Might not break

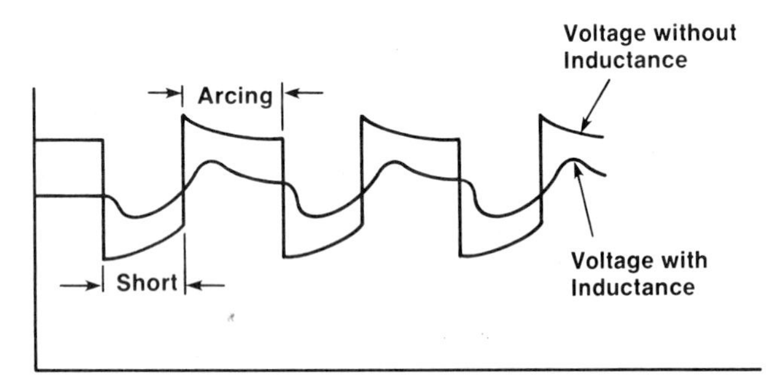

FIGURE 2-24 Effect of inductance on the voltage pattern

higher resistance causes a steeper slope. In short-circuiting machines, higher inductance also increases the slope. The inductance increase retards the current's rate of change when switching between short circuiting and arcing (Figure 2-24). As a result, slope and inductance become synonymous in the context of this explanation. A steep slope has both decreased short-circuit current and a diminished pinch effect. For a flat slope, both the short-circuit current and the pinch effect are increased.

The short-circuiting metal transfer mode is affected by the machine slope more than any other mode. The high current and pinch effect produced by a flat slope cause a violent short and arc restart cycle. As a result, there is increased spatter. If there is a steep slope and there is too little current and pinch effect, the short circuit is not broken as the wire freezes in the puddle and piles up on the work (Table 2-3).

The slope must be adjusted for proper spatter-free metal transfer to occur. Some machines have adjustable slopes. Changing the slope on machines that have a fixed slope is not as easy, but it can be accomplished by varying the contact tube-to-work distance to change the circuit resistance. The filler wire is not large enough to carry the welding current and heats up due to its resistance to the current flow. As the tube-to-work distance increases, the circuit resistance increases and the slope steepens. For smoother short circuiting with less spatter, this distance can be increased or decreased to obtain a proper slope.

CURRENT RATING AND DUTY CYCLE

In the past, the MIG welding machines used in automobile repair work were typically 220-volt machines. Recently, however, a number of 115-volt machines have been produced. When these machines first appeared on the market they were generally unsuited for the tasks they were to perform. Now most of these machines sold today are quality equipment. In fact, the Inter-Industry Conference on Auto Repair (I-CAR) recently reversed its opposition to this type of machine. This indicates the machines are definitely getting better. Nevertheless, it should be noted that these units still have their inherent limitations.

To comprehend the limits of these machines, it is important to understand what machine capacity is. For every 0.001 inch of metal thickness, one ampere of welding current is needed (Table 2-4). Thus, the current rating of the machine basically determines the limit of metal thickness on which the machine can be used.

By using flux-cored wires or by using straight CO_2 shielded gas instead of a mixed gas, penetration is increased enough to weld metal that is one or two gauges thicker than the specified machine capacity.

TABLE 2-4: CURRENT REQUIREMENTS FOR SOLID WIRE/ARGON CO_2 MIXED GAS

Gauge	Thickness	Required Amperage
24	0.024	25
22	0.029	30
20	0.035	35
18	0.047	47
16	0.059	59
14	0.074	74
12	0.104	104
11	0.119	119

Some general guidelines for determining whether a 115-volt welder is suited for a particular application are as follows:

- The 90- to 100-amp machine can weld 14- to 24-gauge with 0.023 solid wire and an argon/CO_2 mixed shielding gas.
- If straight CO_2 shielding gas is used with 0.023 solid wire, the 90- to 100-amp machine can be used to weld 11- and 12-gauge metal.
- If the 0.030-diameter self-shielded flux-cored wire is used, the 90- to 100-amp machine can weld 20-gauge to 3/16-inch-thick material. The joint must be properly prepared and multiple passes will be required on material thicker than 11 gauge.
- The machines rated at 100- to 120-amps can weld 18-gauge to 5/16-inch material with 0.035-diameter flux-cored wire. Again, proper joint preparation is required and multiple passes must be used on material thicker than 11 gauge.
- The 70- to 75-amp machine is limited to 14 gauge if the solid wire and mixed gas are used; 12 gauge if solid wire and straight CO_2 gas are used; and 1/8 inch to 3/16 inch if the 0.030 flux-cored wire is used (as well as proper joint preparation and multiple passes).

Obviously, the gauges of the metals commonly welded in the shop will dictate which machine is most suitable. Note that plug welds on 14 or 16 gauge will require a little more amperage than seam-type welds. Therefore, a little extra capacity in the machine is needed to assure quality full-penetration when plug welding is to be performed. If only exterior sheet metal welding is to be done, then the 70- to 75-amp machine is probably enough. However, if structural-type welding is to be done, then a 100-amp or possibly even a 120-amp machine is required. Ideally, a 115-volt welder can be set up for small-diameter solid or cored wire to complement a 220-volt machine.

The duty cycle is probably one of the most cost-significant factors. The duty cycle of a particular machine basically tells how many minutes out of 10 a welding machine can run without overheating. Figure 2-25 shows the duty cycle chart for a 115-volt welder. When the heat setting is 30 amps, this machine has a 100 percent duty cycle, so it can weld 10 minutes out of 10 minutes. Yet the machine only has a 20 percent duty cycle when the heat setting is 105 amps, and it is only safe to weld 2 minutes out of 10. Thus, the higher the duty cycle the better the machine.

All machines should have a duty cycle displayed either in the owner's manual or on the machine. Some machines also display a welding guide (Figure 2-26). A typical chart identifies the type of base

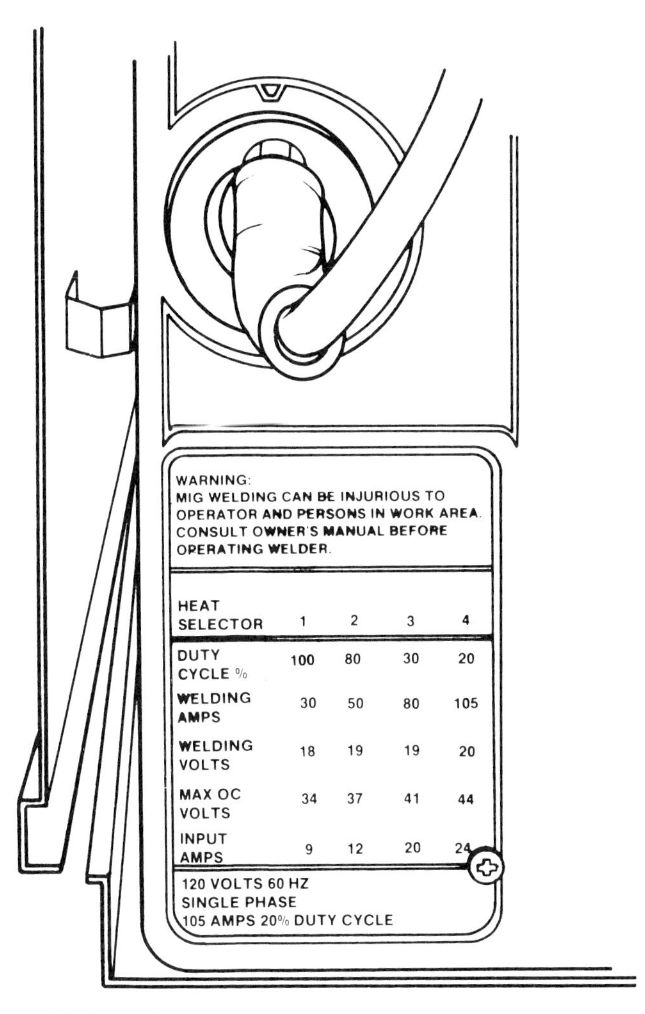

FIGURE 2-25 Duty cycle chart

WELDING GUIDE
Settings are approximate. Adjust as required.

Material	Thickness	Process	Wire		Gas		Polarity	Stickout	Welding Voltage	Wire Speed Control
			Class	Size	Type	Flow				
	24 ga							1/4	1	5.5-6
	18 ga							1/4-5/16	1	6-7
	16 ga			0.024				5/16-1/2	2	6.5-7
	10 ga							5/16-1/2	3 (4)	7 (8)
	3/16"							1/2	4	8
	18 ga				CO_2			5/16	2	5.5-6.5
Carbon Steel	16 ga	GMAW	ER-70S-6 (HB 28)		or	20 CFH	DCEP	5/16-1/2	2	6-6.5
	10 ga			0.030	C_{25}			1/2	3	6.5
	3/16"							1/2	4	7-7.5
	18 ga							5/16	2	5-5.5
	16 ga			0.035				5/16	2	5.5
	10 ga							1/2	3	6
	3/16"							1/2	4	6.5
Stainless Steel	10 ga		ER-308L 308L Stainless	0.030	C_{25}			1/2	4	7.5
	18 ga							1/2-3/4	2	5
Carbon Steel	16 ga	FCAW	E-717-11 Fabshield 21B	0.045	None		DCEN	1/2-3/4	2	5.5
	10 ga							3/4	3	6
	3/16"							3/4	4	6

24 ga = 0.022", 18 ga = 3/64", 16 ga = 1/16", 10 ga = 1/8"
CO_2 = Carbon dioxide
CO_{25} = 25" Carbon dioxide + 75% Argon
DCEP = DC Volts Wire Positive
DCEN = DC Volts Wire Negative

FIGURE 2-26 Typical welding guide

metal, thickness, and the recommended machine settings for solid wires and small-diameter flux-cored wires. It is a great convenience when setting the machine.

ELECTRODE FEED UNIT

The function of the electrode feeder is to provide a steady and reliable supply of wire to the weld. The motor used in a feed unit is usually a DC type on which the supply rate can be adjusted continuously over the desired range. Even slight changes in the wire supply rate have distinct effects on the weld.

Push-Type Feed System

With this system, the wire rollers are clamped securely against the wire. This pressure, which can be adjusted, provides the necessary friction to push the wire through the conduit to the weld. There is a groove in the roller that aids in alignment and lessens the chance of slippage. Most manufacturers provide rollers with V-shaped or U-shaped grooves (Figure 2-27). V-grooved rollers are best suited for hard wires; they might distort the surface of the softer wires. Soft wires, such as aluminum, are best used with smooth U-grooves. The knurled U-groove

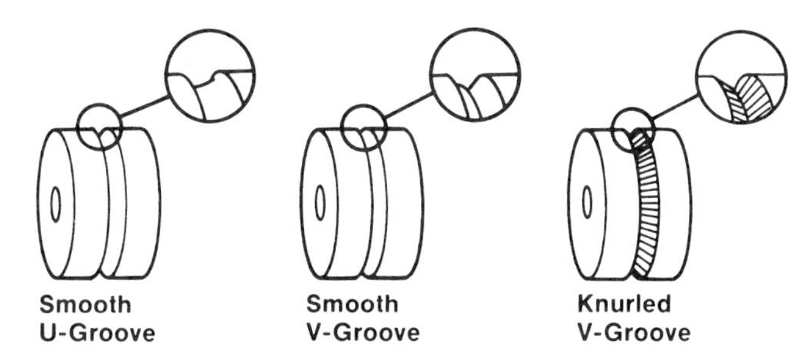

FIGURE 2-27 Feed rollers

rollers have a series of ridges cut into the groove so that the rollers can grip large-diameter wires and push them along more easily. Soft wires should not be used with knurled rollers. The use of the correct size grooves in the rollers is also important.

In the push-type system, the electrode must have enough strength to be pushed through the conduit without kinking. Mild steel and stainless steel can be pushed 15 to 20 feet; aluminum is much harder to push 10 feet.

Pull-Type Feed System

The pull-type system is a variation of the push-type electrode wire feeder. The difference is in the size and location of the drive rollers. In the pull-type system, a smaller but higher speed motor is located in the gun to pull the wire through the conduit. The advantage of this system is its ability to move even soft wire over great distances. However, there are some disadvantages:

- The gun is heavier.
- It is more difficult to use.
- Rethreading the wire takes more time.
- The operating life of the motor is shorter.

Push-Pull-Type Feed System

In this synchronized system, feed motors are located at both ends of the electrode conduit. If a feed roller is periodically installed into the electrode conduit, this system can be used to move any type of wire over long distances. This system is more expensive than the others, but the push-pull-type system also has a number of advantages:

- It moves wire over longer distances.
- It can be rethreaded faster.
- It has prolonged motor life due to the reduced load.

Linear Electrode Feed System

A different method is used to move the wire and change the feed speed in a linear electrode feed system. The linear electrode feed sytem does not use the conventional-type rollers that clamp against the wire. Neither does it use the system of gears used in standard systems between the water and rollers to provide roller speed within the desired range.

In the linear feed system, the wire is fed through the hollow armature shaft of a small motor. The rollers are attached so that they move around the wire. By changing the roller pitch (angle), the wire speed can be changed without changing the motor speed. This system works in the same way that changing the coarseness of the threads on a screw affects the rate that the screw will move through a spinning nut.

Although it has one notable disadvantage (the wire can become twisted as it is moved through the feeder), this system has no bulky system of gears. This feature offers some inherent benefits, including a reduction in weight, size, and wasted power. The motor operates more efficiently since it operates at a constant high speed. Because the system is smaller in size, it can be housed in the gun or within an enclosure in the cable. An extended operating range can be provided by synchronizing several linear wire feeders.

Spool Gun

The compact spool gun consists of a small drive system and a wire supply (Figure 2-28). It is a self-

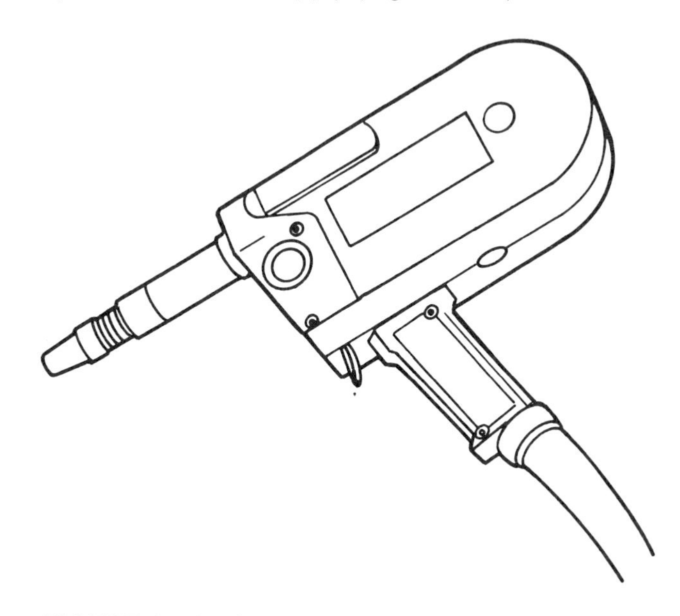

FIGURE 2-28 Spool gun

contained system that allows the welder to move freely around a job with only a power lead and shielding gas hose to manage. The motor and feed rollers are located inside the gun just behind the nozzle and contact tube. A small spool of welding wire is located just behind the feed rollers. Since the distance the wire must be moved is so short, very soft wire can be used. The major control system is usually mounted on the welder.

ELECTRODE CONDUIT

The electrode conduit guides the welding wire from the feed rollers to the gun. It can be encased in a lead that also contains the shielding gas. The power cable and gun switch circuit wires are contained in the conduit. It is made of a tightly wound coil having the needed flexibility and strength. It can also have a nylon or nonstick-coating tube to protect soft, easily scratched metals as they are fed.

The conduit must be firmly attached to both ends of the lead if it is not an integral part of the lead. Otherwise, misalignment can result, which causes additional drag or makes the wire jam completely. If the conduit is too long for the lead, it should be cut off and filed smooth. This will prevent the lead from bending and twisting inside the conduit, which could cause feed problems. If the conduit is not long enough to extend through the lead casing to make a connection, it can be drawn out by tightly coiling the lead. This forces the conduit out so that it can be connected.

WELDING GUN

The welding gun is used by the welder to produce the weld. It has a trigger switch that controls the starting and stopping of the weld cycle. The power cable, electrode conduit, and shielding gas hose are all attached to the welding gun. A contact tube on the gun transfers welding current to the electrode moving through the gun. The gun's gas nozzle directs the shielding gas into the weld.

ELECTRODE

Generally speaking, automotive welding requires fine-diameter welding wire, from 0.015 through 0.030 inch. The 0.015-inch wire is becoming increasingly popular; once a specialty wire, it is now stocked by most manufacturers. These small-diameter wires can be used at low currents and voltages, thus greatly reducing heat input to the base material.

FLUX-CORED ARC WELDING

Flux-cored arc welding (FCAW) is an electric arc welding process that uses a tubular wire with flux inside. With the development of 0.030 self-shielded flux-cored wire, the flux-cored welding process has proven to be valuable for work on high-strength steel (coated or uncoated). The FCAW process uses the same type of constant potential power source as GMAW. It also uses the electrode feed system, contact tube, electrode conduit, welding gun, and many other pieces of equipment that are used in GMAW. Nevertheless, the process itself differs greatly from GMAW.

There is no external shielding gas in FCAW. As the flux within the wire melts in the heat of the arc, the created gases shield the weld puddle, stabilize the arc, help to control penetration, and reduce porosity. The melted flux also mixes with the impurities on the metal surface and brings them to the top of the weld where they solidify as slag. The slag can then be chipped or brushed away.

Two very important advantages of the FCAW process over GMAW are its ability to tolerate surface impurities (thus, it requires less precleaning) and to stabilize the arc. Other beneficial characteristics of the process include the following:

- High deposition rate
- Efficient electrode metal use
- Requires little edge preparation
- Welds in any position
- Welds a wide range of metal thicknesses with one size of electrode
- Produces high-quality welds
- Weld puddle is easily controlled and its surface appearance is smooth and uniform even with minimal operator skill.
- Produces a weld with less porosity than GMAW when welding galvanized steels

SHOP TALK

If the nozzle is removed when using self-shielded wires, visibility is improved.

While the FCAW process has a number of advantages over GMAW, it has the following drawbacks:

- FCAW wires are more expensive than GMAW hard wires. However, the cost is quickly recovered through higher productivity.

- The 0.030 FCAW wire contains toxic fluoride compounds.
- The flux from the wire changes to slag as it cools. Until it does cool, the slag is sharp and hot and should be considered an eye and skin hazard. Once it cools, this slag must be removed prior to the application of fillers, seam sealouts, primers, or paint.
- Spatter is worse when using flux-cored wires. Use nozzle gel and keep the nozzle scraped clean. Spatter buildup in the gun nozzle can jam the wire in the contact tip; it can also fall off during welding and mix with the molten puddle, diminishing the quality of the weld.
- Excessive tension on the drive rollers or using the incorrect style of drive rollers can collapse the tubular wire. Check the owner's manual for flux-cored wire requirements.
- The open butt joint is difficult to weld with even penetration.
- Only ferrous metals can be welded.

FLUX

The flux is poured into the electrode, which has already been formed into a U-shaped channel (Figure 2-29). The thin sheet of U-shaped metal is then passed through a series of dies to size it and squeeze it shut, further compacting the flux (Figure 2-30).

Basically, the fluxes used are rutile or lime based. The purpose of the fluxes is to provide deoxiders, slag formers, arc stabilizers, alloying elements, and a shielding gas. Rutile-based wires produce a fine drop transfer, a relatively low level of fume, and a slag that is easy to remove. Lime-based wires produce a more globular transfer, more spatter, more fume, and a slag that is harder to remove.

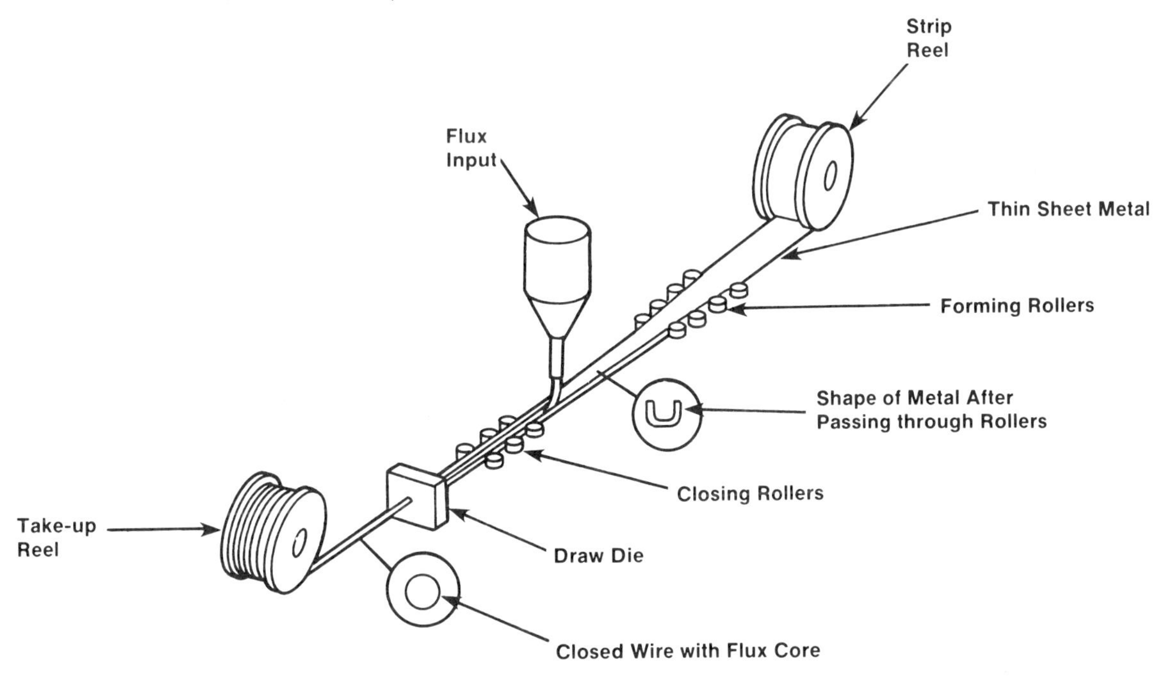

FIGURE 2-29 Pouring the flux into the U-shaped channel

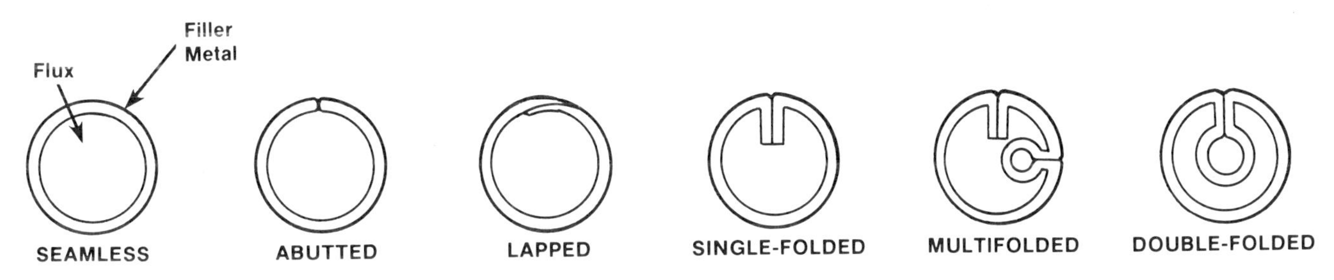

FIGURE 2-30 While the abutted shape is the most common by far, there are other variations that allow a higher percentage of filler metal-to-flux without losing the necessary flexibility.

However, when it is necessary to deposit very tough weld metal, these characteristics are tolerated. Lime-based wires are also used for welding materials having a low tolerance for hydrogen.

Be careful to use the cored wires with the recommended gases. In fact, it is usually best not to use gas at all with self-shielded wires. Defective welds can result from using a shielding gas with a self-shielding flux-cored wire, because the gas prevents the proper disintegration of much of the deoxidizers. The deoxidizers, unable to disintegrate, are transferred across the arc to the weld. When deoxidizers are present in high concentrations, they can produce slag that gets trapped in the welds, resulting in defects. In lower concentrations, only brittleness results. In both cases, the possibility of weld failure is increased.

While some fluxes can be used on both single- and multiple-pass welds, others are limited to single-pass welds only. If a single-pass welding electrode is used for multipass welds, the result could be an excessive amount of manganese. When making large, single-pass welds, the manganese is necessary to retain strength. However, the lower dilution associated with multipass techniques causes the manganese to strengthen the weld metal too much and reduce its ductility. This problem can be controlled in some instances by making small welds that deeply penetrate the base metal.

Flux composition can also place a limit on welding position capabilities. Although some electrodes can be used in all positions, others are limited in their out-of-position welding ability. The out-of-position capabilities of some electrodes are improved by the use of the correct shielding gas or gas mixture.

In Table 2-5, the welding features of the seven classifications of fluxes are detailed. The letter "G" indicates an unspecified classification. This means that the American Welding Society has not classified the electrode. Composition, current, and self-shielding/nonshielding characteristics are mutually agreed upon by the manufacturer and user. The tensile strength alone is specified.

SETUP AND USE

Like GMAW, FCAW has certain variables that must be carefully controlled to deliver quality welds. These include:

- Welding voltage (heat)
- Welding current (wire speed)
- Travel speed

TABLE 2-5: WELDING CHARACTERISTICS OF SEVEN FLUX CLASSIFICATIONS

Classifications	Characteristics	Shielding Gas
T-1	Requires clean surfaces and produces little spatter. It can be used for single- and multiple-pass welds in the flat (1G and 1F) and horizontal (2F) positions.	Carbon dioxide (CO_2)
T-2	Requires clean surfaces and produces little spatter. It can be used for single-pass welds in the flat (1G and 1F) and horizontal (2F) positions only.	Carbon dioxide (CO_2)
T-3	Used on thin-gauge steel for single-pass welds in the flat (1G and 1F) and horizontal (2F) positions only.	None
T-4	Low penetration and moderate tendency to crack for single- and multiple-pass welds in the flat (1G and 1F) and horizontal (2F) positions.	None
T-5	Low penetration and a thin, easily removed slag, used for single- and multiple-pass welds in the flat (1G and 1F) position only.	With or without carbon dioxide (CO_2)
T-6	Similar to T-5 without externally applied shielding gas.	None
T-G	The composition and classification of this electrode is not given in the preceding classes. It may be used for single- or multiple-pass welds.	With or without shielding

- Gun angle
- Contact tip height (electrical stickout)

Optimum voltage and current value for 0.030 flux-cored wire is approximately 100 amps at 16 volts. With the variety of machines in use today, it is best to check the owner's manual for setting recommendations. Settings for 0.030 hard wire are similar to settings for 0.030 flux-cored wire. Remember that voltage and current are determined primarily by the thickness of the base metal. Secondary considerations are the type of joint, the position in which the welding is to be done, and the skill of the operator.

MIG WELDING METHODS

Match the welding power unit in the MIG machine to the available input voltage, following the procedure outlined on the machine or in the owner's manual (Figure 2-31). This should always be done before attempting to hook up the equipment.

Handle the shielding gas with care. The cylinder might be pressurized to more than 2000 psi. Chain or strap it to a support sturdy enough to hold it securely to the MIG machine's running gear (Figure 2-32), (if so equipped), or to a wall, post, or the like. Standing to one side of the cylinder, quickly crack the valve to blow out any dirt before the flowmeter regulator is attached. Install the regulator, making sure to observe the recommended safety precautions (Figure 2-33). Attach the correct hose from the regulator to the "gas-in" connection on the electrode feed unit or machine.

Install the electrode reel on the holder and secure it (Figure 2-34). Check the roller size to ensure that it matches the wire size (Figure 2-35). The conduit liner size should be checked to be sure that it is compatible with the wire size. Connect the conduit to the feed unit. The conduit or an extension should be aligned with the groove in the roller and set as close to the roller as possible without touching (Figure 2-36). Misalignment at this point can contribute to bird nesting (Figure 2-37).

By attaching the clamp to clean metal on the vehicle (Figure 2-38) near the weld site, the welding circuit from the machine to the work and back to the machine is completed. This clamp is not referred to as a ground cable or ground clamp. The ground connection is for safety purposes and is usually made from the machine's case to the building ground through the third wire in the electric input

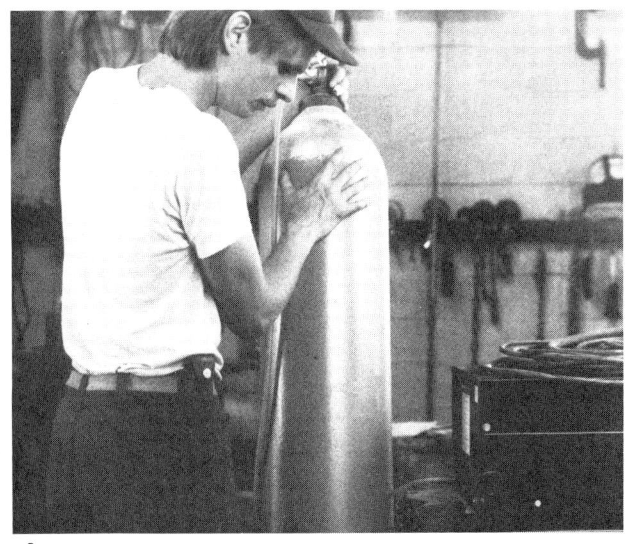

A

B

FIGURE 2-32 (A) Install the shielding gas cylinder with care and (B) chain or strap it in place.

FIGURE 2-31 Check the owner's manual before hooking up the equipment.

FIGURE 2-33 Installing the regulator on the cylinder

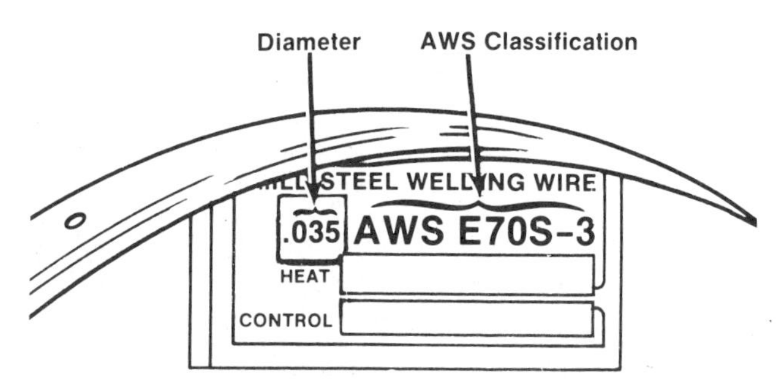

FIGURE 2-34 When installing the reel, be sure the wire is the correct type and size.

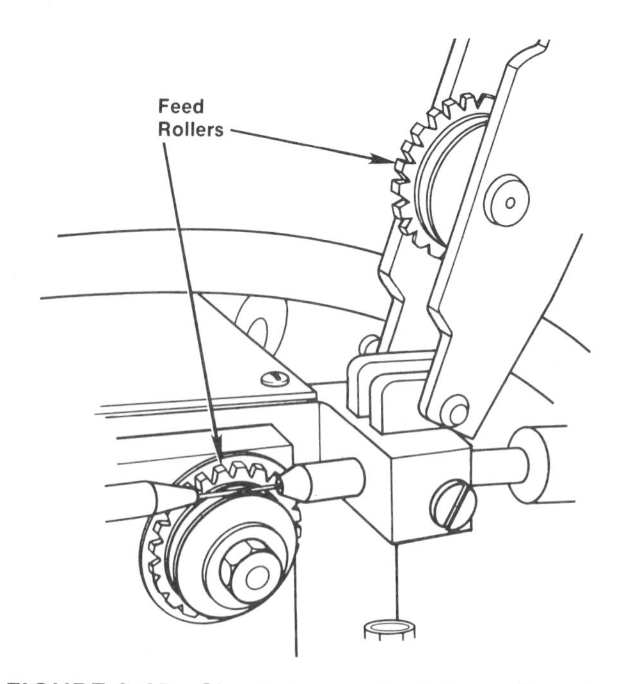

FIGURE 2-35 Check to see that the roller size matches the wire size.

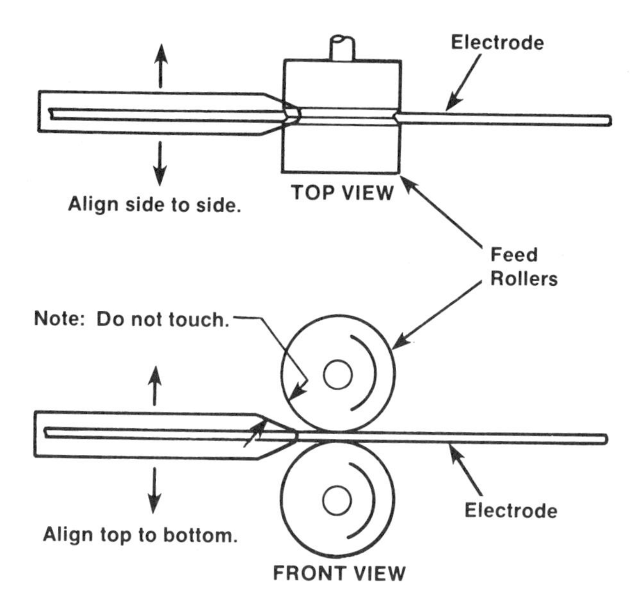

FIGURE 2-36 Feed roller and conduit alignment

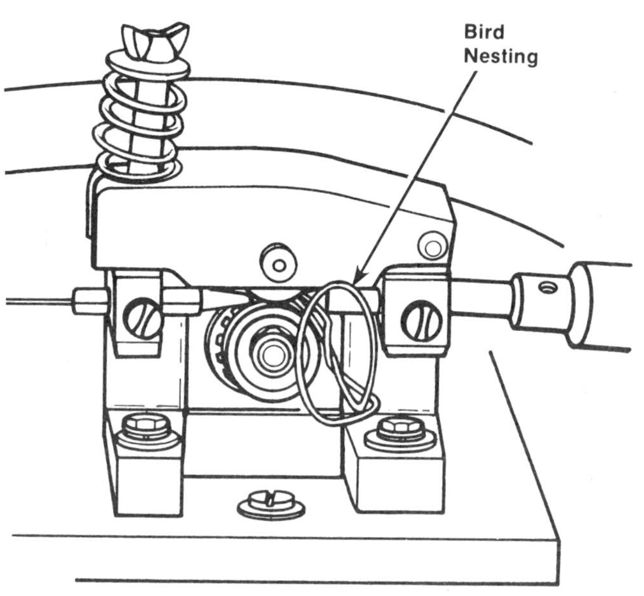

FIGURE 2-37 Bird nesting can result from misalignment.

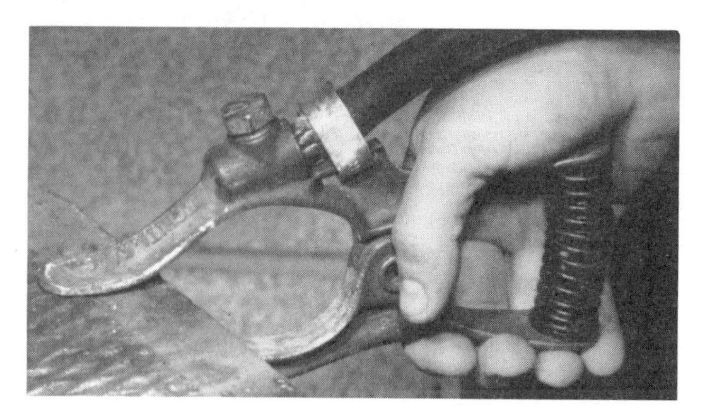

FIGURE 2-38 Attaching the clamp to clean metal

cable. Attach the shielding "gas-out" side of the solenoid to the gun lead. If a separate splice is required from the gun switch circuit to the feed unit, it should be connected at this time. Be sure the welding contactor circuit is connected from the feed unit to the power source. The welding gun should be permanently attached to the main lead cable and conduit (Figure 2-39). There should be a gas diffuser attached to the end of the conduit liner to ensure proper alignment. Install a contact tube of the correct size to match the electrode wire size (Figure 2-40). Attach a shielding gas nozzle also.

Recheck all fittings and connections for tightness. Loose fittings can leak; loose connections can cause added resistance, thus reducing the welding efficiency.

Next, the GMAW wire must be threaded and the wire feeder adjusted. Consult the owner's manual to see that the unit is assembled correctly according to the manufacturer's specifications. Turn on the power and check the gun switch circuit by depressing the switch. The power source relays, feed relays, gas solenoid, and feed motor should all activate.

Cut the end of the electrode wire free and hold it tightly so that it does not unwind. The wire has a natural curve that is known as its cast. The cast helps the wire make a good electrical contact as it passes through the contact tube (Figure 2-41). However, the cast can be a problem when threading the system. To make threading easier, straighten about 12 inches of the end of the wire and remove any kinks by cutting.

Separate the wire feed rollers and push the wire first through the guides, then between the rollers and into the conduit liner (Figure 2-42). Reset the rollers so there is a slight amount of compression on the wire (Figure 2-43). Set the wire feed speed con-

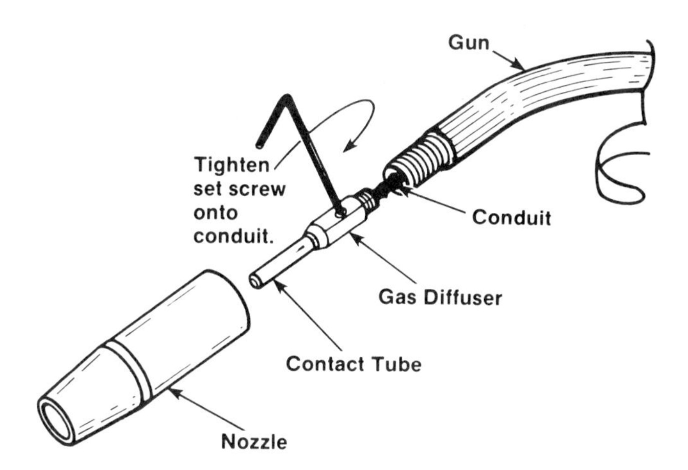

FIGURE 2-39 The welding gun is attached to the main lead cable and conduit.

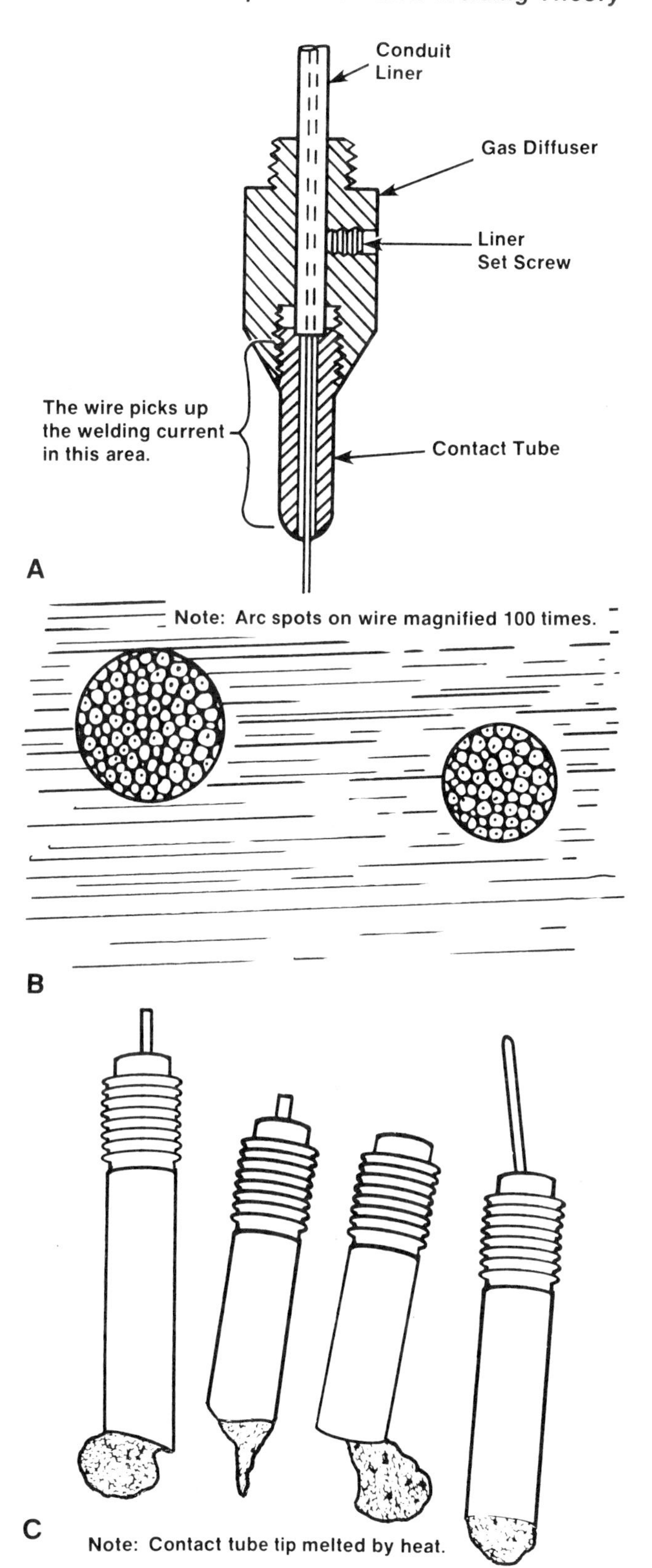

FIGURE 2-40 (A) The contact tube must be the correct size. Too small a tube will cause the wire to stick. (B) Too large a tube can cause arcing to occur between the wire and tube. (C) Heat from arcing can damage the tube.

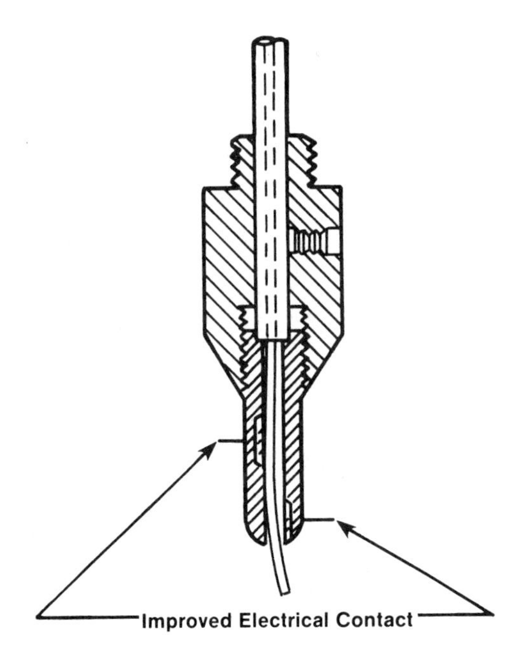

FIGURE 2-41 The cast helps the wire make good electrical contact.

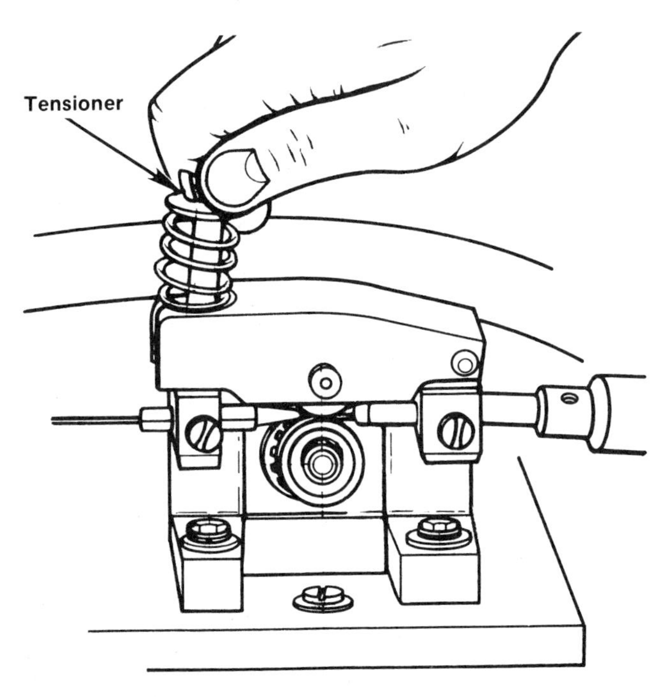

FIGURE 2-43 Adjust the wire feed tensioner.

trol to a slow speed. Hold the welding gun so that the electrode conduit and cable are as straight as possible.

Press the gun switch. The wire should start feeding into the liner. Watch to make certain that the wire feeds smoothly and release the gun switch as soon as the end comes through the contact tube.

If the wire stops feeding before it reaches the end of the contact tube, stop and check the system. If no obvious problem can be found, mark the wire with tape and remove it from the gun. It then can be held next to the system to determine the location of the problem.

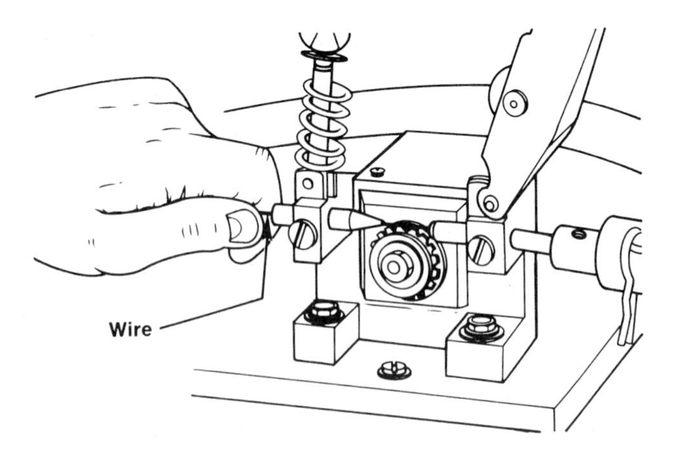

FIGURE 2-42 Push the wire through the guides by hand.

Adjust the drive rollers so that they apply just enough pressure on the wire to pull it off the spool and through the gun/cable assembly (Figure 2-44). The tension must be such that the wire will slip at the rollers when it is stopped at the nozzle but is tight enough to withstand a 30- to 40-degree deflection. If too much pressure is applied, the wire will be deformed, creating a spiral effect through the liner and an erratic feed.

Stopping the wire at the tip with excessive pressure will also cause the wire to bird-nest between the rollers and cable entrance. The tension on the spool spindle should also be set so that the wire can be pulled off easily yet still be tight enough to stop the spool from freewheeling when the trigger is released (Figure 2-45).

Proper handling of the welding equipment is essential to successful welding. When tuning the MIG welder for a welding job, the operator has to deal with a number of variables: input voltage to the welding equipment, welding current, arc voltage, tip-to-base metal distance, torch angle, welding direction, shield gas flow volume, welding speed, and wire speed. Most manufacturers of MIG welders provide tables that show the variable control parameters that apply to their machines.

WELDING CURRENT

The welding current affects the base metal penetration depth (Figure 2-46), the speed at which

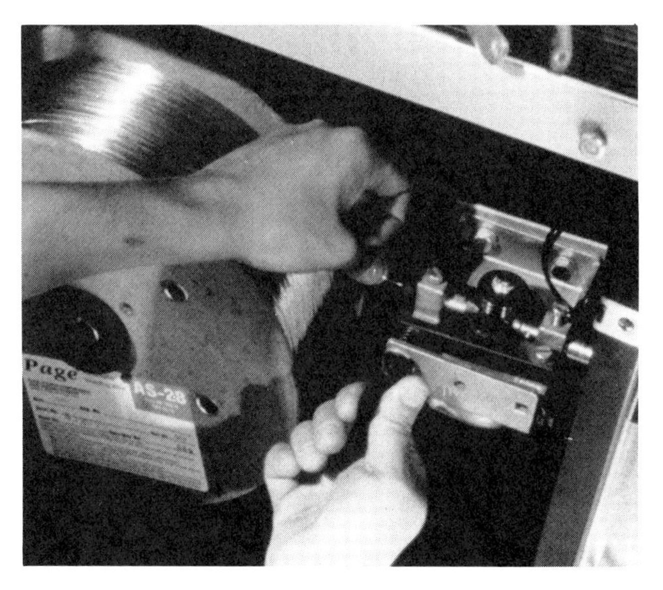

FIGURE 2-44 Always be sure the drive roll grooves, wire guides, cable liner, and gun contact tube correspond with the wire size being used.

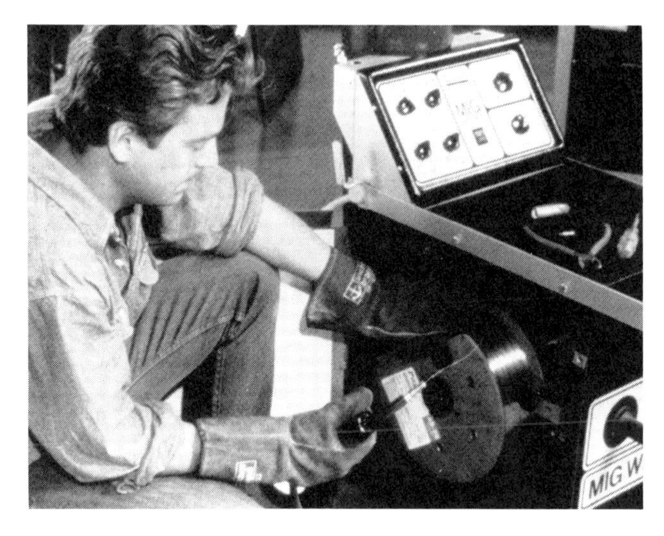

FIGURE 2-45 The tension should be light enough so the wire slips at the rollers when it is stopped at the gun nozzle.

the wire is melted, arc stability, and the amount of weld spatter. As the current is increased, the penetration depth, excess metal height, and bead width also increase (Table 2-6).

ARC VOLTAGE

Good welding results depend on a proper arc length. The length of the arc is determined by the arc voltage. When the voltage is set properly, a continuous light hissing or cracking sound is emitted from the welding area. When the voltage is high, the arc length increases, the penetration is shallow, and the bead is wide and flat. When the voltage is low, the arc length decreases, penetration is deep, and the bead is narrow and dome shaped.

Since the length of the arc depends on the amount of voltage, voltage that is too high will result in an overly long arc and an increase in the amount of weld spatter (Figure 2-47). A sputtering sound and no arc means that the voltage is too low.

TIP-TO-BASE METAL DISTANCE

The tip-to-base metal distance (Figure 2-48) is also an important factor in obtaining good welding results. The standard distance is approximately 1/4 to 5/8 inch.

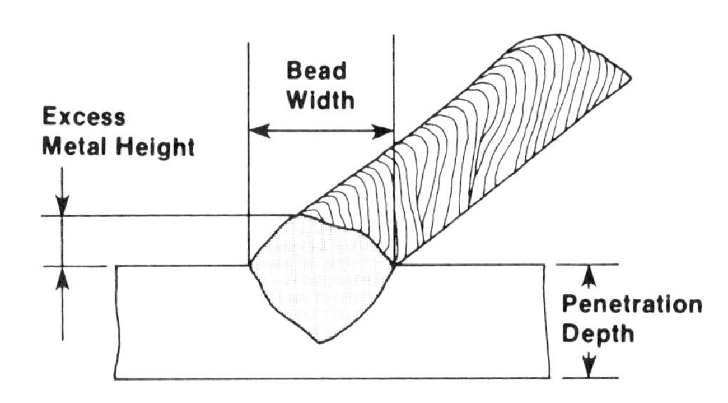

FIGURE 2-46 Penetration depth, excess metal height, and bead width

TABLE 2-6: RELATIONSHIP BETWEEN WIRE DIAMETER, PANEL THICKNESS, AND WELDING CURRENT

	Panel Thickness						
Wire Diameter	**1/64″**	**1/32″**	**Less Than 3/64″**	**3/64″**	**1/16″**	**3/32″**	**1/8″**
1/64″	20-30A	30-40A	40-50A	50-60A	—	—	—
1/32″	—	—	40-50A	50-60A	60-90A	100-120A	—
More Than 1/32″	—	—	—	—	60-90A	100-120A	120-150A

If the tip-to-base metal distance is too long, the length of wire protruding from the end of the gun increases and becomes preheated, which increases the melting speed of the wire. Also, the shield gas effect will be reduced. If the tip-to-base metal distance is too short, it is difficult to see the progress of the weld because it will be hidden behind the tip of the gun.

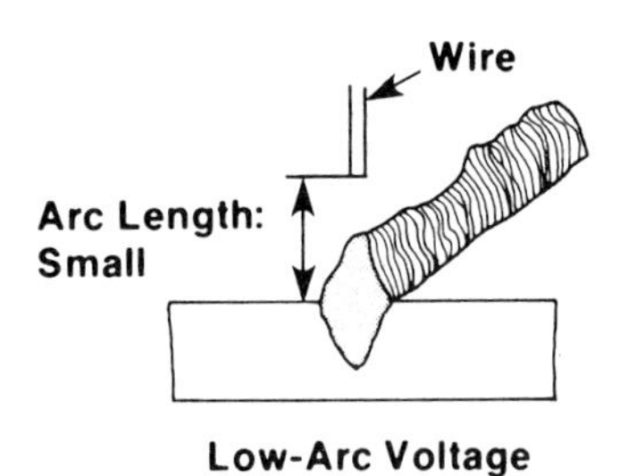

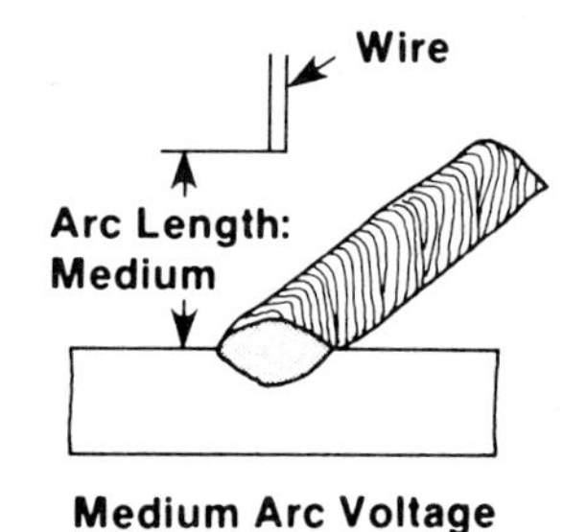

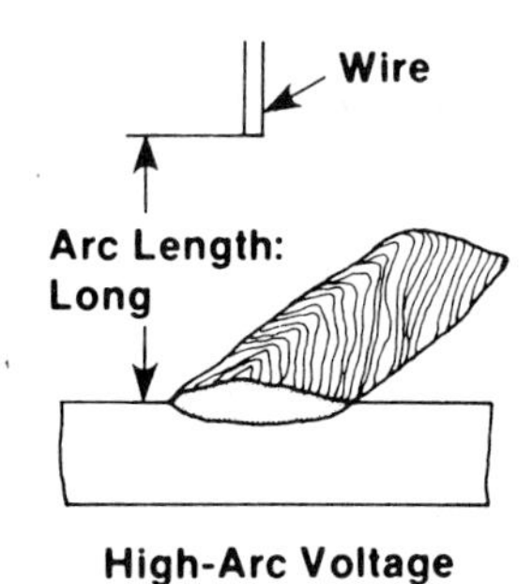

FIGURE 2-47 Arc voltage and bead shape

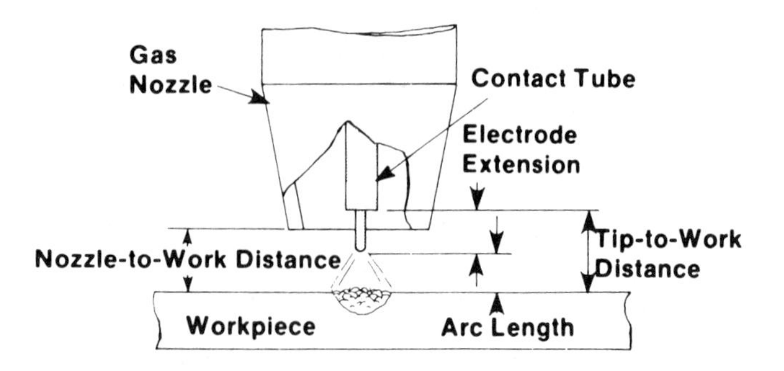

FIGURE 2-48 Tip-to-base metal distance

GUN ANGLE

There are two basic MIG welding methods: the forward or forehand method and the reverse or backhand method (Figure 2-49). With the forward method, the penetration depth is shallow and the bead is flat. With the reverse method, the penetration is deep and a large amount of metal is deposited (Figure 2-50). The gun angle for both methods should be between 10 and 30 degrees.

SHIELD GAS FLOW VOLUME

Precise gas flow is essential for a good weld. If the volume of gas is too high, it will flow in eddies and reduce the shield effect. If there is not enough gas, the shield effect will also be reduced. Adjustment is made in accordance with the distance between the nozzle and the base metal, the welding current, welding speed, and welding environment (nearby air currents). The standard flow volume is approximately 1-3/8 to 1-1/2 cubic inches per minute.

WELDING SPEED

If the operator welds at a rapid pace, the penetration depth and bead width will decrease, and the bead will be dome shaped. If the speed is increased even more, undercutting can occur. Welding at too slow of a speed can cause burn-through holes. Ordinarily, welding speed is determined by base metal panel thickness and/or voltage of the welding machine (Table 2-7).

WIRE SPEED

An even, high-pitched buzzing sound indicates the correct wire-to-heat ratio producing a temperature in the 9000 degrees Fahrenheit range. A steady, reflected light should be observed; it will start to fade in intensity as the arc is shortened and wire speed is increased.

If the wire speed is too slow (Figure 2-51), a hissing and plopping sound will be heard as the wire melts away from the puddle and deposits the molten glob. There will be a much brighter reflected light. Too much wire speed will choke the arc; more wire is being deposited than the heat and puddle can absorb. The result is that the wire melts into tiny balls of molten metal that fly away from the weld. There is a strobe light arc effect.

Before this critical wire-to-heat ratio can be obtained, an understanding of what produces these signals is essential. Using wire and C-25 gas, and a

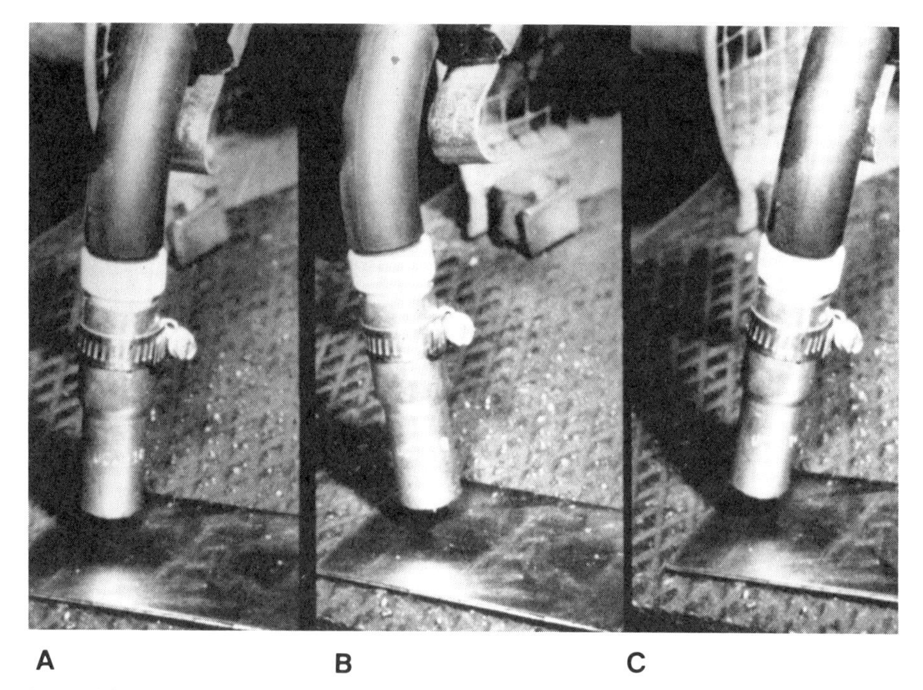

FIGURE 2-49 (A) Hold the torch at a transverse angle of 90 degrees directly over the center of the joint and find the longitudinal angle by experimentation. (B) The trailing or dragging torch angle—where the torch is pointing opposite or reverse to the direction of travel—should be tried at about 10 degrees perpendicular. (C) A leading or pushing torch angle—where the torch is pointing forward in the direction of travel—should be used at the same inclination.

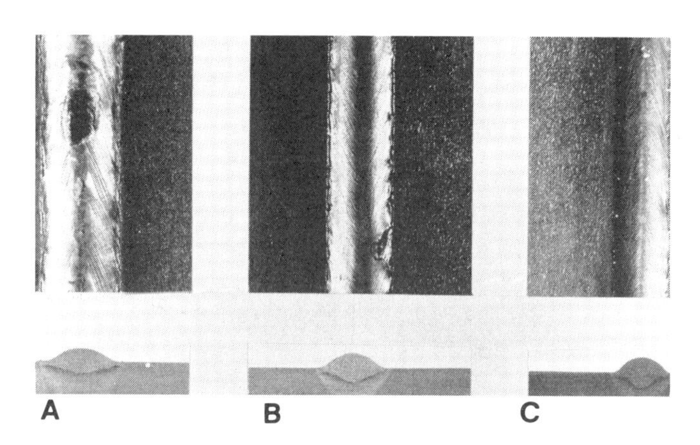

FIGURE 2-50 The penetration and weld pattern for the three travel methods: (A) forehand; (B) transverse angle of 90 degrees; (C) backhand.

TABLE 2-7: WELDING SPEED

Panel Thickness	Welding Speed (in./min.)
1/32"	41-11/32-45-9/32
More Than 1/32"	39-3/8
3/64"	35-7/16-39-3/8
1/16"	31-1/2-33-15/32

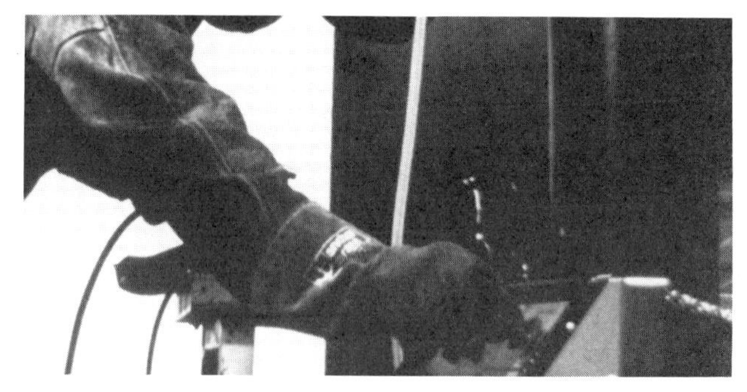

FIGURE 2-51 If the wire speed is too slow, increase the dial setting on the welder.

heat setting producing approximately 24 volts and 45 to 50 amperes, approximately 200 to 230 short circuits and deposits take place per second.

When the trigger is first activated, a solid steel wire makes its initial contact with a solid steel plate. The wire has been charged with current and the gas flow has been started prior to contact (Figure 2-52). The first contact produces tiny oxide sparks being burned off the wire and base metal.

Immediately after the sparks, tiny molten balls form as the wire melts. Once the heat creates the puddle, the balls stop forming. Oxide sparks are present as they burn off the wire and base metal during the weld process.

After the arc transfer has been started, an on-off action occurs in slow motion. It is off approximately 1 to 230 deposits per second. Every time the metal is deposited a plop is heard; when it pulls away, a hiss is heard. Speeding it up to approximately 200 plops and hisses per second creates a smooth buzzing sound.

Increasing and decreasing the wire speed causes the ball to burn farther back at 150 transfers than at 230. Similarly the sound is a lower buzz at 150 and the light is brighter, compared to a high-pitched buzz and a much dimmer light at 230. The more time the ball has to burn back, the more time the arc has to heat the molten puddle. Therefore, a flatter weld is produced with a slower wire speed.

When welding overhead, the danger of having too large a puddle and balls is obvious. The balls are pulled by gravity down onto the contact tip or into the gas nozzle, where they can create serious problems. Therefore, overhead welding should always be done at a higher wire speed, with the arc and balls kept small and close together. Pressing the gas nozzle against the work will keep the wire in the puddle. If it is moved out, the balls are produced by melting wire until a new puddle is formed to absorb them. Normal buildup of oxide sparks in the gas nozzle area must always be removed before it falls inside and shorts out the nozzle. Balls produced by too slow a wire speed must also be removed before a short occurs.

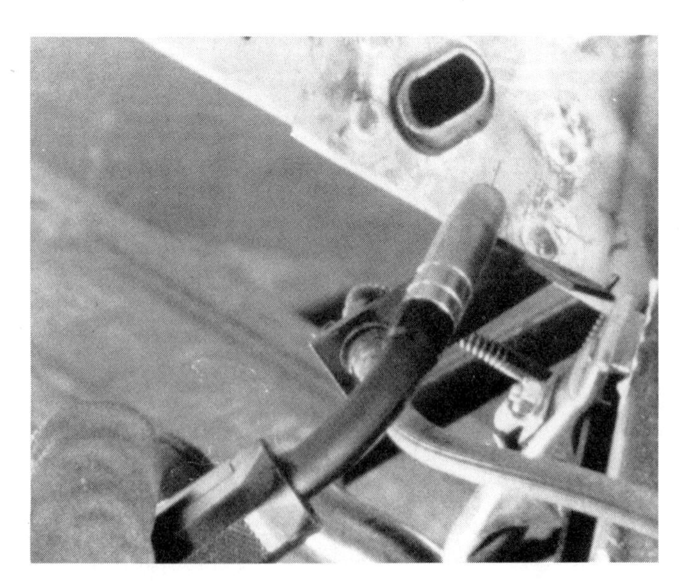

FIGURE 2-52 Prior to contact, the wire has been charged and the gas flow has started.

In summary, Table 2-8 outlines the effects of several welding variables and the changes necessary to alter the weld characteristics.

GUN NOZZLE ADJUSTMENT

The guns used on MIG welders serve two main functions:

- Provide proper gas protection
- Allow pressure to be applied to the work area, thus preventing the wire from moving out of the puddle

If the insulation is bypassed, the power intended for the wire is transferred to the gas nozzle, causing the wire to glow and sputter and the nozzle to be burned up. Welding on dirty or rusty material can cause heavy bombardment into the nozzle. To successfully weld on a rusty surface, slow the wire speed to approximately 130 transfers per second. Set the burnback control to its maximum and tap the trigger, floating the ball on and off the material.

Of the four main components in a MIG welder, the nozzle area and wire feed delivery are the most crucial. A clogged or damaged liner will cause erratic wire speed and produce molten balls that, in turn, will short out the gas nozzle.

The basic adjustment procedure of a gas nozzle is as follows:

- *Arc Generation.* Position the tip of the gun near the base metal. When the gun switch is activated (Figure 2-53), the wire is fed at the same time as the shield gas. Bring the end of the wire in contact with the base metal and create an arc. If the distance between the tip and the base metal is shortened a little, it will be easier to create the arc (Figure 2-54). If the end of the wire forms a large ball, it will be difficult to generate an arc, so quickly cut off the end of the wire with a pair of wire cutters (Figure 2-55).

CAUTION: Hold the tip of the gun away from your face when cutting off the end of the wire.

- *Spatter Treatment.* Remove weld spatter promptly. If it adheres to the end of the nozzle, the shield gas will not flow properly and a poor weld will result. Antispatter compounds are available that reduce the amount of spatter on the nozzle (Figure 2-56). Weld spatter

TABLE 2-8: ADJUSTMENTS IN WELDING VARIABLES AND TECHNIQUES

Welding Variables to Change	Desired Changes							
	Penetration		Deposition Rate		Bead Size		Bead Width	
	Increase	Decrease	Increase	Decrease	Increase	Decrease	Increase	Decrease
Current and Wire Feed Speed	Increase	Decrease	Increase	Decrease	Increase	Decrease	No effect	No effect
Voltage	Little effect	Little effect	No effect	No effect	No effect	No effect	Increase	Decrease
Travel Speed	Little effect	Little effect	No effect	No effect	Decrease	Increase	Increase	Decrease
Stickout	Decrease	Increase	Increase	Decrease	Increase	Decrease	Decrease	Increase
Wire Diameter	Decrease	Increase	Decrease	Increase	No effect	No effect	No effect	No effect
Shield Gas Percent CO_2	Increase	Decrease	No effect	No effect	No effect	No effect	Increase	Decrease
Torch Angle	Backhand to 25°	Forehand	No effect	No effect	No effect	No effect	Backhand	Forehand

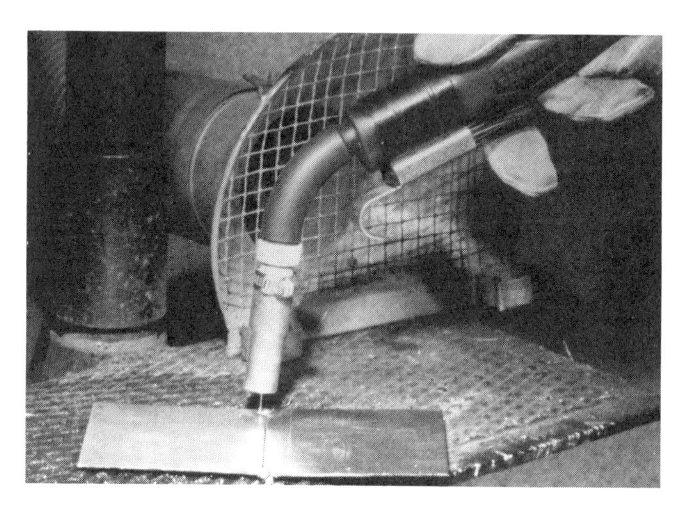

FIGURE 2-53 Activating the gun switch starts the wire feed.

on the tip will prevent the wire from moving freely. If the wire feed switch is turned on and the wire is not able to move freely through the tip, the wire will become twisted inside the welder. Use a suitable tool, such as a file, to remove spatter from the tip, then check to see that the wire comes out smoothly.

- *Contact Tip Conditions.* To ensure a stable arc, the tip should be replaced if it has become worn. For a good current flow and sta-

A

B

FIGURE 2-54 (A) Typically, the contact tip should be flush with the nozzle to 1/8 inch inside it. (B) The wire should extend 3/16 to 5/16 inch beyond the contact tip.

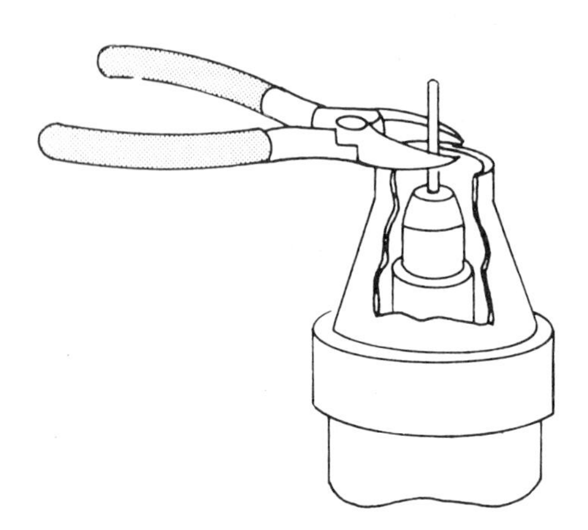

FIGURE 2-55 Cutting the end of the wire with wire cutters

FIGURE 2-56 Using antispatter compound

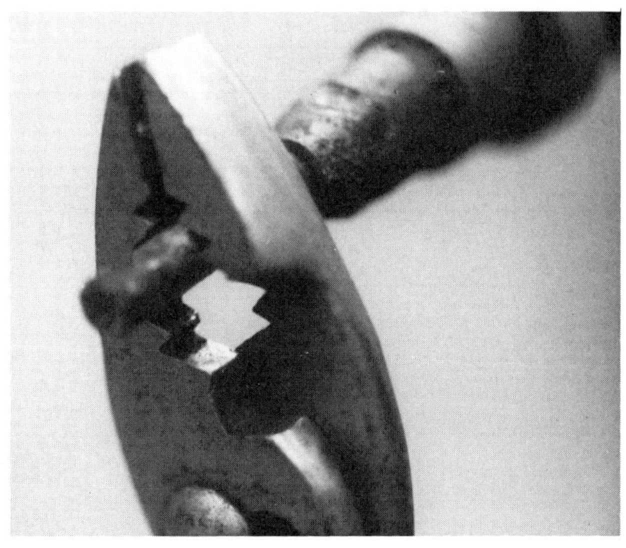

A

B

FIGURE 2-57 (A) Keep the tip properly tightened and (B) check its condition regularly.

ble arc, keep the tip properly tightened (Figure 2-57).

CLAMPING TOOLS

Locking jaw (vise) pliers, C-clamps, sheet metal screws, and special clamps are all necessary tools for good welding. Clamping panels together correctly will require close attention to detail (Figure 2-58). As shown in Figure 2-59, a hammer and dolly can often be used to fit panels closely together in places that cannot be clamped.

Clamping both sides of a panel is not always possible. In these cases, a simple technique using sheet metal screws can be employed. To hold panels together with sheet metal screws, the panels should have holes punched and drilled through the panel closest to the operator. In the case of plug welding, every other hole is filled with a screw. The empty holes are then plug welded. After the original holes are plug welded, the screws are removed and the remaining holes are then plug welded.

Fixtures can also be used in some cases to hold panels in proper alignment. Fixtures alone, however, should not be depended upon to maintain tight clamping force at the welded joint. Some additional clamping will be required to make sure that the pan-

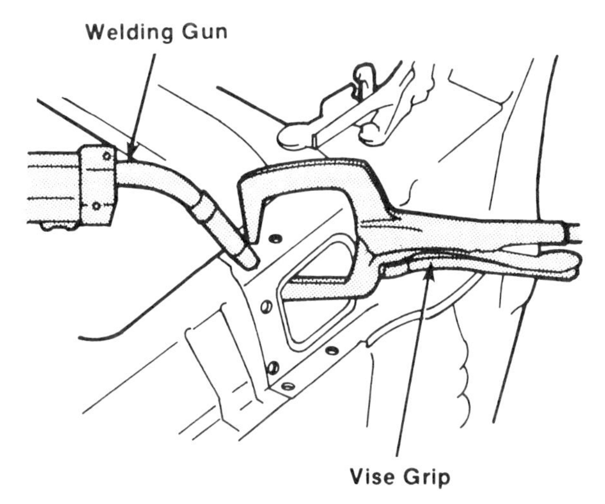

FIGURE 2-58 Clamping for front fender apron welding

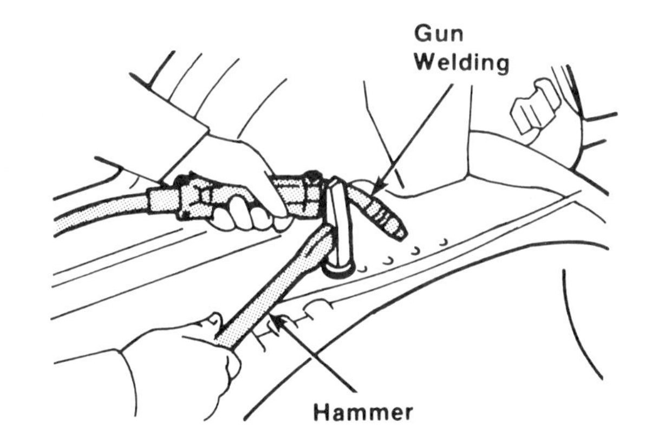

FIGURE 2-59 Clamping for floor pan welding

els are tight together and not just held in proper alignment.

WELDING POSITION

In vehicle repair, the welding position is usually dictated by the location of the weld in the structure of the car. Both the heat and wire speed parameters can be affected by the welding position (Figure 2-60).

Flat welding is generally easier and faster and allows for the best penetration (Figure 2-61). When welding a member that is off the vehicle, try to place it so that it can be welded in the flat position.

When welding a horizontal joint (Figure 2-62), angle the gun upward to hold the weld puddle in place against the pull of gravity.

To weld a vertical joint (Figure 2-63), the best procedure is usually to start the arc at the top of the joint and pull downward with a steady drag.

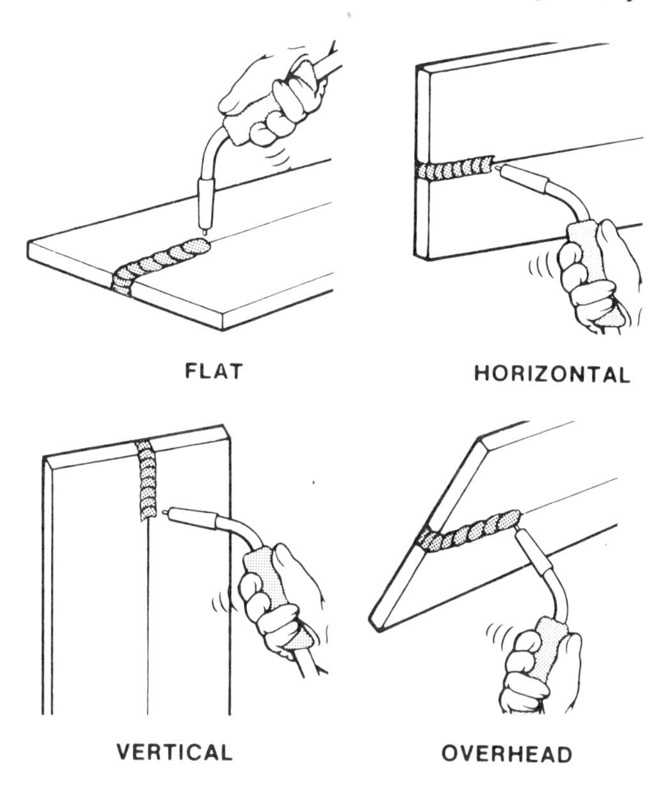

FIGURE 2-60 Typical welding positions

FIGURE 2-61 Flat welding position

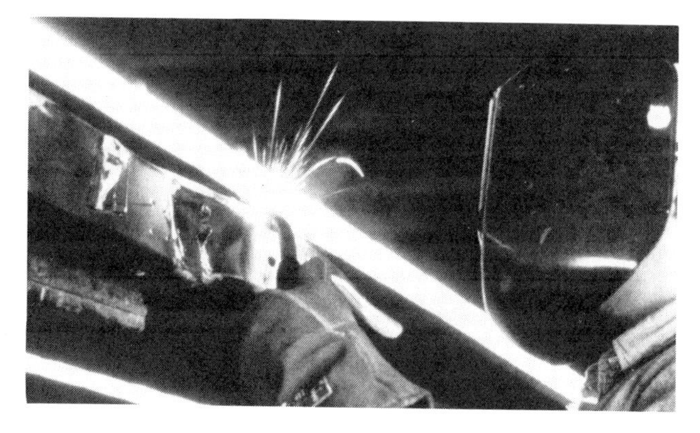

FIGURE 2-62 Horizontal welding position

FIGURE 2-63 Vertical welding position

FIGURE 2-64 Overhead welding position

Overhead welding, as mentioned previously, is the most difficult. In this position (Figure 2-64), the danger of having too large a puddle is obvious; some of the molten metal can fall down into the nozzle, where it can create problems. Always do overhead welding at a lower voltage, while keeping the arc as short as possible and the weld puddle as small as possible. Press the nozzle against the work to ensure that the wire is not moved away from the puddle, and pull the gun along the joint with a steady drag.

REVIEW QUESTIONS

1. In which of the following categories are the pieces of metal heated to the melting point, joined together (usually with a filler rod), and allowed to cool?
   a. pressure welding
   b. fusion welding
   c. braze welding
   d. all of the above

2. Which of the following is not characteristic of welding?
   a. Air and water tightness are excellent.
   b. Production efficiency is very high.
   c. The strength of a welded joint is not greatly influenced by the skill level of the welder.
   d. None of the above

3. How much fusion in the base metals does MIG welding produce?
   a. 25 percent
   b. 50 percent
   c. 75 percent
   d. 100 percent

4. Which type of gas is noted for cleaning the work area?
   a. argon
   b. helium
   c. CO_2
   d. all of the above

5. Which gas is preferred with short-circuiting metal transfer?
   a. argon
   b. argon helium
   c. carbon dioxide
   d. argon/carbon dioxide

6. Using straight polarity for MIG welding results in ____________.
   a. panel warping
   b. greater penetration
   c. a high, rope-like bead
   d. none of the above

7. Which of the following statements is true?
   a. A steeper slope increases pinch effect.
   b. A flat slope decreases the short-circuit current.
   c. An inductance increase decreases the slope.
   d. A higher resistance causes a steeper slope.
   e. None of the above

8. One ampere of current is needed for every ____________.
   a. 0.1 square inch of metal thickness
   b. 0.01 square inch of metal thickness
   c. 0.001 square inch of metal thickness
   d. none of the above

9. The duty cycle tells how many minutes out of 10 a welding machine can run without ____________.
   a. having to replace the electrode spool
   b. overheating

c. a slope adjustment
d. none of the above

10. Which rollers are best suited for large-diameter wires?
    a. V-shaped grooves
    b. smooth U-shaped grooves
    c. knurled U-shaped grooves
    d. all of the above

11. Which of the following indicates the correct wire-to-heat ratio?
    a. an even, high-pitched buzzing sound
    b. a steady, reflected light
    c. both a and b
    d. neither a nor b

12. Which of the following statements is true?
    a. FCAW is less tolerant than GMAW to surface impurities.
    b. FCAW has greater arc stability than GMAW.
    c. FCAW wires are less expensive than GMAW.
    d. None of the above

13. Welding current affects which of the following?
    a. base metal penetration depth
    b. arc stability
    c. amount of weld spatter
    d. all of the above

14. Welder A uses a forward gun angle to achieve a deep penetration in the metal. Welder B uses the reverse gun angle to achieve a flat bead. Who is right?
    a. Welder A
    b. Welder B
    c. Both A and B
    d. Neither A nor B

15. Welder A says that the main function of the gun nozzle is to provide proper gas protection. Welder B says the main function of the gun nozzle is to allow pressure to be applied to the work area. Who is right?
    a. Welder A
    b. Welder B
    c. Both A and B
    d. Neither A nor B

16. In which of the following categories is metal with a melting point lower than that of the pieces of base metal melted over the joint of the pieces?
    a. pressure welding
    b. fusion welding
    c. braze welding
    d. all of the above

17. Welder A says that all steels can be MIG welded with one common type of weld wire. Welder B says that metals of different thicknesses can be MIG welded with the same diameter wire. Who is right?
    a. Welder A
    b. Welder B
    c. Both A and B
    d. Neither A nor B

18. Which gas causes a more concentrated arc than any of the other commonly used gases employed with the gas metal arc process?
    a. argon
    b. helium
    c. carbon dioxide
    d. none of the above

19. Which of the following must the welder set before starting to MIG weld?
    a. voltage for the arc
    b. wire speed
    c. shielding gas flow rate
    d. all of the above

20. Welder A says that by using flux-cored wires, penetration is increased enough to weld metal that is one or two gauges thicker than the specified machine capacity. Welder B says this amount of increased penetration can be achieved by using straight CO_2 shielded gas instead of a mixed gas. Who is right?
    a. Welder A
    b. Welder B
    c. Both A and B
    d. Neither A nor B

21. Which of the following is an advantage of a pull-type feed system?
    a. The gun is lighter.
    b. It has the ability to move even soft wire over great distances.
    c. The operating life of the motor is longer.
    d. None of the above

22. Which of the following is characteristic of FCAW?
    a. high deposition rating
    b. wire contains toxic fluoride compounds
    c. requires little edge preparation
    d. all of the above

23. When the arc voltage is high, the ______________.
    a. arc voltage increases
    b. penetration is deep
    c. bead is narrow and dome shaped
    d. none of the above

CHAPTER THREE

MIG WELDING TECHNIQUES

Objectives

After reading this chapter, you should be able to

- Identify the major differences between mild and high-strength steel.
- Name the three types of high-strength steel used most often in automotive applications and describe the uses of each.
- Name the six basic welding techniques used with MIG equipment and describe how each is accomplished.
- Identify butt, lap or flange, plug, spot, and stitch welds.
- Describe the problems unique to MIG welding galvanized metals.
- Distinguish between weld discontinuities and weld defects.
- Describe how a penetrant inspection and an ultrasonic inspection are performed.
- Identify common MIG weld defects and their causes.

With the revolutionary changes in automobile construction in recent years, the importance of employing correct MIG welding techniques cannot be overstated. The popular Chrysler New Yorker is a good case in point. In 1965, this vehicle weighed approximately 4600 pounds. With a typical payload of 1000 pounds of passengers and cargo, this produced a vehicle-to-occupancy ratio of 4 to 1. Obviously, this meant that a rather large margin for error existed during a collision repair. By comparison, a 1989 Chrysler New Yorker weighs approximately 2600 pounds; with the same cargo, a 2-to-1 vehicle-to-occupancy ratio is produced. The end result is a very close tolerance of vehicle weight to passenger cargo, less margin for error, and a much more delicate situation when working with today's modern high-strength steels, which are so sensitive to heat and repair techniques. Without a doubt, the changes in the automobile industry have placed a tremendous responsibility on the collision repair technician.

And remember, even the most reliable method and the best equipment are of little use if improper procedures are used—or if the proper procedures are carried out in a hurried or slipshod manner. The basic MIG welding techniques and joints discussed in this chapter can only be mastered through diligent study and repeated practice. There are certainly no shortcuts to obtaining high-quality, long-lasting MIG welds.

WORKING WITH HIGH-STRENGTH STEEL

For many years, low carbon or "mild" steel was used in most automotive structural applications. This particular type of steel has a carbon content of 0.30 percent or less, making it extremely weldable. (The higher the carbon content of the steel, the more difficult it is to weld.) However, by the mid-1970s carmakers began to design smaller, unitized vehicles in an effort to reduce vehicle weight and thereby improve fuel economy. Low carbon steel was now inadequate for the structural members, which were forced to handle far greater load-carrying and energy-absorption requirements (Figure 3-1). Stronger, lighter, high-strength steels filled these needs while also improving crash worthiness.

One of the things that complicates structural repairs and even some cosmetic repairs to external sheet metal is not knowing exactly where high-strength steels are apt to be found. Lists and diagrams are available from the vehicle manufacturers and aftermarket sources that specify locations of

FIGURE 3-1 Low carbon steel was inadequate for the structural members of the smaller, unitized cars of the mid-1970s. *(Courtesy of American Best Car-Parts, Inc.)*

special steels on various makes and models (Figure 3-2). It can be a time-consuming job to find the information needed. One good way to approach the problem is to treat all late-model thin-gauge panels and structural members as if they were made from high-strength steel.

MILD VERSUS HIGH-STRENGTH STEEL

It is important to treat high-strength steels differently than ordinary mild steel for two reasons. First and foremost is heat sensitivity. Although some high-strength steels can withstand temperatures of 1200 degrees Fahrenheit for up to 3 minutes without weakening significantly, others have temperature

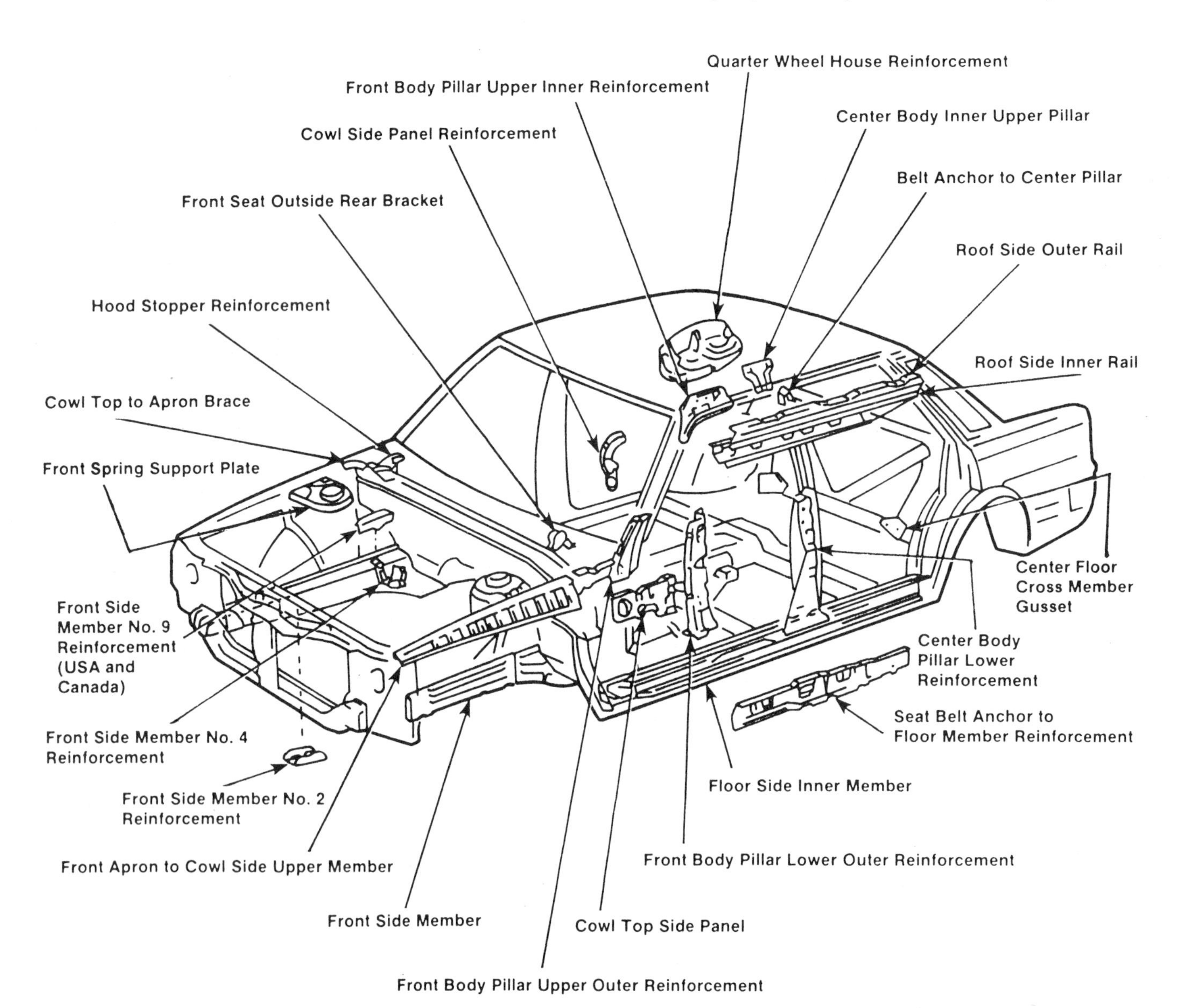

FIGURE 3-2 Because it can be difficult to find high-strength steel, charts like this one are often necessary.

limits as low as 700 to 900 degrees Fahrenheit. Most, but not all, high-strength steels are heat sensitive and will be weakened if heated excessively. Naturally, this presents a unique challenge to the auto repair industry. Such steels cannot be welded with an oxy-acetylene torch. The recommended method of welding in this case is with a MIG welder.

What many body technicians do not realize is that when highly sensitive steel is subjected to temperatures in excess of 1200 degrees Fahrenheit, it is, for all intents and purposes, converted to mild steel. What this means is that a weld or section of a car, which was originally designed with a yield strength of 65,000 psi and sections of 20-gauge high-strength steel, is now much weaker than anticipated. Upon visual inspection and subjecting the panel to hammering and prying, it may very well appear undamaged and perfectly capable of meeting original specifications. However, the 20-gauge high-strength steel has now been converted to 20-gauge mild steel and is no longer capable of withstanding the high stresses required. This is now a potentially dangerous situation that can cause serious damage to the car and injury to the occupants if the car should be involved in another collision, or even hit a large pothole. Simply put, this section of the car will now yield much earlier than its engineers originally intended in its design.

The second reason for treating high-strength steels differently is because most are brittle. A hard steel will not pull like a mild steel; the harder the steel, the more it tends to tear or crack if straightened. With some of the new high-strength steels, it is safe to use heat to relax stress when pulling and straightening, but with other steels the heat will weaken the metal too much.

TYPES OF HIGH-STRENGTH STEEL

In spite of the fact that most types of steel look alike to the naked eye, there can be considerable differences in chemical makeup and crystalline structure that affect strength and sensitivity to heat. There is a variety of high-strength steels, all of which have unique properties that relate back to the way in which they can or cannot be repaired.

The following types of steel can be found in this class:

- Chromium alloy
- Carbon molybdenum
- Chromium molybdenum
- Chromium vanadium
- Manganese alloy
- Nickel molybdenum
- Manganese molybdenum
- Nickel chromium
- Nickel chromium molybdenum
- Nickel copper

Other high-strength, low-alloy steels are available under a variety of trade names and are listed in the Metals Handbook, published by the American Society for Metals.

Two primary numbering systems have been developed to classify standard grades of steel, including both carbon and alloy steels. These systems classify the types according to their basic chemical composition. One classification system was developed by the Society of Automotive Engineers (SAE); the other is sponsored by the American Iron and Steel Institute (AISI). The so-called Unified Numbering System currently being promoted for all metals may eventually replace these systems.

The numbers used in both systems are now just about the same. However, the AISI system uses a letter before the number to indicate the method used in the manufacture of the steel. Both systems usually have a four-digit series of numbers. In some cases, a five-digit series is used for certain alloy steels. The entire number is a code for the approximate composition of the steel.

In both classification systems, the first number usually refers to the basic type of steel, as follows:

1XXX Carbon
2XXX Nickel
3XXX Nickel chrome
4XXX Molybdenum
5XXX Chromium
6XXX Chromium vanadium
7XXX Tungsten
8XXX Nickel chromium vanadium
9XXX Silicomanganese

The first two digits together indicate the series within the basic alloy group. There may be several series within a group, depending upon the amount of principle alloying elements. The last two or three digits refer to the approximate permissible range of carbon content.

The letters in the AISI system, if used, indicate the manufacturing process as follows:

- C—Basic open-hearth or electric furnace steel and basic oxygen furnace steel
- E—Electric furnace alloy steel

Table 3-1 shows the AISI and SAE numerical designations of alloy steels. Note that the elements in the table are expressed in percent.

Strength

Before understanding the difference between the types of high-strength steel, *strength* must be defined. There are four basic kinds of strength, all of which relate to the ability of the steel to resist permanent deformation (Figure 3-3).

- *Tensile Strength.* This is the property of a material which resists forces applied to pull it apart. Tension includes both yield stress and ultimate strength. Yield stress is the amount of strain needed to permanently deform a test specimen; ultimate strength is a measure of the load that breaks a specimen. The tensile strength of a metal can be determined by a tensile testing machine.
- *Compressive Strength.* This is the property of a material to resist being crushed.
- *Shear Strength.* This is a measure of how well a material can withstand forces acting to cut or slice it apart.
- *Torsional Strength.* This is the property of a material that withstands a twisting force.

Strength is expressed in pounds per square inch (psi) or kilograms per millimeter squared (kg/mm^2). Heat treatment, cold rolling, and chemical additives are among the manufacturing procedures used to increase the strength of steel. Any further heating beyond the set time and temperature limits can alter the strength significantly and permanently. These limits are determined by the type and extent of the heat and chemical treatment.

TABLE 3-1: AISI AND ASE ALLOY STEEL DESIGNATIONS*

13XX	Manganese 1.75
23XX**	Nickel 3.50
25XX**	Nickel 5.00
31XX	Nickel 1.25; chromium 0.65
E33XX	Nickel 3.50; chromium 1.55; electric furnace
40XX	Molybdenum 0.25
41XX	Chromium 0.50 or 0.95; molybdenum 0.12 or 0.20
43XX	Nickel 1.80; chromium 0.50 or 0.80; molybdenum 0.25
E43XX	Same as above, produced in basic electric furnace
44XX	Manganese 0.80; molybdenum 0.40
45XX	Nickel 1.85; molybdenum 0.25
47XX	Nickel 1.05; chromium 0.45; molybdenum 0.20 or 0.35
50XX	Chromium 0.28 or 0.40
51XX	Chromium 0.80, 0.88, 0.93, 0.95, or 1.00
E5XXXX	High carbon; high chromium; electric furnace bearing steel
E50100	Carbon 1.00; chromium 0.50
E51100	Carbon 1.00; chromium 1.00
E52100	Carbon 1.00; chromium 1.45
61XX	Chromium 0.60, 0.80, or 0.95; vanadium 0.12, or 0.10, or .015 minimum
7140	Carbon 0.40; chromium 1.60; molybdenum 0.35; aluminum 1.15
81XX	Nickel 0.30; chromium 0.40; molybdenum 0.12
86XX	Nickel 0.55; chromium 0.50; molybdenum 0.20
87XX	Nickel 0.55; chromium 0.50; molybdenum 0.25
88XX	Nickel 0.55; chromium 0.50; molybdenum 0.35
92XX	Manganese 0.85; silicon 2.00; 9262 chromium 0.25 to 0.40
93XX	Nickel 3.25; chromium 1.20; molybdenum 0.12
98XX	Nickel 1.00; chromium 0.80; molybdenum 0.25
14BXX	Boron
50BXX	Chromium 0.50 or 0.28; boron
51BXX	Chromium 0.80; boron
81BXX	Nickel 0.33; chromium 0.45; molybdenum 0.12; boron
86BXX	Nickel 0.55; chromium 0.50; molybdenum 0.20; boron
94BXX	Nickel 0.45; chromium 0.40; molybdenum 0.12; boron

*Consult current AISI and SAE publications for the latest revisions.

**Nonstandard steel

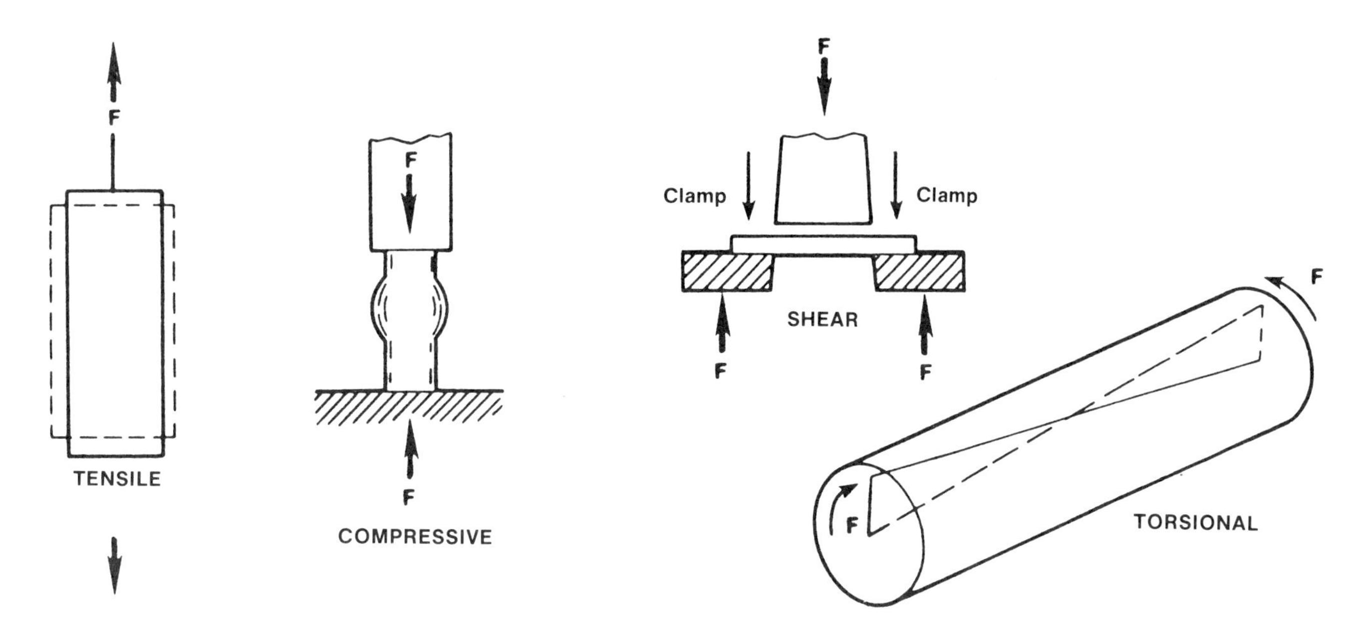

FIGURE 3-3 Types of strength and how they react to force

HIGH-STRENGTH STEEL CLASSIFICATIONS

There are several types of steel within the classification of high strength, all of which must be treated with extreme care when heating (Figure 3-4). Those most often used in automotive applications are:

- High tensile strength steel (HSS)
- High-strength, low-alloy steel (HSLA)
- Martensitic steel

Whenever possible, check with the automobile manufacturer regarding the acceptable heat ranges for specific components. Also, be sure to follow the manufacturer's instructions for restoring corrosion protection after any repairs are completed.

HSS Steel

High tensile strength, or HSS, steel is used in the structural components and some exterior panels of many Japanese imports; its strength is derived from heat treatment. Because HSS has a yield strength of more than 35,000 psi and a tensile strength of at least 45,000 psi, conventional heating and welding methods do not weaken it. When high tensile strength steel is deformed (Figure 3-5) it experiences an increase in stress that exceeds its yield strength. When heat is applied to assist in straightening it, the stresses resulting from the collision are decreased and the strength of the material is restored to precollision levels. HSS steel will tear or fracture if the collision stresses exceed the tensile strength.

Normal MIG welding procedures are used to restore HSS components. Keep in mind that door guard beams cannot be straightened and should instead be replaced; the same is true of some bumper reinforcements. All replacement parts should be MIG welded using AWS-E-705-6 wire because it possesses the same strength level as HSS.

HSLA Steel

High-strength, low-alloy steel, or HSLA, is more often than not found in domestic vehicles. HSLA is

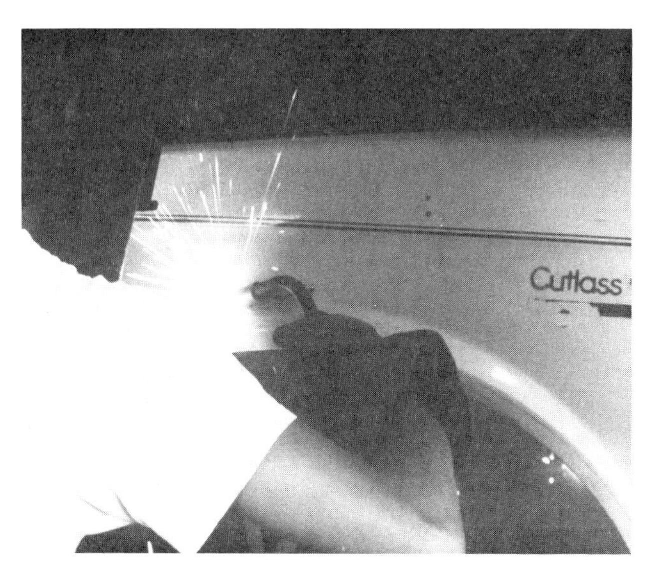

FIGURE 3-4 All heating of high-strength steel must be done with care.

FIGURE 3-5 An example of HSS steel deformed in a collision

used in structural components, including front and rear rails, rocker panels (Figure 3-6), bumper face bars, bumper reinforcements, door hinges, and lock pillars. Special chemical elements added to the steel give HSLA its high strength. After exposure to temperatures of 1200 degrees Fahrenheit or higher for several minutes (the temperature limit can be lower for some parts), the special hardening elements are absorbed by large softer elements in the heated area. The result is reduced strength, which in turn substantially reduces the ability of the part to react to the normal loads and collision forces. Use AWS-E-705-6 wire when MIG welding HSLA steel.

Martensitic Steel

In some domestic models, ultra high-strength Martensitic steel is used in door guard beams, bumper face bars, and reinforcements. Because it is very resistant to atmospheric corrosion, this type of steel is ideal wherever high-strength and wear-resistance properties are required. Martensitic steel is so heat sensitive that heat cannot be applied to it for purposes of repair. Therefore, any damaged parts must be replaced; they cannot be straightened.

FIGURE 3-6 Rocker panels are frequently made from HSLA steel.

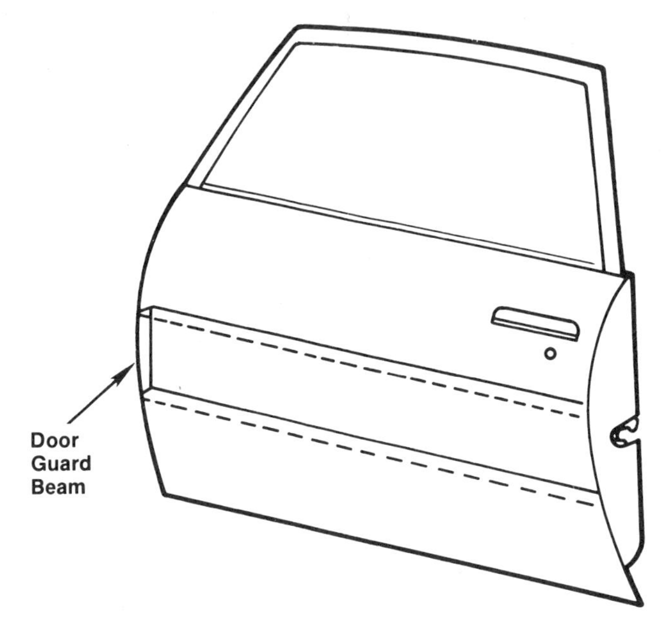

FIGURE 3-7 When a door guard beam sustains an impact, it must be replaced only if the damage affects door alignment or function.

Replacement parts should be installed by MIG plug welding with AWS-705-6 wire.

REPAIR OR REPLACEMENT

In the event of relatively minor damage to high-strength steel, a judgment must be made regarding the question of repair versus replacement. For example, if damage to a door guard beam (Figure 3-7) does not interfere with door alignment or function, it can be ignored. However, if the corrugations in the beam are damaged, the door beam must be replaced. Bumper face bar reinforcements that are covered by separate fascias do not require replacement in the event of minor damage, provided the operation of the bumper is unaffected. Plastic filler and sealer can be applied to shallow dents to restore the original contour.

MIG WELDING TECHNIQUES

There are six basic welding techniques employed with MIG equipment, as shown in Figure 3-8.

- *Tack Weld.* The tack weld is exactly that: a tack, or relatively small, temporary MIG spot weld that is used instead of a clamp or sheet metal screw to hold the fit-up in place while proceeding to make a permanent weld. And like the clamp or sheet metal screw, a tack weld is always and only a temporary device.

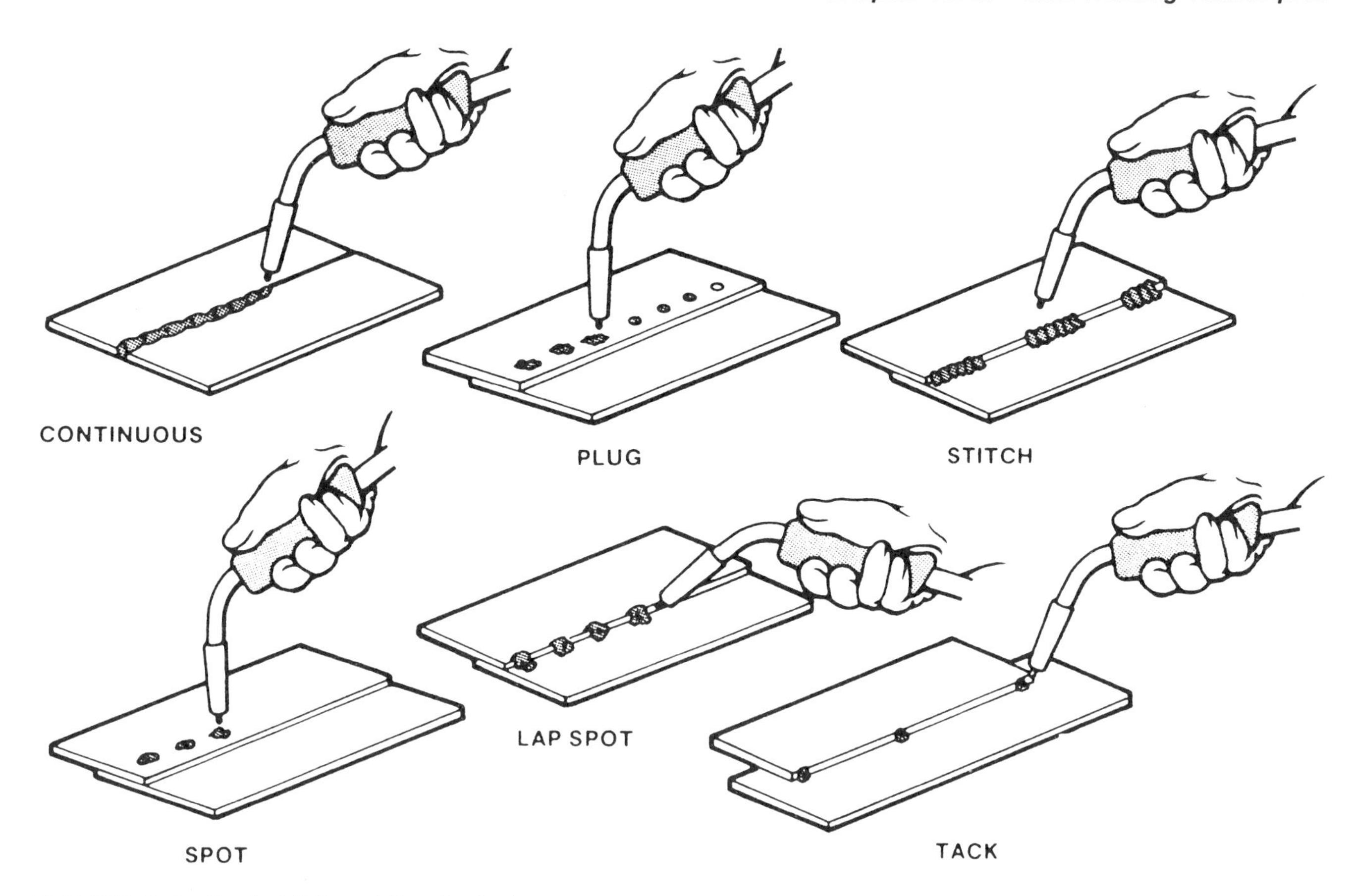

FIGURE 3-8 Basic MIG welding techniques

The length of the tack weld is determined by the thickness of the panel. Ordinarily, a length of 15 to 30 times the thickness of the panel is appropriate (Figure 3-9). Temporary welds are very important in maintaining proper panel alignment and therefore must be done accurately.

- *Continuous Weld.* In a continuous weld, an uninterrupted seam or bead is laid down in a slow, steady, ongoing movement. The gun must be supported securely so it does not wobble. Use the forward method, moving the gun continuously at a constant speed, looking frequently at the welding bead. The gun should be inclined between 10 to 15 degrees to obtain the optimum bead shape, welding line, and shield effect (Figure 3-10).

  Maintain proper tip-to-base metal distance and correct gun angle when making a continuous weld. If the weld is not progressing well, the problem might be that the wire length is too long. If this is the case, penetration of the metal will not be adequate. For proper penetration and a better weld, bring the gun closer to the base metal. If the gun handling is smooth and even, the bead will be

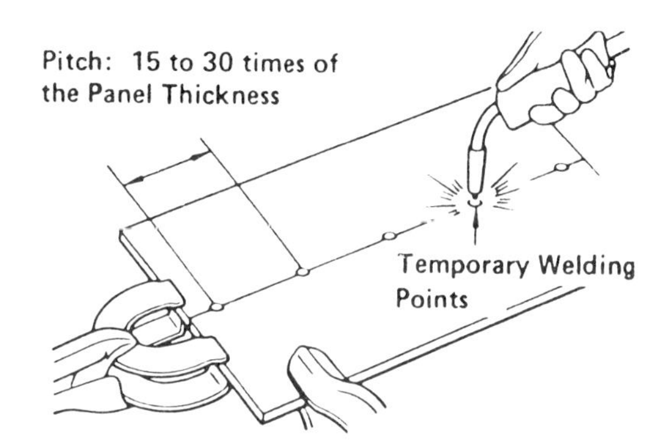

FIGURE 3-9 Temporary or tack welding

of consistent height and width, with a uniform, closely spaced ripple.

- *Plug Weld.* To do a MIG plug weld, a hole is drilled or punched through the outside piece (or pieces) of metal, the arc is directed through the hole to penetrate the inside piece, and the hole is filled with molten metal.
- *Spot Weld.* In a MIG spot weld, the arc is directed to penetrate both pieces of metal, while triggering a timed impulse of wire feed.
- *Lap Spot Weld.* In the MIG lap spot technique, the arc is directed to penetrate the

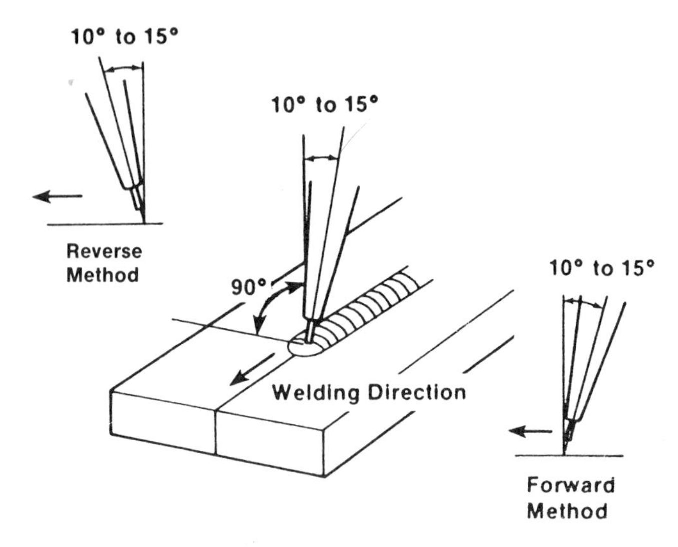

FIGURE 3-10 Continuous welding *(Courtesy of Toyota Motor Corp.)*

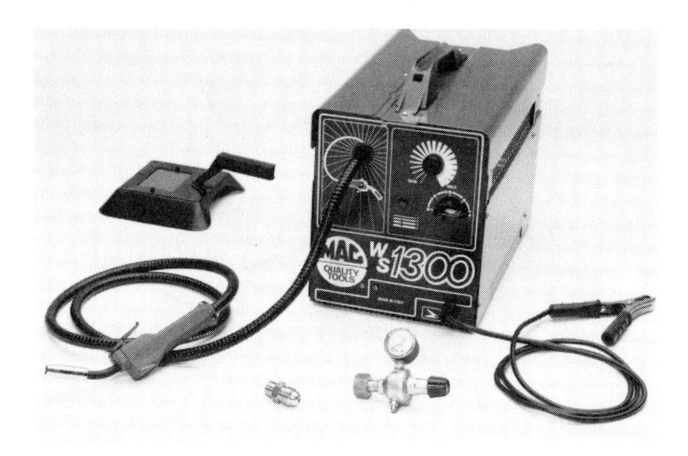

FIGURE 3-11 Small welders are suitable for welding single layers of sheet metal.

bottom piece of metal; the puddle is allowed to flow into the edge of the top piece.

- *Stitch Weld.* A stitch weld is a series of connecting or overlapping spot welds, creating a continuous seam.

CHOOSING THE PROPER EQUIPMENT

Many collision repair "experts" fail to select the proper size MIG welder for a given welding situation. The number one reason for this is that they underestimate the thickness of the workpiece and the overall requirements of the job at hand. While small, inexpensive, do-it-yourself type welders (Figure 3-11) might be fine for welding single layers of standard 18- to 22-gauge sheet metal—as is the case when attaching small patches on a door skin or lower quarter panel during a rust repair—many procedures require a greater capacity welder. For example, if a quarter panel is being joined or attached, plug welds and laminated sections consisting of two, three, and sometimes four layers of 20- to 22-gauge sheet metal are used. Obviously, a plug weld through two or three layers of sheet metal requires far more energy than a small, do-it-yourself type welder can possibly put out. The same can be said for installing an upper or lower structural rail; this entails plug welding through a 3/16- or 5/16-inch hole consisting often of three layers of sheet metal.

Using the proper capacity welder also promotes rapid fusion of the panels, thus reducing the chances of the adjoining metal being affected by the heat zone. Keep in mind that most welds in structural areas are performed on a flanged, U-shaped panel; the factory makes all the original spot welds on the outer flanges of this panel. The strength is derived from the 90-degree bends in the panel and the welds attached to the outer flanges. The role of the welds is to retain the rigid panel in position and maintain structural integrity, both during normal driving conditions and in the event of a collision impact. It is therefore crucial that the heat be contained in the flange and not allowed to enter the bent areas of the panel, since this is where the structural strength is obtained. To do this, heat control is crucial; the shorter the duration of the weld, the better the heat is contained.

SHOP TALK

Always disconnect the battery terminals before doing any MIG or arc welding on the vehicle. Also, remove the computer module (if applicable) and store it outside the immediate welding area. The magnetic fields created by welding machines have been known to affect on-board computers.

BASIC WELDING JOINTS

The MIG welds used for the repair or reattachment of damaged or replacement sections are butt welds, lap welds, flange welds, and plug or spot welds. Each type of joint can be welded by several different techniques, depending mainly on the given situation and parameters: the thickness or thinness of the metal; the condition of the metal; the amount of gap, if any, between the pieces to be welded; the welding position; and so on. For instance, the butt joint can be welded with the continuous technique, or the stitch technique. And it can be tack welded at various points along the joint to hold the fit-up in place while completing the joint with a permanent continuous weld or a stitch weld. Lap and flange joints can be made using all six of the previously discussed welding techniques.

Butt Welds

Butt welds are formed by fitting two edges of adjacent panels together and welding along the mating or butting edges. When butt welding, especially on thin panels, it is wise not to weld more than 3/4 of an inch at one time. Closely watch the melting of the panel, welding wire, and the continuity of the bead. At the same time, be sure the end of the wire does not wander away from the butted portion of the panels.

If the butt weld is to be long, it is a good idea to tack weld the panels in several locations to prevent panel warpage (Figure 3-12). Figure 3-13 illustrates how to generate an arc a short distance ahead of the point where the weld ends and then immediately move the gun to the point where the bead should begin. The bead width and height should be consistent at this time, with a uniform, closely spaced ripple—provided the gun handling is smooth and even.

When butt welding, a sequence should be established to allow the weld area to cool naturally before the next area is welded (Figure 3-14). While butt welds of outer panels are far less sensitive, the same sequencing procedure should be used to prevent warpage and distortion from temperature buildup. To fill the spaces between intermittently placed beads, first grind the beads along the surface of the panel using a sander or grinder, then fill the space with metal (Figure 3-15). If weld metal is placed without grinding the surface of the beads, blowholes can be produced.

When butt welding panels that are 1/32 of an inch or less, an intermittent or stitch welding tech-

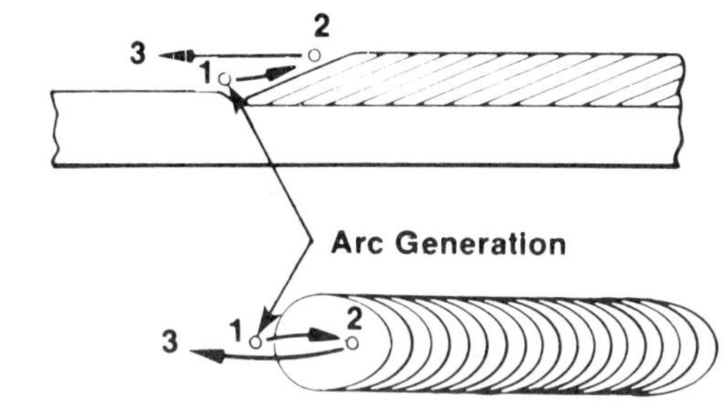

FIGURE 3-13 If the gun handling is smooth and even, the bead will be of consistent height and width, with a uniform, closely spaced ripple.

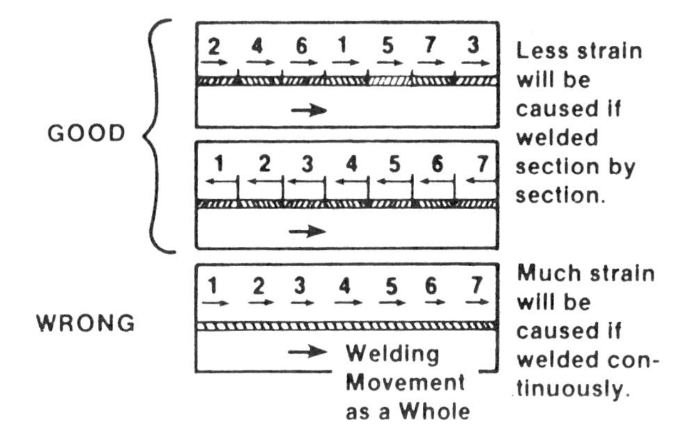

FIGURE 3-14 Right and wrong welding sequences *(Courtesy of Nissan Motor Corp.)*

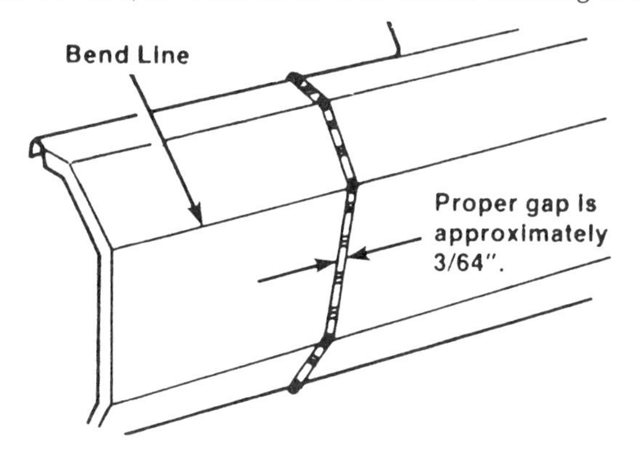

FIGURE 3-12 Tack welding panels prevents warpage. *(Courtesy of Nissan Motor Corp.)*

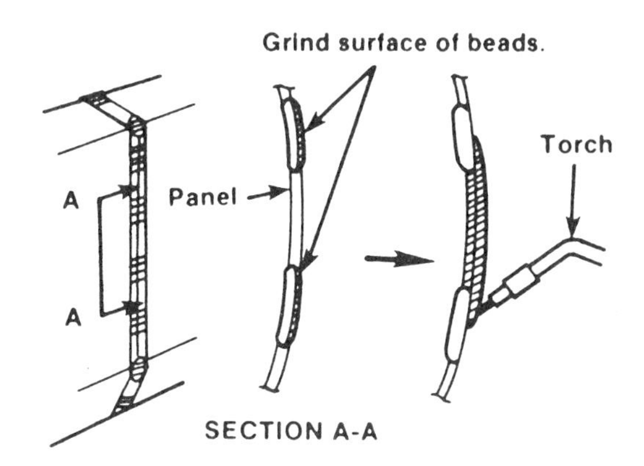

FIGURE 3-15 Filling the space between intermittently spaced beads

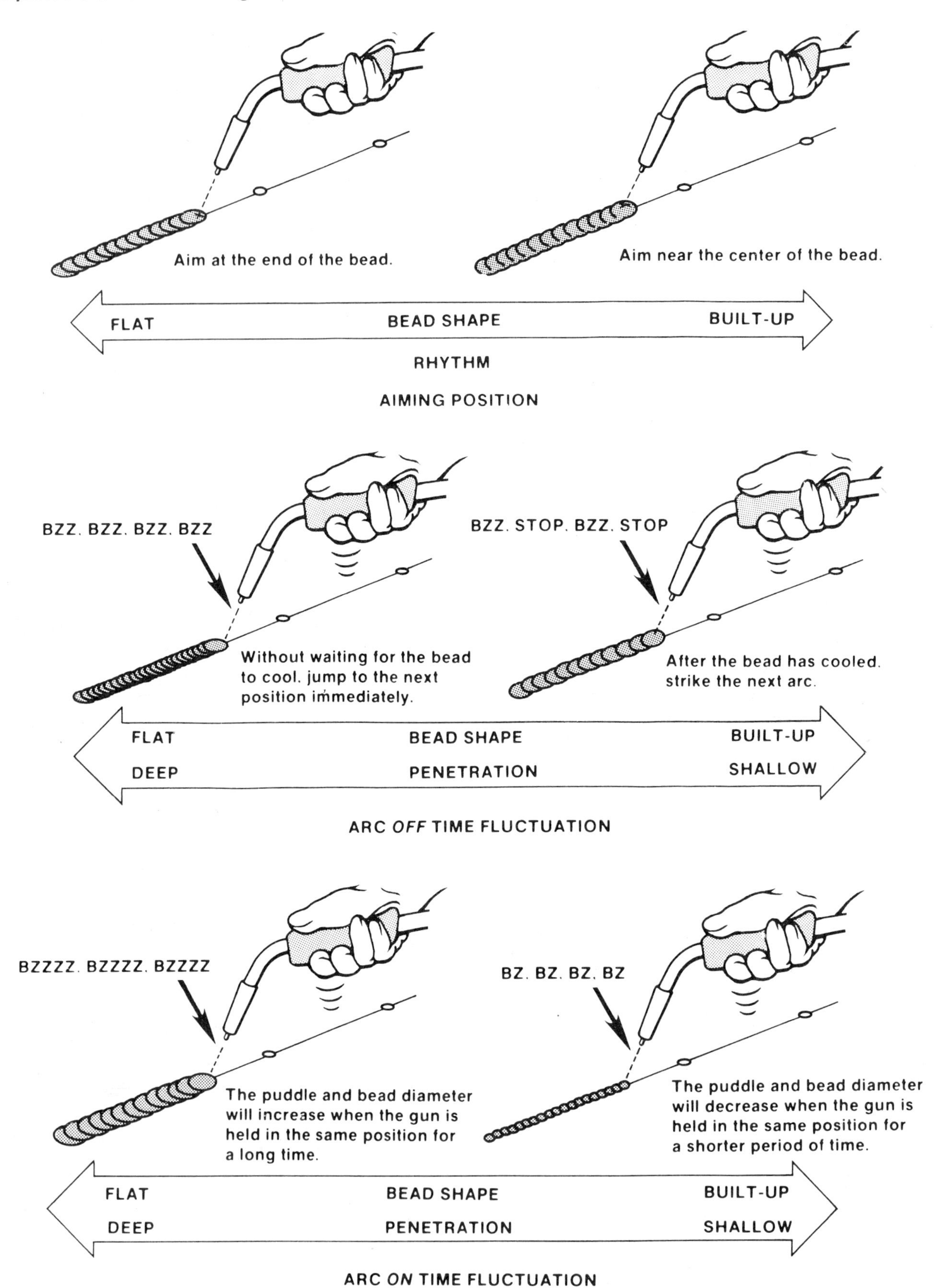

FIGURE 3–16 Steps in achieving a proper weld bead *(Courtesy of Toyota Motor Corp.)*

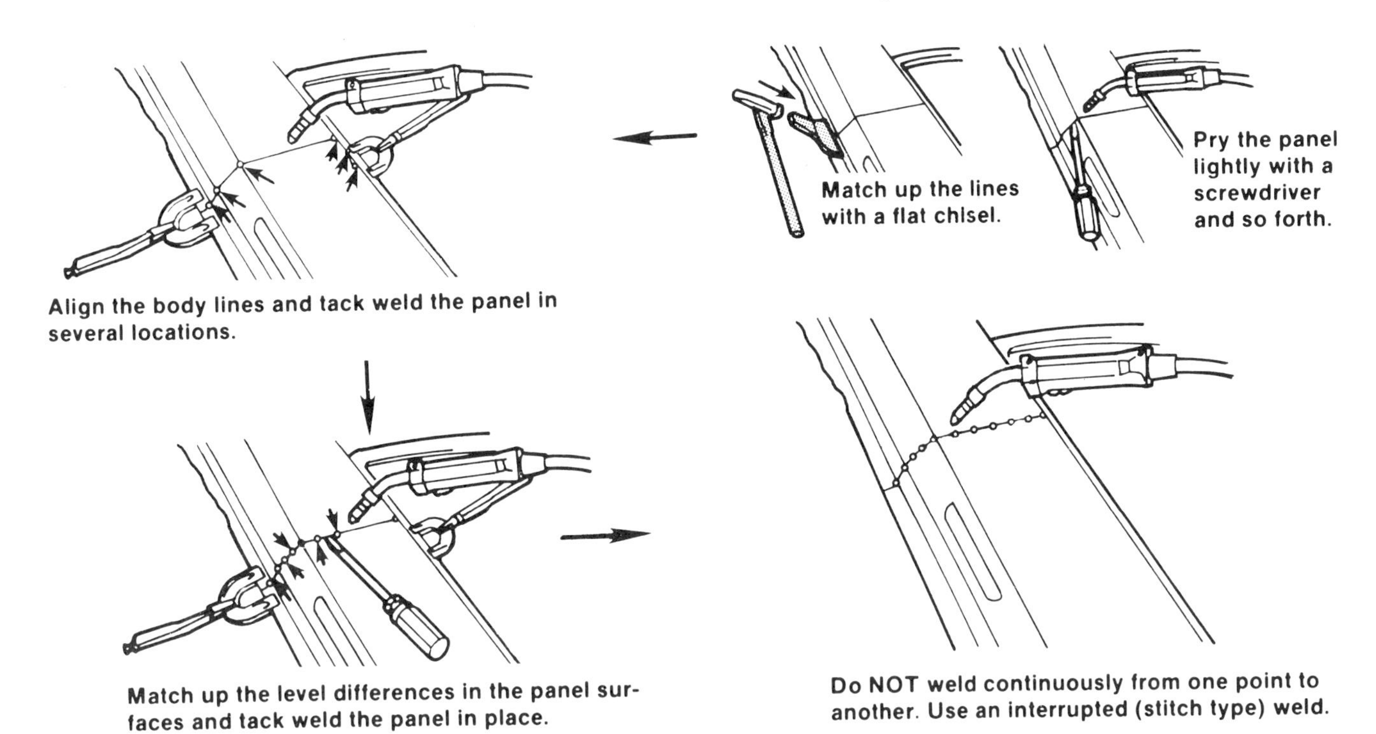

FIGURE 3-17 Procedure for butt welding replacement panels *(Courtesy of Toyota Motor Corp.)*

nique is a must to prevent burn-through. The combination of the proper gun angle and correct cycling techniques will achieve a satisfactory weld bead (Figure 3-16). The reverse welding method should be used for moving the gun because it is easier to aim at the bead.

Figure 3-17 shows a typical butt welding procedure for installing a replacement panel. If the desired results are not obtained using this method, the cause for the problem may be that the distance between the tip of the gun and the base metal is too great. Remember that weld penetration decreases as the distance between the tip and the base metal increases. Try holding the tip of the gun at several different distances away from the base metal until the proper distance that gives the desired results is found (Figure 3-18).

Moving the gun too fast or too slowly (Figure 3-19) will yield poor results (even if the speed of the wire feed is constant). A gun speed that is too slow will cause melt through. Conversely, a gun speed that is too fast will cause shallow penetration and inferior weld strength.

Even if a proper bead is formed during butt welding, panel warpage can result if the weld is started at or near the edge of the metal (Figure 3-20A). Therefore, to prevent panel warpage, disperse the heat into the base metal by starting the weld in the center and frequently changing the location of the weld

INCORRECT

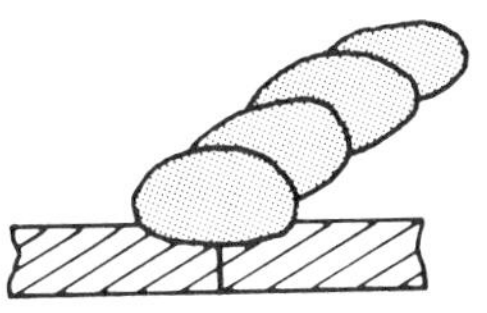

Insufficient penetration. Weld strength is poor and the panel could separate when the panel is finished with a grinder.

INCORRECT

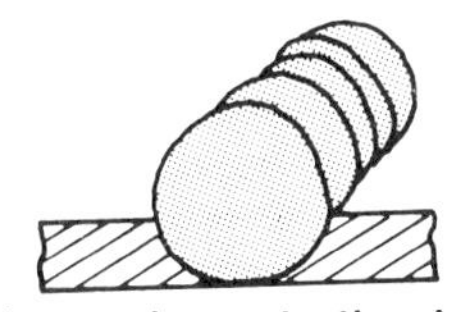

There is good penetration but finish grinding will be both difficult and time consuming.

CORRECT

There is good penetration and the bead will be easy to grind.

Good Penetration and Easy to Grind

FIGURE 3-18 Analyzing bead cutaway shapes

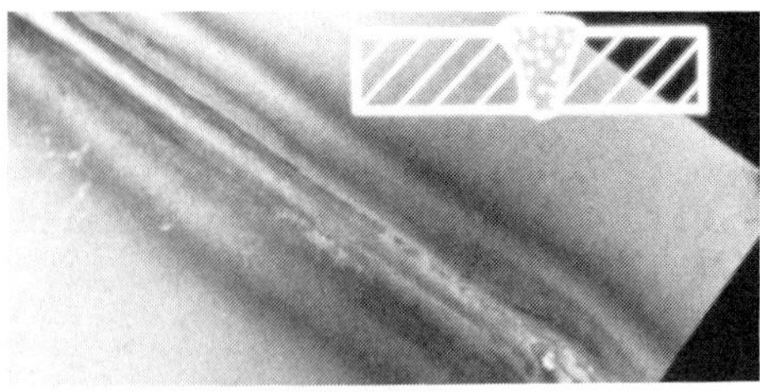

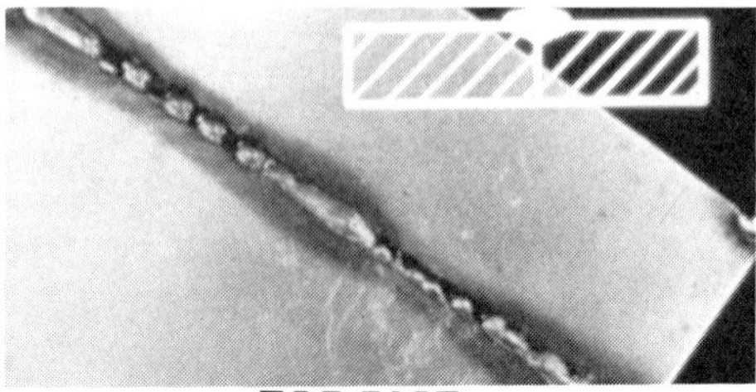

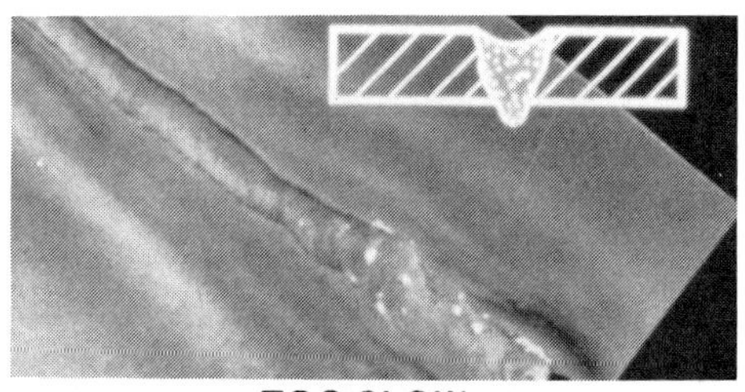

FIGURE 3-19 The relation of gun speed to bead shape

area (Figure 3-20B). A good rule of thumb is this: the thinner the panel thickness, the shorter the bead length.

When welding a butt joint, be sure the weld penetrates all the way through to the backside. Where the metal thickness at a butt joint is 1/16 inch or more, a gap should be left to assure full penetration. If it is not practical to leave a gap, grind a V-groove in the joint (Figure 3-21) so the weld can penetrate to the backside.

When butt welding vertical sections of structural panels, the weld cannot be made in a straight line on the panels. In addition, do not reinforce butt welds. Reinforcement can create a weaker condition than nonreinforced welds because of stress or structural buildup at the area of reinforcement.

Lap and Flange Welds

Both lap and flange welds (Figure 3-22) are made with identical MIG techniques. They are formed by melting the two surfaces to be joined at the edge of the top of the two overlapping surfaces. This is similar to butt welding, except only the top surface has an edge. Lap and flange welds should only be made in repairs where they replace identical original factory welds, or where outer panels and not structural panels are involved. These welds should not be used to join more than two thicknesses of material together. The same technique used for temperature control in butt welding should be followed for lap and flange welding. Welds should never be made continuously, but should be sequenced to allow for natural cooling; this will prevent temperature buildup in the weld area.

A flat lap or flange joint can be easily welded if some basic measures are taken. When heating the two plates, caution must be exercised to ensure that they start melting at the same time. Otherwise, the heat is not distributed uniformly in the joint, as seen in Figure 3-23.

Because of this difference in heating rate, the gun must be directed on the bottom plate (Figure 3-24). The filler metal should be added to the top plate. Gravity will pull the molten puddle down, so it is not necessary to put metal on the bottom plate.

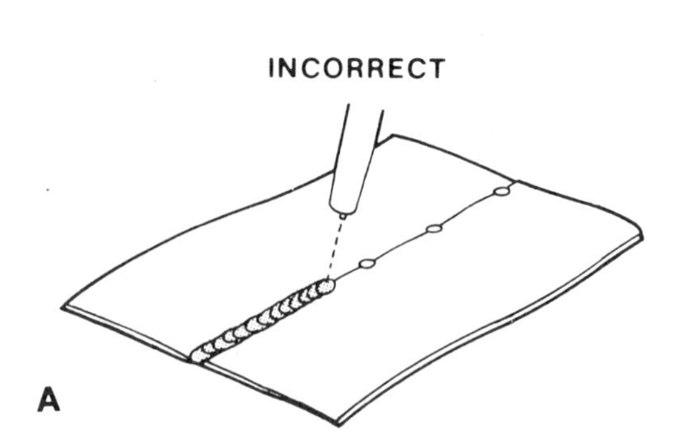

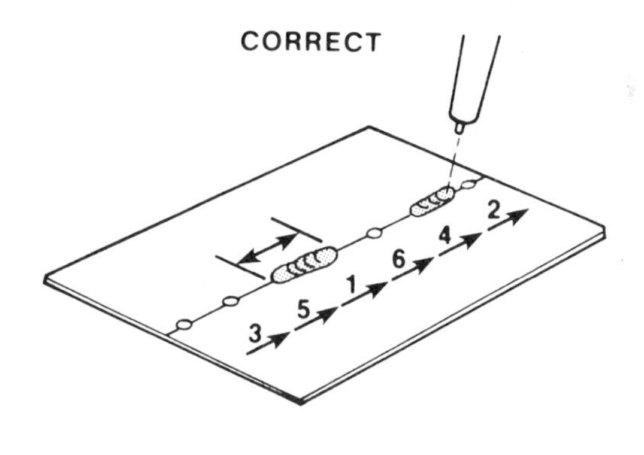

FIGURE 3-20 (A) Starting a butt weld at or near the edge of the panel causes warpage. (B) Instead, always start in the center and frequently change the location of the weld area. *(Courtesy of Toyota Motor Corp.)*

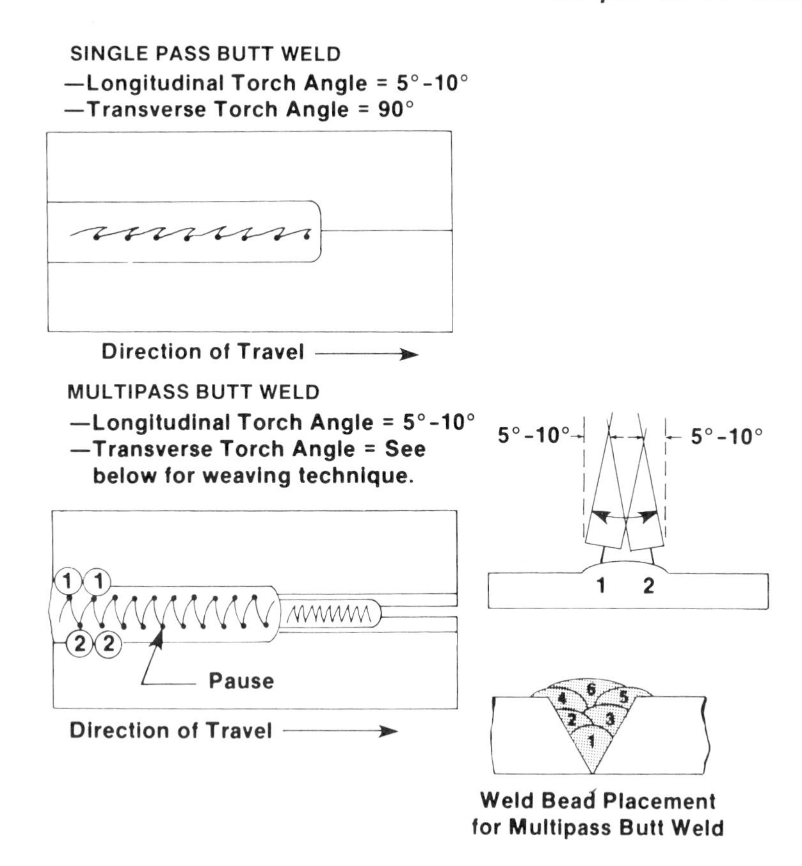

FIGURE 3-21 Gun positions for butt welding

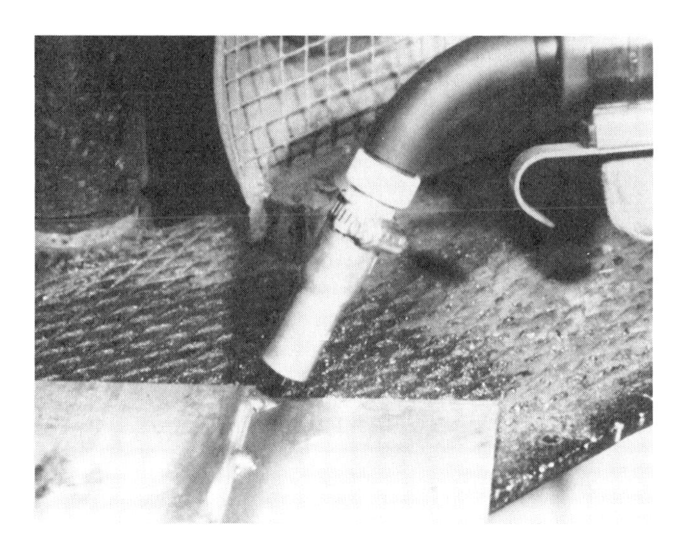

FIGURE 3-22 Welding a lap or flange joint

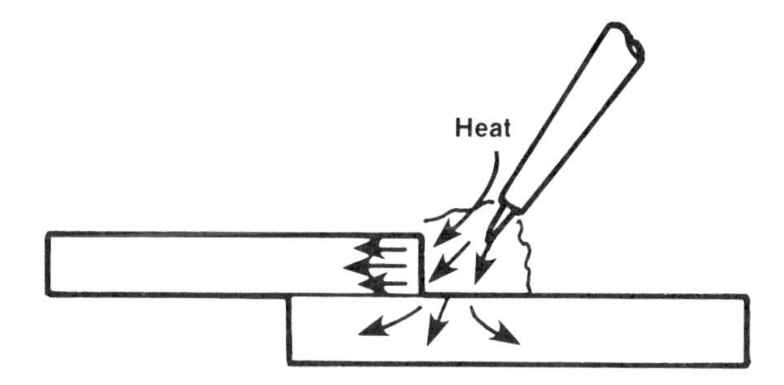

FIGURE 3-23 Heat is conducted away faster in the bottom plate, resulting in the top plate melting more quickly.

Keep in mind that if the filler metal is not added to the top plate, or if it is not added fast enough, surface tension will pull the puddle back from the joint. When this happens, the filler metal should be inserted directly into the notch to close it. The weld appearance and strength will not be affected.

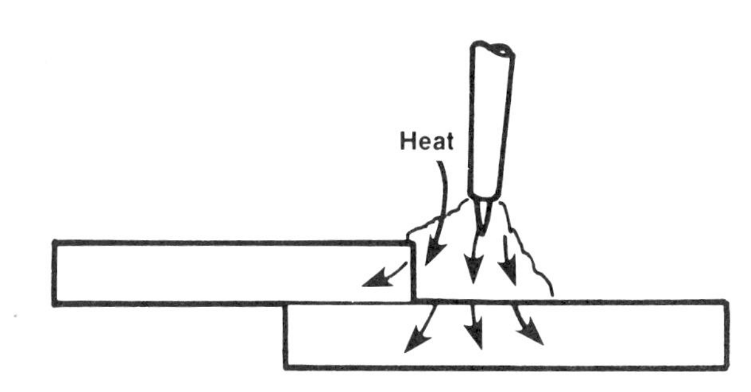

FIGURE 3-24 Direct the heat at the bottom of the plate to compensate for thermal conductivity.

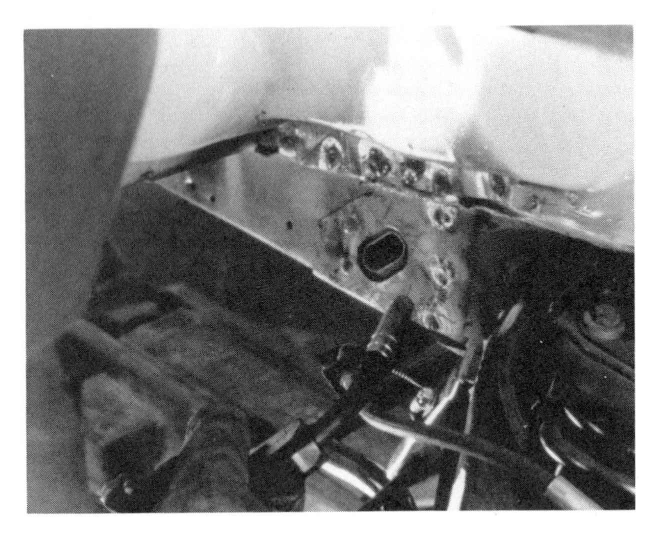

FIGURE 3-25 Plug welding is similar to spot welding.

Plug Welds

The plug weld is the body shop alternative to the OEM resistance spot welds made at the factory. It can be used anywhere in the body structure where a resistance spot weld was used. A plug weld has ample strength for welding load bearing structural members, and it can also be used on cosmetic body skins and other thin-gauge sheet metal.

Plug welding (Figure 3-25) is a form of spot welding; it is basically spot welding through a hole. That is, a plug weld is formed by drilling or punching (Figure 3-26) a hole in the outer panel being joined (Figure 2-27). The materials should be tightly clamped together. Holding the torch at right angles to the surface, put the electrode wire in the hole, trigger the arc briefly, then release the trigger. The puddle fills the hole and solidifies (Figure 3-28).

When plug welding, try to duplicate both the number and the size of the original factory spot welds. The hole that is punched or drilled should not be larger in diameter than the factory weld nugget. The 3/16-inch hole customarily used for plug welding cosmetic panels is not sufficient for plug welding

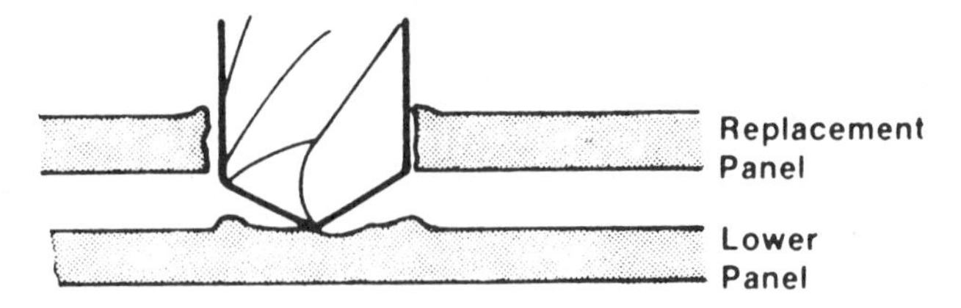

FIGURE 3-26 Plug welds are formed by drilling or punching a hole in the outer panel.

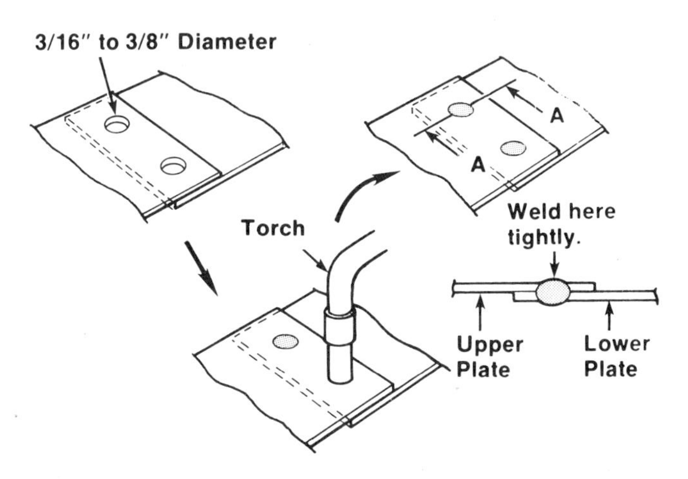

FIGURE 3-27 Steps in making a plug weld *(Courtesy of Nissan Motor Corp.)*

FIGURE 3-28 Plugging the weld hole

structural members. Most structural members require a 5/16- to 3/8-inch hole to achieve acceptable weld strength. When filling a larger hole, move the gun slowly in a circular motion around the edges of the hole (Figure 3-29), filling in the cavity. For smaller holes, it is best to aim the gun at the center of the hole and keep it stationary. A flat, gently sloping bead gives a nice appearance and reduces the grinding or sanding operations.

Proper wire length is an important factor in obtaining a good plug weld. If the wire protruding out of the end of the gun is too long, the wire will not melt properly, causing inferior weld penetration. The weld will improve if the gun is held closer to the base metal. Be sure the weld penetrates into the lower panel; round, dome-shaped protrusions on the underside of the metal are good indicators of proper weld penetration.

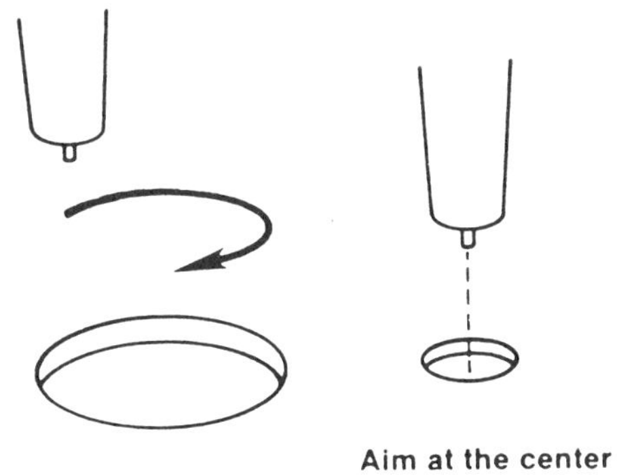

FIGURE 3-29 Slowly move the gun around the edges of the hole to fill the plug weld.

SHOP TALK

Intermittent plug welding leads to the generation of oxide film on the surface, which in turn causes blowholes. If this occurs, remove the oxide film with a wire brush (Figure 3-30).

The weld area should be allowed to cool naturally before any adjacent welds are made. Areas around the weld should not be force cooled using water or air; it is important that they are allowed to cool naturally. Slow, natural cooling without the use of water or air will minimize any panel distortion and maintain the strength designed into the panels.

FIGURE 3-30 Removing oxide film with a wire brush

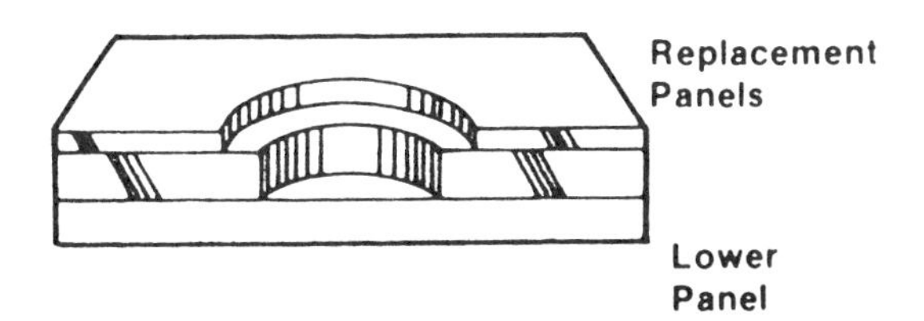

FIGURE 3-31 Welding two or more panels using the plug welding technique

Plug welds can also be used to join more than two panels together. When this is being done, a hole is punched in every panel except the lower one (Figure 3-31). The diameter of the plug weld hole in each additional panel being joined should be smaller than the diameter of the plug weld hole on top. Likewise, if panels of different thicknesses are being joined, a larger hole is punched in the thinner panel to assure that the thicker panel is melted into the weld first. In addition, make sure the thinner panel is on top.

A plug weld can be accomplished automatically and in a minimum amount of time using a MIG welder, creating less temperature buildup in adjacent panels. And while adjacent welds should not be made immediately, the area being welded will cool in a very short period of time.

Spot Welds

It is possible to spot weld (Figure 3-32) with MIG equipment. In fact, most of the better MIG machines now available and designed for collision repair work have built-in timers that shut off the wire feed and welding arc after the time required to weld one spot (Figure 3-33). Some MIG equipment also has a burnback time setting that is adjusted to prevent the

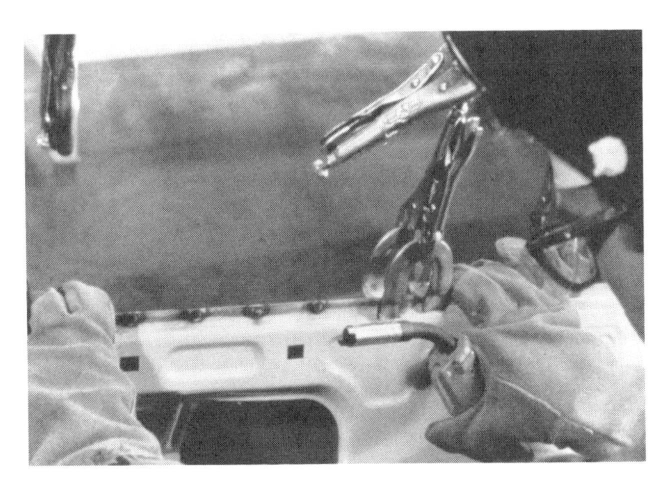

FIGURE 3-32 Most MIG welders can be used for spot welding.

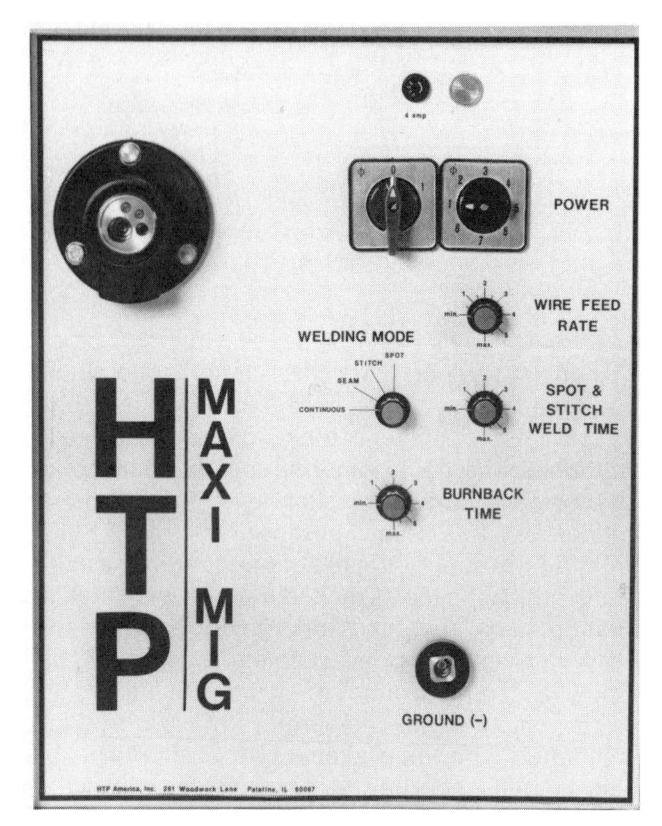

FIGURE 3-33 Controls on a typical MIG welder *(Courtesy of HTP America, Inc.)*

wire from sticking in the puddle. The setting of these timers depends on the thickness of the workpiece. This information can be found in the machine's owner's manual.

Since spot welding usually involves more heat than running a continuous bead, make a test weld first to select the right welder settings. Check the penetration of the spot weld by pulling the test weld apart. A good weld will tear a small hole out of the bottom piece, while a weak weld will break off at the surface. To get more penetration, increase the weld time or heat. To reduce penetration, reduce the time or heat.

For MIG spot welding, a special nozzle (Figure 3-34) must replace the standard nozzle. Once the gun is in place and the spot timing, welding heat, and backburn times are set for the given situation, the spot nozzle is held against the weld site and the gun triggered. For a very brief period of time, the timed pulses of wire feed and welding current are activated, and the arc melts through the outer layer and penetrates the inner layer (Figure 3-35). Then the automatic shut-off goes into action and no matter how long the trigger is squeezed, nothing will happen. However, when the trigger is released and then squeezed again, the next spot pulse is obtained.

FIGURE 3-34 Special spot welding nozzle

FIGURE 3-35 The arc melts through the outer layer and penetrates the inner layer.

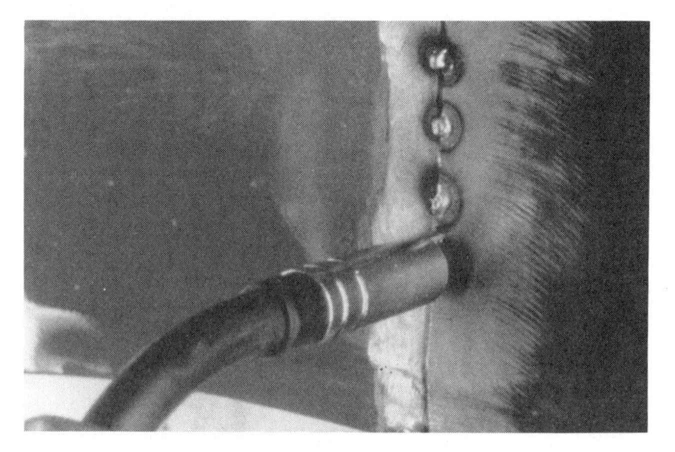

FIGURE 3-36 The lap spot technique is effective for quick welding of lap joints and flanges on thin-gauge nonstructural sheets and skins.

Because of varying conditions, the quality of a MIG spot weld is difficult to ascertain. On load bearing members, therefore, the plug weld or resistance spot welding technique described later is the preferred method. However, the MIG lap spot technique is effective for quick welding of lap joints and flanges on thin-gauge nonstructural sheets and skins (Figure 3-36). Here again the spot timer is set, but this time the nozzle is positioned over the edge of the outer sheet, at an angle slightly off 90 degrees. This will allow contact with both pieces of metal at the same time. The arc melts into the edge and penetrates the lower sheet.

Stitch Welds

In MIG stitch welding, the standard nozzle is used, not the spot nozzle. To make a stitch weld, combine the spot welding process with the continuous welding technique and gun travel (Figure 3-37). Set the automatic timer—either a shut-off or pulsed interval timer—depending on the MIG machine (Figure 3-38). The spot weld pulses and shut-offs recur with automatic regularity: weld-stop-weld-stop-weld-stop as long as the trigger is held in.

FIGURE 3-37 The results of combining spot welding with the continuous welding technique and gun travel

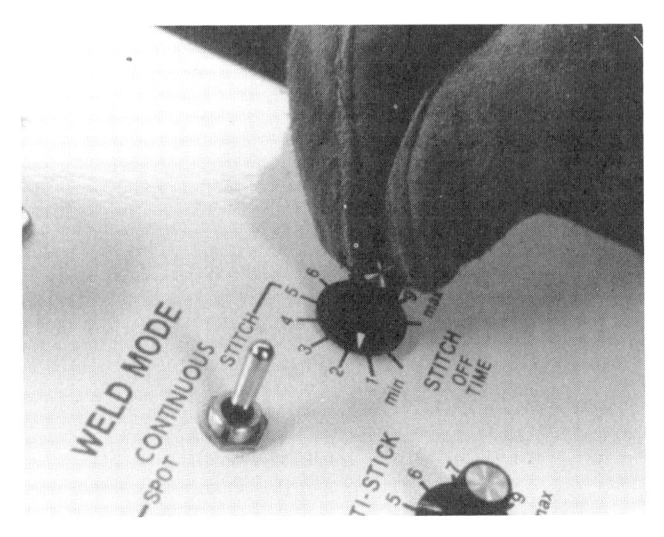

FIGURE 3-38 Setting the shutoff, or interval, timer

Another way to look at this is weld-cool-weld-cool-weld-cool and so on, because the arc-off period allows the last spot to cool slightly and start to solidify before the next spot is deposited. This intermittent technique means less distortion and less melt through or burn through. These characteristics make the stitch weld preferable to the continuous weld for working thin-gauge cosmetic panels.

The intermittent cooling and solidifying of the stitch weld also makes it preferable to continuous welding on vertical joints where distortion is a problem (Figure 3-39). The operator does not have to contend with a continuous puddle that gravity is trying to pull down the joint ahead of the arc. By the same token, stitch welding is also preferable in the overhead position (Figure 3-40), because there is virtually no puddle for gravity to pull.

If the MIG machine does not have the automatic stitch modes, the spot and stitch welds can be made manually. The operator merely has to be capable of triggering the gun on-off, on-off, on-off—the same way the automatic system does (Figure 3-41). When finish grinding is required after welding, use light pressure and short grinds to keep the heat from building up. Otherwise, clean the weld area with a wire brush.

USING NOZZLE SPRAY

Welding spray or paste should never be present in a vaporous state in the nozzle when welds are

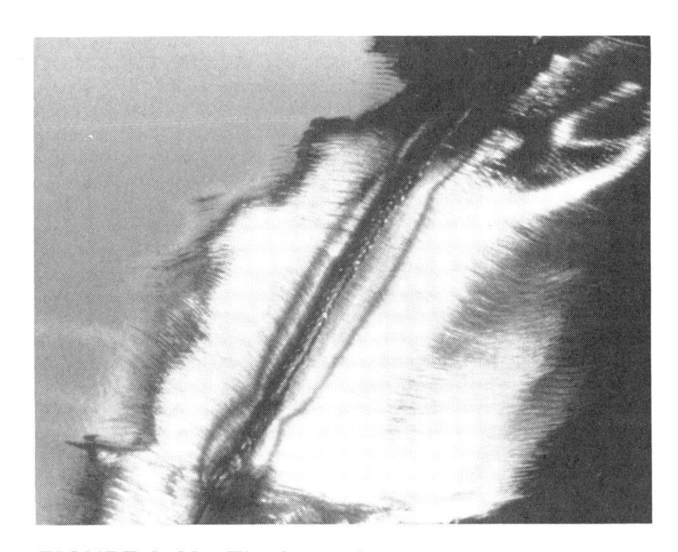

FIGURE 3-39 The intermittent cooling and solidifying of the stitch weld makes it preferable to continuous welding on vertical joints.

FIGURE 3-40 Stitch welding is preferable in the overhead position.

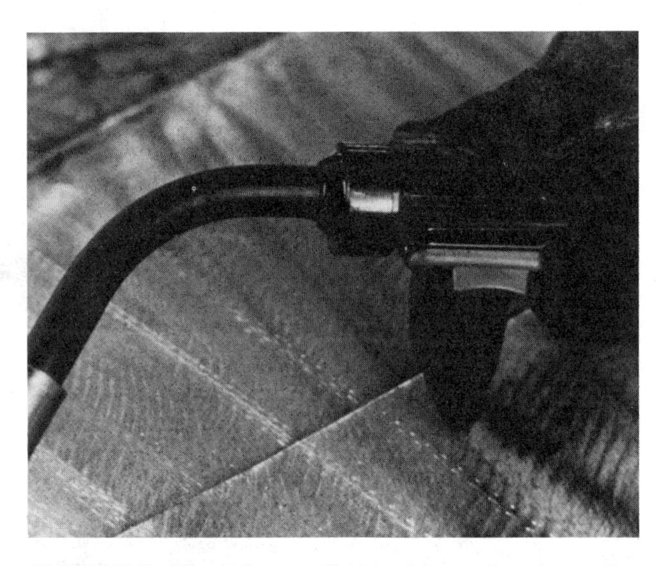

FIGURE 3-41 It is possible to manually trigger the gun on-off, on-off, on-off, and so on.

being made. Instead, nozzle spray is highly preferred because it does not leave a heavy residue inside the nozzle. Nozzle spray should never be put into the nozzle while it is attached to the welding gun. Proper application is with the nozzle off, allowing the spray to dry or "flash off" prior to reinstalling the nozzle on the gun. A straight blade screwdriver makes an excellent tool for scraping the nozzle.

Care should also be taken never to put the nozzle spray into the contact tip; this can form a coating inside the tip, thus preventing proper electrical contact of the wire with the tip. As the operator learns to control the wire transfer by adjusting the heat and wire speed, it will be possible to achieve the desired penetration and welding speed in clean, uncoated metal as well as in zinc-plated, primed, and painted material.

MIG WELDING GALVANIZED METALS

Since there is little question as to the superior durability of coated steels when compared to uncoated steels, it is very important that the auto body technician is able to weld them. But although the name might change (galvanized, zinc-metallized, zinc-coated), the basic problem remains the same: coated steels just do not weld very well. In fact, the weldability is so poor that some technicians grind off the zinc near the weld seam to avoid dealing with it. Of course, this results in premature corrosion of the steel unless a well-through primer is used.

Some of the problems associated with welding coated steels compared to uncoated include the following:

- Zinc fumes
- Erratic arc
- Reduced penetration
- Increased spatter
- Porosity

Although these problems cannot be totally eliminated, they can at least be controlled to a degree.

DEALING WITH FUMES AND SPATTER

The fumes given off while welding galvanized metals are due to the fact that zinc melts at 786 degrees Fahrenheit and has a boiling point of 1665 degrees Fahrenheit. The melting point of steel is about 2600 degrees Fahrenheit. For this reason, it is easy to understand why welding zinc-coated steels yields fumes in the form of a white powdered smoke. The zinc melts at a temperature so much lower than that of steel that as soon as the welding arc is initiated, zinc fumes appear. Besides being hazardous to one's health, these fumes cause the weld puddle to become very erratic, which results in porosity, spatter, and reduced penetration.

The first considerations for anyone welding galvanized metals are safety and health. Inhaling substantial amounts of zinc fumes will usually make a person sick. Always provide adequate ventilation and wear a respirator. A small fan set off to one side will help clear out the fumes. Do not direct the fan right at the work or a porous, weak weld could result.

When welding zinc-coated steels, sometimes the weld puddle seems to literally explode, sending globules of hot molten metal everywhere. These globules can cause severe burns as well as ignite combustibles. Make sure to dress appropriately for welding—leather cape sleeves are a great help in protecting the upper body and arms from burns. Use a good quality welding helmet and wear earplugs and a skull cap.

Make sure to protect nearby glass, upholstery, and trim from the spatter and globules. Make sure that combustible materials are moved away from the weld site. A portable welding screen (Figure 3-42) offers excellent protection for walls and objects that cannot be easily moved out of the general welding area. These screens are flame retardant, so weld spatter falls right off them.

ERRATIC ARCS

As already noted, zinc melts at a lower temperature than steel, and this causes white smoke or residue. When the zinc fumes become vaporized in the heat of the arc, the weld puddle becomes quite erratic. It is very hard trying to weld thin-gauge metals with an arc that seems to have a mind of its own, but there are some things the technician can do to stabilize the arc.

First, increase the amperage by five or ten and increase the volts by one. On most MIG machines, the increased amperage means an increase in wire speed of two or three short marks on the wire speed adjustment dial. This change in machine setting to a "hotter arc" will help vaporize the zinc fumes somewhat.

Next, move the gun somewhat slower than normal. This will give the leading edge of the weld puddle time to vaporize the zinc before the bead actually passes over that area. Pulling the gun concentrates the heat of the arc, which aids in burning out the zinc. Pushing the gun angle seems to spread the heat of the arc sideways, which vaporizes more zinc than necessary.

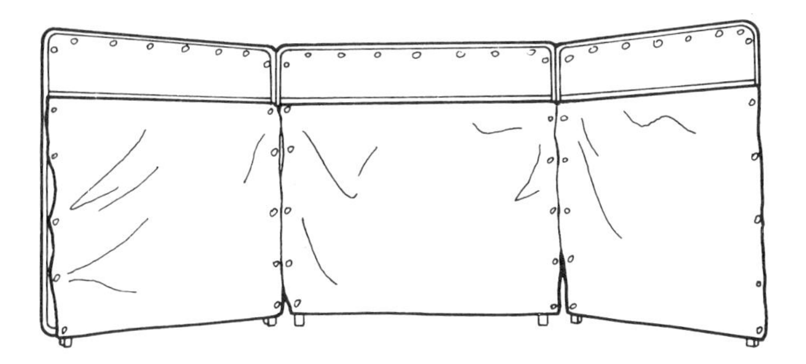

FIGURE 3-42 A flame-retardant welding screen

HEAT INPUT

Because a "hotter" arc is recommended for welding galvanized metals, heat input is very important. Heat input is in direct relationship to the machine settings, travel speed, and how quickly the welded joints must be made to complete a repair. Because the "hotter" settings and slow travel speed work better on zinc-coated steels, use a skip welding technique, weld short beads (no more than 1-1/2 inches at a time), and allow the heat to dissipate before continuing to weld. (Simply pause several minutes between welds.)

Penetration is reduced slightly when welding zinc-coated steels. Wider gaps are usually required for butt joints, which makes penetration control difficult. For this reason, the I-CAR procedure, which uses a backup piece when making splices, is the ideal way to weld a butt joint. When the gap is opened up to compensate for the reduced penetration on an open butt joint, the gap becomes so wide that burn-through problems are created. Avoid the open butt joint whenever possible and use the I-CAR method of inserts on all zinc-coated steels.

The most common shielding gas used for welding zinc-coated steel is 75 percent argon and 25 percent CO_2. The exception is that a flow of 40 cfh seems to work a bit better on coated steels than a flow of 20 or 30 cfh. Without a doubt, no other shielding gas works as well as 75/25. As for the wire, the ER-705-6 is recommended for MIG welding all zinc-coated steel.

DEALING WITH POROSITY

One of the biggest faults with the weld bead made on coated steels is porosity. If the holes are no larger than 1/32 inch in diameter and are no more frequent than one hole for every 1 inch of weld, porosity will probably not impair the quality of the weld or the strength of the joint. If there are three or four holes for every inch of weld, grind them out, then reweld the bad portion. The extent of porosity will depend on amperage, volts, welding speed, and thickness of the coating. Remember to increase amperage slightly, increase the volts by one, and travel slower than when welding uncoated steels (Figure 3-43). The problem of porosity is explored in further detail later in this chapter.

Welding on coated steels will result in some loss of zinc immediately on either side of the welded seam. This loss usually extends 1/8 to 3/16 inch from the weld on either side. Extending out from the weld for a distance of about 1/2 inch, sometimes it is

FIGURE 3-43 When welding galvanized metals, use a slower gun travel speed.

possible to see where the surface was heated enough to melt and redistribute the zinc. This area will create no special problems if proper rustproofing methods are employed to make the repair corrosion resistant.

When grinding the weld, be careful not to remove any more coating than is necessary. Also, avoid grinding away too much surface metal; this can result in a reduction in the thickness of the metal.

TESTING MIG WELDS

To ensure the quality, durability, and strength of a MIG weld, it is important to periodically test the welding results. Naturally, the extent of any testing depends on the type of service the welded part will be required to withstand. For example, a structural member must be welded to much more exacting standards than purely cosmetic parts of the vehicle. Results acceptable in one welding application might not meet the needs of another.

Discontinuity is the term used for an interruption of the typical structure of a weld. Ideally, a finished weld should have no discontinuities, but this is virtually impossible. When inspecting a weld, note the type, size, and location of the discontinuity. The most common weld discontinuities are:

- Porosity
- Inclusions
- Inadequate joint penetration
- Incomplete fusion
- Overlap
- Undercut
- Cracks
- Laminations
- Delaminations
- Lamellar tears
- Arc strikes

POROSITY

Porosity results from gas trapped during solidification of the weld metal. It is most often caused by improper welding techniques, contamination, or an improper chemical balance between the filler and the base metal. Porosity is either spherical or cylindrical in shape. Cylindrical porosity is also known as wormholes; the rounded edges tend to reduce the stresses around them. Therefore, unless this type of porosity is extensive, there is little or no loss in strength.

The intense heat of a weld can decompose paint, dirt, oil, rust, or other oxides, producing hydrogen. This gas, like nitrogen, can also become trapped in the solidifying weld pool, producing porosity. When it causes porosity, hydrogen can also diffuse into the heat-affected zone, producing underbead cracking in some steels.

Porosity can be grouped into the following major types:

- Uniformly scattered porosity (Figure 3-44) is most frequently caused by poor welding techniques or faulty materials.
- Clustered porosity (Figure 3-45) is most often caused by improper starting and stopping techniques.

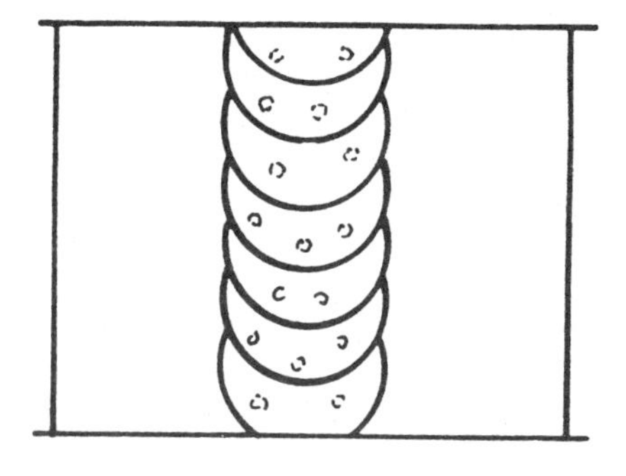

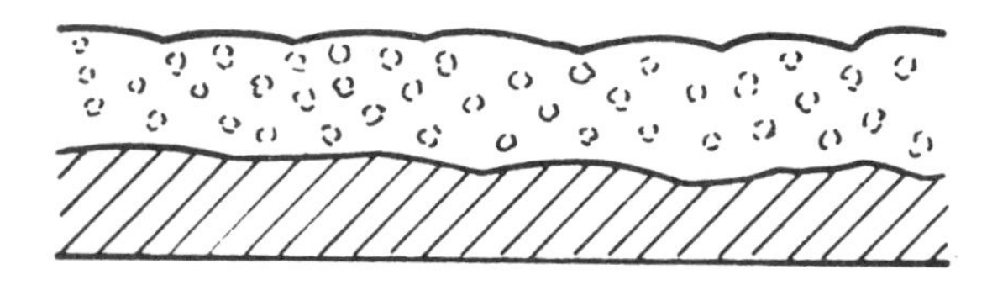

FIGURE 3-44 Uniformly scattered porosities

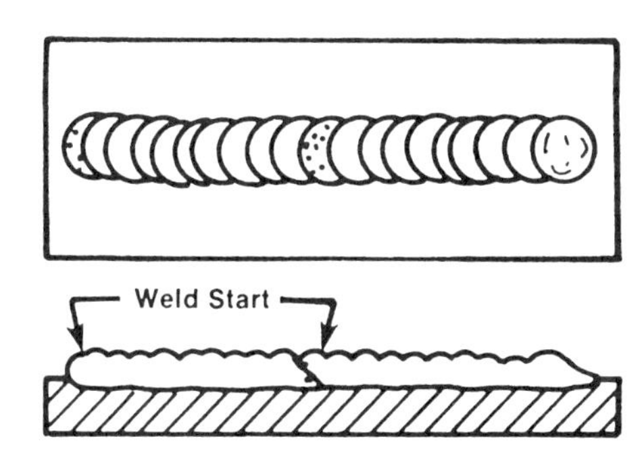

FIGURE 3-45 A clustered porosity

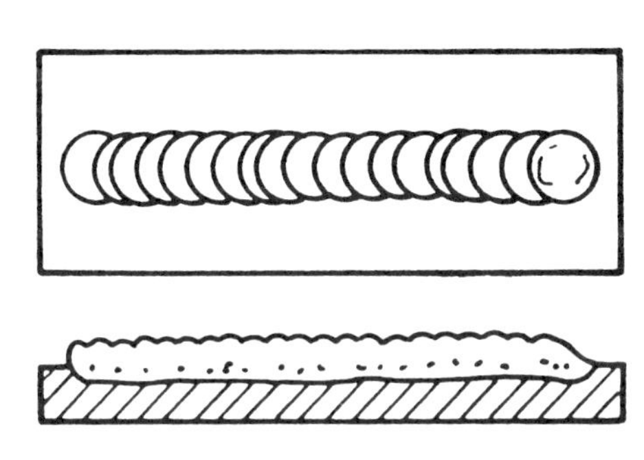

FIGURE 3-46 A linear porosity

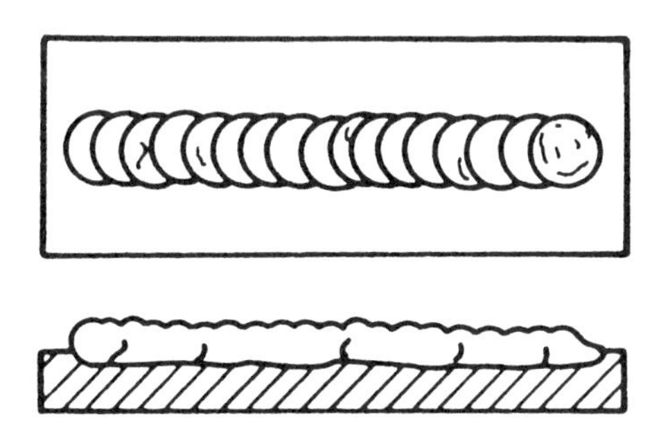

FIGURE 3-47 A piping porosity

- Linear porosity (Figure 3-46) is most frequently caused by contamination within the joint, root, or interbead boundaries.
- Piping porosity (Figure 3-47) is most often caused by contamination. This porosity is unique because its formation depends on the gas escaping from the weld pool at the same rate as the pool is solidifying.

INCLUSIONS

Inclusions are nonmetallic materials (such as slag and oxides) that are trapped in the weld metal, between weld beads, or between the weld and the base metal. Inclusions are sometimes jagged and irregularly shaped and can form in a continuous line. This reduces the structural integrity of the weld.

Inclusions are often found when the following conditions exist:

- Slag and/or oxides do not have enough time to float to the surface of the molten puddle.
- Sharp notches are present between weld beads or between the bead and the base metal that trap the material so that it cannot float out.
- The joint was designed with insufficient room for the correct manipulation of the puddle.

INADEQUATE JOINT PREPARATION

Inadequate joint penetration means that the depth that the weld penetrates the joint is less than is needed to fuse through the plate or into the preceding weld (Figure 3-48). A defect results that could reduce the required cross-sectional area of the joint or become a source of stress concentration that leads to fatigue failure. The importance of such defects depends on the safety factor to which the weld has been designed. Generally, if proper welding procedures were followed, such defects have only minimal effects on service.

The major causes of inadequate joint penetration are improper welding technique, insufficient welding current, improper joint fit-up, and improper joint design. The welding technique might require that both starting and runout tabs be used, so that the puddle is well estalished before it reaches the joint. Metals that are thick or have a high thermal conductivity can be preheated so that the weld heat is not drawn away so quickly by the metal that it cannot penetrate the joint. A root gap that is too small or a root face that is too large will also keep the weld from penetrating adequately. When joints are accessible from both sides, back gouging is often used to ensure 100 percent root fusion.

FIGURE 3-48 Examples of inadequate joint preparation

INCOMPLETE FUSION

Incomplete fusion is the failure of the molten filler metal and the base metal to mix completely. Even if the base metal melts, a thin layer of oxide can still prevent fusion from occurring. This can be caused by improper welding techniques, improper edge preparation, or improper joint design. Incomplete fusion can also result from insufficient heat to melt the base metal, inadequate agitation to break up oxide layers, or too little space allowed for correct puddle manipulation.

OVERLAP

When the weld metal protrudes beyond the toe, face, or root of the weld without actually being fused to it, the resulting discontinuity is known as *overlap.* It can occur as a result of lack of control of the welding process, improper selection of welding materials, or improper preparation of the materials prior to welding. If there are tightly adhering oxides on the base metal that interfere with fusion, overlap will result. Overlap is a surface-connected discontinuity that forms a severe notch.

UNDERCUT

This discontinuity is generally associated with either improper welding techniques or excessive current—or both. It is generally located at the junction of the weld and the base metal (the toe or root). If examined carefully, all welds have some undercut. Often it may be seen only when examined under magnification. When undercut is controlled within the limits of the specifications and does not constitute a sharp or deep notch, it is not usually considered a weld defect.

CRACKS

If localized stress exceeds the strength of a weld, cracks occur. Cracks are an excellent indicator of discontinuities present in either the weld or base metal. Welding-related cracks are generally brittle in nature, usually due to the high residual stresses present.

Cracks are classified as either hot or cold, depending upon the temperatures at which they form. Hot cracks develop at elevated temperatures and commonly form during the solidification of the metal at temperatures near the melting point. Cold cracks develop after the solidification of the metal is complete and are sometimes called *delayed cracks.* Hydrogen embrittlement is usually associated with the formation of cold cracks. Hot cracks form between the grains, while cold cracks form both between and through the grains.

LAMINATIONS

Laminations consist of flat, elongated discontinuities in the base metal. They may be internal and thus detectable only through ultrasonic testing. In other cases, laminations can extend to an edge of the component, where they are visible at the surface. Metals containing laminations often cannot be relied on to carry tensile stress.

DELAMINATIONS

The separation of a lamination under stress is known as a *delamination.* The stress can be residual from welding or it can be applied stress. Delaminations can be found visually at the edges of pieces or ultrasonically by testing with a normal beam search unit. A delamination discontinuity cannot be relied on to transmit tensile loads.

LAMELLAR TEARS

Lamellar tears are terrace-like separations in the base metal typically caused by shrinkage stresses resulting from welding. Lamellar tearing can extend over long distances. The tears are roughly parallel to the surface and generally initiate in regions of the base metal having a high incidence of inclusions or high residual stresses, or a combination of the two. The fracture usually forms at one lamellar plane and moves to another by shearing action.

ARC STRIKES

Arc strikes are small, localized points where surface melting occurs away from the joint. These spots can be caused by accidentally striking the arc in the wrong place and/or by faulty ground connections. Even though arc strikes can be ground smooth, they cannot be removed. Keep in mind that these spots can be the starting point for cracking.

DESTRUCTIVE TESTING

Most of the destructive testing methods used in industrial welding situations are simply not feasible in an automotive application. Tensile, fatigue, and impact testing are more conducive to evaluating welds on small, individual specimens; they are of little use on vehicles weighing more than 2000

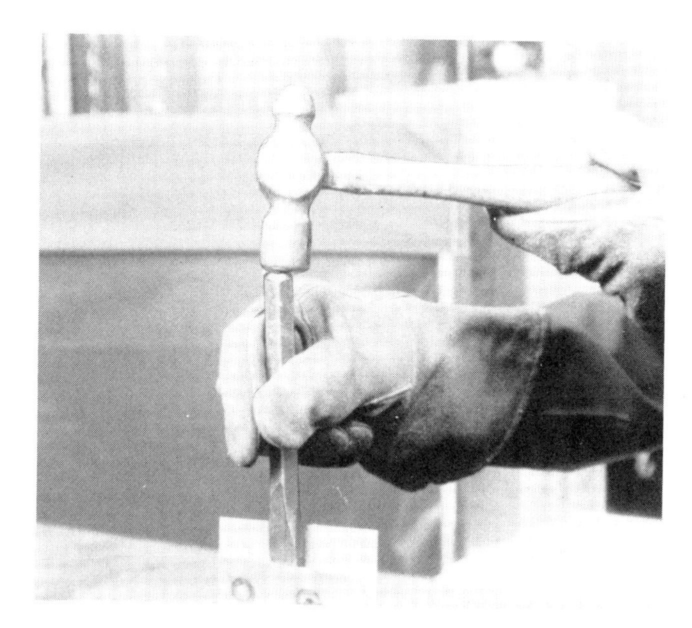

FIGURE 3-49 Checking the quality of a MIG weld

pounds. Probably the simplest, yet most effective, destructive test for automotive welds is illustrated in Figure 3-49. Using a hammer and chisel, try to break the weld apart. If reasonable force does not harm it, the weld integrity is good.

In lieu of destructive testing, it is good practice to make repair welds from time to time on each job. Before making any welds on the vehicle, secure test panels of the same gauge steel that is to be repaired on the vehicle. Make a few welds on the test panels; if the proper settings on the MIG welder are obtained on these pieces, the quality of the welds on the vehicle can be assured.

NONDESTRUCTIVE TESTING

Nondestructive testing is an effective method of checking for surface defects, including cracks, undercuts, arc burns, and lack of penetration. The three methods of nondestructive testing that will be discussed here include visual inspection, penetrant inspection, and ultrasonic inspection.

Visual Inspection

Visual inspection is the most frequently used nondestructive testing method. Many times, a weld receives only a visual inspection. If the weld looks good, it passes; if it looks bad, it is rejected. This procedure can be easily overlooked when more sophisticated nondestructive testing methods are used, but it should not be. In fact, a visual inspection should be done before any other tests to eliminate the obvious problem welds.

A visual inspection schedule can reduce the finished weld rejection rate by more than 75 percent. Visual inspection can easily be used to check for fit-up, interpass acceptance, welder technique, and other variables that ultimately affect weld quality. Minor problems can be identified and corrected before the weld is completed, thus eliminating costly repairs or rejection.

Penetrant Inspection

Penetrant inspection is used to locate very small surface cracks, as well as porosity. There are two types of penetrants now in use, the color-contrast and the fluorescent versions. Color-contrast penetrants contain a colored dye that is visible under ordinary white light. Fluorescent penetrants contain a more effective fluorescent dye that is only visible under black light.

When using a penetrant, proceed as follows:

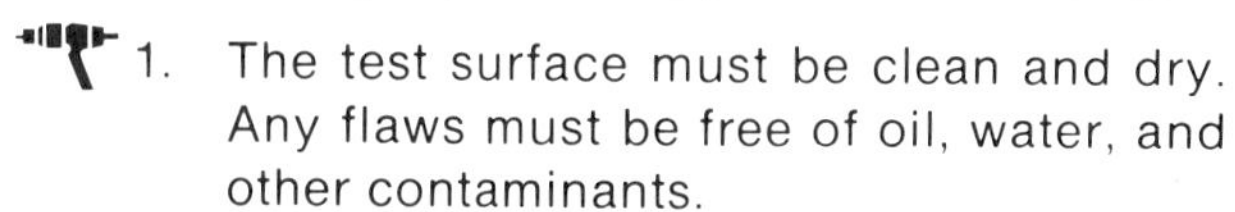

1. The test surface must be clean and dry. Any flaws must be free of oil, water, and other contaminants.
2. Cover the test surface with a film of penetrant. This can be accomplished by dipping, bathing, spraying, or brushing.
3. Wipe, wash, or rinse the test surfaces free of excess penetrant. Dry with cloths or hot air.
4. Apply developing powder to the test surface. It will act like a blotter and speed the penetrant out of any flaws on the test surface.
5. Depending on the type of penetrant applied, make the inspection under ordinary white light or near-ultraviolet black light. When viewed under the latter, the penetrant fluoresces to a yellow-green color which clearly defines the defect.

Ultrasonic Inspection

Ultrasonics is a fast and relatively low-cost nondestructive testing method. It employs electronically produced, high-frequency sound waves (roughly 25 million cycles per second), which penetrate metal at speeds of several thousand feet per second. The two types of ultrasonic equipment are pulse and resonance. The pulse-echo system, most often employed with welding, uses sound generated in short bursts or pulses. Since high-frequency sound has

TABLE 3-2: MIG WELDING DEFECTS AND THEIR CAUSES

Defect	Defect Condition	Remarks	Possible Causes
Pores/Pits	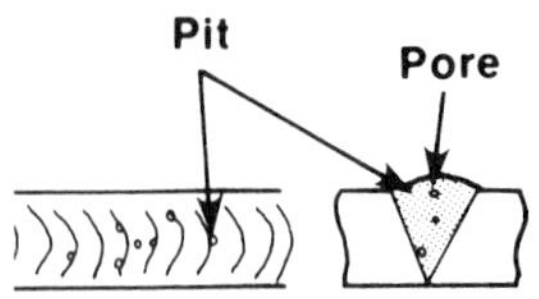 Pit Pore	There is a hole made when gas is trapped in the weld metal.	There is rust or dirt on the base metal. There is rust or moisture adhering to the wire. Improper shielding action (the nozzle is blocked or wind or the gas flow volume is low). Weld is cooling off too fast. Arc length is too long. Wrong wire is elected. Gas is sealed improperly. Weld joint surface is not clean.
Undercut	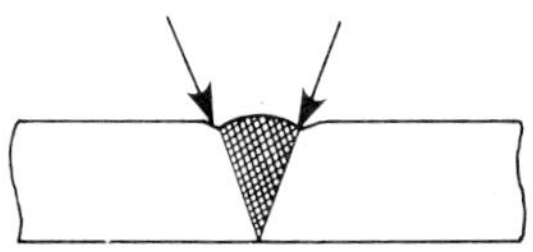	Undercut is a condition where the overmelted base metal has made grooves or an indentation. The base metal's section is made smaller and, therefore, the weld zone's strength is severely lowered.	Arc length is too long. Gun angle is improper. Welding speed is too fast. Current is too large. Torch feed is too fast. Torch angle is tilted.
Improper Fusion	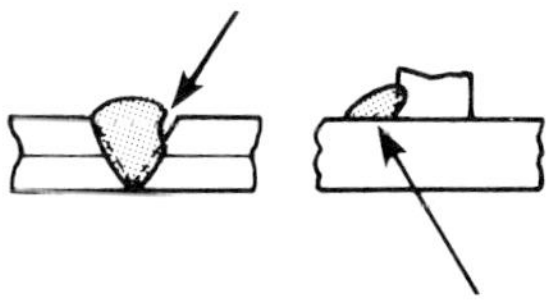	This is an unfused condition between weld metal and base metal or between deposited metals.	Check torch feed operation. Is voltage lowered? Weld area is not clean.
Overlap	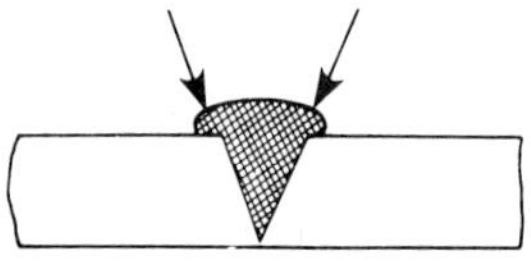	Overlap is apt to occur in fillet weld rather than in butt weld. Overlap causes stress concentration and results in premature corrosion.	Welding speed is too slow. Arc length is too short. Torch feed is too slow. Current is too low.
Insufficient Penetration	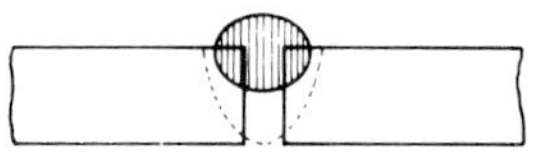	This is a condition in which there is insufficient deposition made under the panel.	Welding current is too low. Arch length is too long. The end of the wire is not aligned with the butted portion of the panels. Groove face is to small.
Excess Weld Spatter	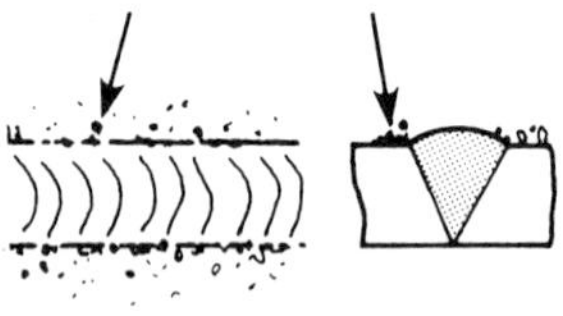	Excess weld spatter occurs as speckles and bumps along either side of the weld bead.	Arc length is too long. Rust is on the base metal. Gun angle is too severe.
Spatter (short throat)		Spatter is prone to occur in fillet welds.	Current is too great. Wrong wire is selected.

TABLE 3-2: MIG WELDING DEFECTS AND THEIR CAUSES (CONTINUED)

Defect	Defect Condition	Remarks	Possible Causes
Vertical Crack		Cracks usually occur on top surface only.	There are stains on welded surface (paint, oil, rust).
The Bead Is Not Uniform.		This is a condition in which the weld bead is misshapen and uneven rather than streamlined and even.	The contact tip hole is worn or deformed and the wire is oscillating as it comes out of the tip. The gun is not steady during welding.
Burn Through		Burn through is the condition of holes in the weld bead.	The welding current is too high. The gap between the metal is too wide. The speed of the gun is too slow. The gun-to-base metal distance is too short.

little ability to travel through air, it must be conducted into the component through a medium such as oil or water.

Sound is directed into the component with a probe held in a preselected angle or direction so that any defects will reflect energy back to the probe. Ultrasonic devices operate very much like depth sounders or "fish finders." The speed of sound through a material is a known quantity; ultrasonic devices measure the time required for a pulse to return from a reflective surface. Internal computers calculate the distance and present the information on a cathode ray tube where the results are interpreted. Sound not reflected by defects continues into the component. Defect size is determined by plotting the length, height, width, and shape.

Figure 3-50 shows the path of the sound beam used in butt welding testing. The operator must know the exit point of the sound beam, the exact angle of the refracted beam, and the thickness of the workpiece.

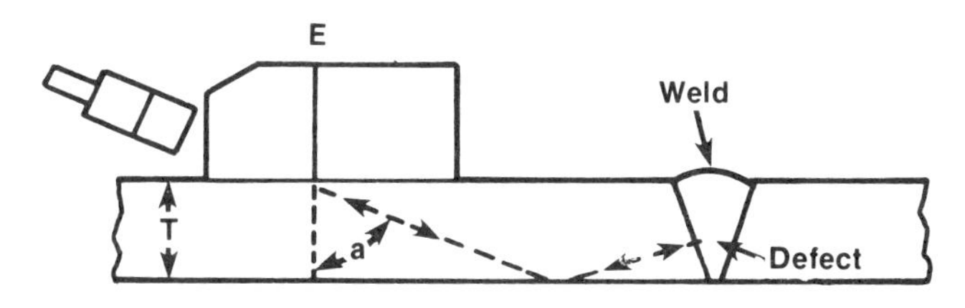

FIGURE 3-50 Ultrasonic testing of a butt weld

MIG WELD DEFECTS

A defect is officially defined as "one or more discontinuities that render a component unable to meet minimum accepted standards or specifications." While a defect may or may not be cause for rejection of a weld, it is almost always cause for repair. It is most important to become familiar with the various types of MIG welding defects in order to better recognize, repair, and, most importantly, avoid them. Table 3-2 summarizes common defects and their causes. Keep in mind that it is not intended to be all-inclusive; there are others not mentioned here, and those that are mentioned will not always conform to these exact descriptions. When defects do occur, always think of ways that the method of operation can be changed to prevent making the same mistake again.

REVIEW QUESTIONS

1. The higher the carbon content of steel, the ____________.
   a. more difficult it is to weld
   b. easier it is to weld
   c. less brittle it is
   d. both b and c

2. The property of a material that resists forces applied to pull it apart is known as ______________.
   a. tensile strength
   b. torsional strength
   c. compressive strength
   d. none of the above

3. Welder A says that normal MIG welding procedures can be used to restore HSS components. Welder B says that the MIG process should not be used on HSS steel. Who is correct?
   a. Welder A
   b. Welder B
   c. Both A and B
   d. Neither A nor B

4. HSLA steel is used in ______________.
   a. rocker panels
   b. bumper reinforcements
   c. front and rear rails
   d. all of the above

5. What type of steel cannot have heat applied to it for repairs?
   a. HSS
   b. HSLA
   c. martensitic
   d. all of the above

6. Which of the following welds is intended only as a temporary device?
   a. tack
   b. plug
   c. spot
   d. lap spot

7. Which of the following statements is incorrect?
   a. The shorter the duration of the weld, the better the heat is contained.
   b. Using the proper capacity welder promotes rapid fusion of the panels.
   c. When butt welding on thin panels, it is not good practice to weld more than 3/4 of an inch at a time.
   d. Weld penetration increases as the distance between the tip of the gun and the base metal increases.

8. Welder A starts a butt weld in the center of the metal. Welder B starts a butt weld at the edge of the metal. Who is correct?
   a. Welder A
   b. Welder B
   c. Both A and B
   d. Neither A nor B

9. Which of the following welds is the body shop alternative to the OEM resistance spot welds made at the factory?
   a. spot
   b. plug
   c. stitch
   d. all of the above

10. Where is the stitch weld preferable to the continuous weld?
    a. on vertical joints where distortion is a problem
    b. on thin-gauge cosmetic panels
    c. both a and b
    d. neither a nor b

11. Which of the following is not a characteristic of welding coated steels?
    a. superior penetration
    b. increased spatter
    c. porosity
    d. erratic arc

12. The most common shielding gas for welding zinc coated steel is ______________.
    a. 75 percent argon and 25 percent CO_2
    b. 25 percent argon and 75 percent CO_2
    c. 50 percent argon and 50 percent CO_2
    d. 50 percent argon and 50 percent helium

13. What results from gas trapped during solidification of the weld metal?
    a. incomplete fusion
    b. porosity
    c. overlap
    d. undercut

14. When the molten filler metal and the base metal fail to mix completely, the result is ______________.
    a. inclusions
    b. lamellar tears
    c. arc strikes
    d. incomplete fusion

15. Welder A uses a fluorescent penetrant to inspect a MIG weld. Welder B uses a color-contrast penetrant to inspect a MIG weld. Who is correct?
    a. Welder A
    b. Welder B
    c. Both A and B
    d. Neither A nor B

16. How long can some high-strength steels withstand temperatures of 1200 degrees Fahrenheit without weakening significantly?
   a. indefinitely
   b. for up to 1 hour
   c. for up to 3 minutes
   d. none of the above

17. In both the SAE and AISI steel classification systems, the first two digits together indicate the ____________.
   a. basic type of steel
   b. series within the basic alloy group
   c. approximate permissible range of carbon content
   d. approximate permissible range of iron content

18. Strength is expressed in ____________.
   a. kilograms per millimeter squared
   b. pounds per square inch
   c. both a and b
   d. neither a nor b

19. In some domestic cars, where is Martensitic steel used?
   a. door guard beams
   b. lock pillars
   c. front and rear rails
   d. all of the above

20. Welder A says that the length of a tack weld is determined by the thickness of the panel. Welder B says the length is up to the preference of the welder. Who is right?
   a. Welder A
   b. Welder B
   c. Both A and B
   d. Neither A nor B

21. Welder A MIG welds a butt joint using the stitch technique. Welder B MIG welds a butt joint using the continuous technique. Who is right?
   a. Welder A
   b. Welder B
   c. Both A and B
   d. Neither A nor B

22. When weld metal is used to fill the spaces between intermittently placed beads without first grinding the surface of the beads, what can result?
   a. inclusions
   b. linear porosity
   c. piping porosity
   d. blowholes

23. A gun speed that is too fast will cause ____________.
   a. inferior weld strength
   b. shallow penetration
   c. melt through
   d. both a and b

24. What type of welds are made with identical MIG techniques?
   a. lap and plug
   b. lap and flange
   c. flange and butt
   d. plug and stitch

25. When MIG welding a flat lap joint, Welder A directs the gun on the top plate; Welder B directs it on the bottom plate. Who is right?
   a. Welder A
   b. Welder B
   c. Both A and B
   d. Neither A nor B

26. Which of the following statements concerning plug welds is incorrect?
   a. Most structural members require a 5/16- to 3/8-inch hole to achieve acceptable weld strength.
   b. Plug welding is essentially spot welding through a hole.
   c. Areas around the weld must be allowed to cool naturally.
   d. Plug welds are not sufficiently strong for welding load bearing structural members.

27. The first consideration when welding galvanized metals should be ____________.
   a. porosity
   b. accuracy
   c. safety and health
   d. repair versus replacement

28. When welding thin-gauge galvanized metals, Welder A moves the gun slower than normal and pushes it. Welder B moves the gun faster than normal and pulls it. Who is right?
   a. Welder A
   b. Welder B
   c. Both A and B
   d. Neither A nor B

29. Which of the following is not a major cause of inadequate joint penetration?
    a. improper joint design
    b. improper welding technique
    c. insufficient welding current
    d. contamination within the joint

30. What type of crack forms between and through the grains of a weld?
    a. cold
    b. hot
    c. tensile
    d. lamellar

CHAPTER FOUR

RESTORING A VEHICLE'S STRUCTURAL INTEGRITY

Objectives

After reading this chapter, you should be able to

- List the general guidelines used to determine whether a part should be repaired or replaced.
- Name the two basic procedures used for part replacement.
- Explain how to remove structural panels.
- Describe how to install new panels.
- Identify caution areas to be avoided when making cuts for a section.
- Identify the basic types of sectioning joints.
- Explain what is involved in the preparation for sectioning.
- Explain the sectioning methods used for different structural members.
- Define a full body section.

Most late-model vehicles are unibodies. Instead of having a separate frame, these vehicles have a structure that integrates the frame and body. The load-bearing components of a unitized frame are called structural members. All structural members of a collision-damaged vehicle must be restored to their original integrity. Otherwise, the vehicle's passengers will not receive the protection that the vehicle was designed to provide. If another accident occurred that resulted in serious injury or death to anyone in the car, the shop where the repairs were made could be held liable due to improper repair techniques. The competent welder will not allow this possibility to arise but will instead follow proven techniques to make quality repairs. These techniques are presented in this chapter.

REPAIR VERSUS REPLACEMENT

The first decision that must be made when surveying damage is whether the damaged component should be repaired or replaced. In some cases, the original unibody components of a vehicle damaged in a collision can be restored to their original condition using a variety of repair operations. Bent structural members can be pulled and realigned. Bulges, dents, and creases can be eliminated by proper hammering, pulling, and shrinking procedures. Some components, however, might be so badly damaged that replacement is the only practical and effective alternative. There are some general criteria to be used when deciding whether to repair or replace a part. A repair should be made whenever it is practical. Parts that are bent should be repaired. On the other hand, parts that are kinked and parts that crack when pulled should be replaced. A structural member that has any cracks in it must be replaced. Even small cracks will grow and lead to component failure due to the normal flexing of the vehicle once it is on the road again. Obviously, these guidelines are very general and the decision to repair or replace will also have to be based upon one's own judgment and experience.

Today's welder will see more unrepairable metal since automobile manufacturers are using more high-strength steels and thinner gauge metals. These metals are more prone to crack during pulling and straightening. High-strength steels are less ductile than mild steels, so they will not straighten as easily.

REPLACEMENT METHODS

When a component must be replaced, there are two basic procedures that can be used to accomplish a quality restoration. The first procedure is to replace the structural member at the factory seams with a new or used part. The second procedure is called *sectioning,* which is the replacement of a component at practical locations other than the factory seams. Sectioning is often more cost effective than replacement at factory seams because it avoids a number of problems frequently encountered in the latter. Replacement at factory seams often involves the following:

- extensive dismantling
- destruction of numerous factory spot welds
- destruction of factory corrosion protection

Sectioning is a replacement method that enables the welder to sidestep these difficulties.

REPLACEMENT AT FACTORY SEAMS

Structural panels should be replaced at factory seams whenever it is possible and practical. This straightforward method is still the first choice when replacing a structural panel. Body repair manuals are available from many automobile manufacturers. These manuals usually contain the instructions for panel replacement. Use them whenever they are available. This whole operation can be broken down into two segments: removal of the damaged panels and installation of the new.

REMOVING STRUCTURAL PANELS

Structural body panels are joined together in the factory by spot welding. Therefore, removing panels mainly involves the separation of spot welds. Spot welds can be drilled out, blown out with a plasma torch, chiseled out, or ground out with a high-speed grinding wheel. The best method for removing a spot-welded panel is determined by the number and arrangement of the mating panels and the accessibility of the weld.

Some spot weld areas have a number of layers of sheet metal. The removal tool is determined by the position of the weld and the arrangement of the panels.

Determining Spot Weld Positions

It is usually necessary to remove the paint film, undercoat, sealer, or other coatings covering the joint area to find the locations of spot welds. To do this, scorch the paint film with an oxyacetylene or propane torch and brush it off with a wire brush (Figure 4-1). Propane has a cooler flame than oxyacetylene, subjecting the metal to less heat stress. A

FIGURE 4-1 Removing paint to determine spot weld locations

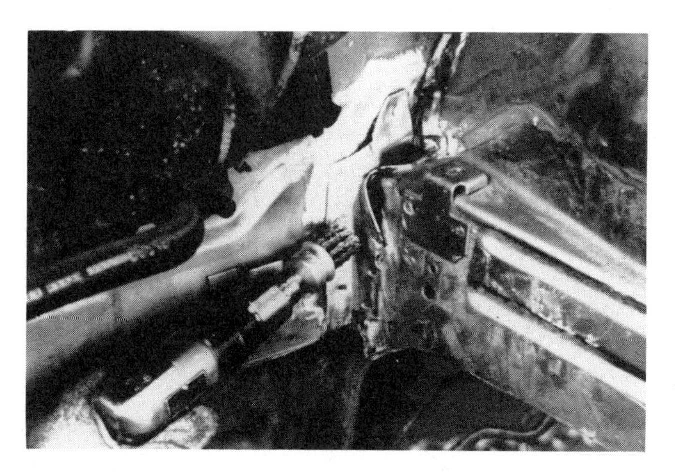

FIGURE 4-2 Removing paint with a wire brush

coarse wire wheel or brush attached to a drill can also be used to grind off the paint (Figure 4-2).

Scrape off thick portions of undercoating or wax sealer before scorching the paint. Do not burn through the paint film so that the sheet metal panel begins to turn color. Heat the area just enough to soften the paint and then brush or scrape it off. It is not necessary to remove paint from areas where the spot welds are visible through the paint film.

In areas where the spot weld positions are not visible after the paint is removed, drive a chisel between the panels as shown in Figure 4-3. Doing so will cause the outline of the spot welds to appear.

Separating Spot Welds

After the spot welds have been located, the welds can be drilled out, using a spot weld cutter (Figure 4-4). Two types of cutting bits can be used: a drill type or a hole saw type (Figure 4-5). Table 4-1 shows when each type should be used. Regardless of which is used, be careful not to cut into the lower

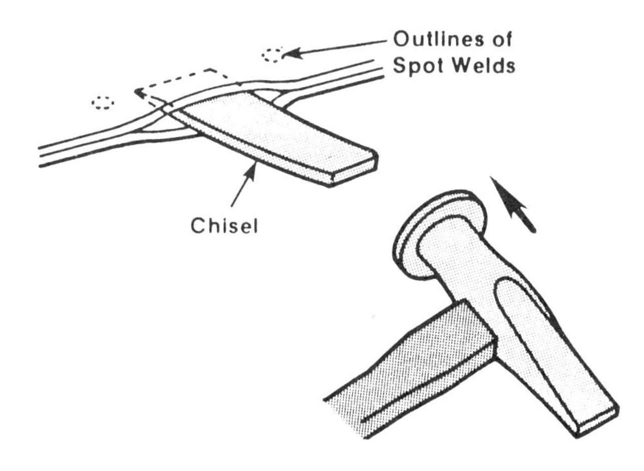

FIGURE 4-3 Determining spot weld locations with a chisel *(Courtesy of Toyota Motor Corp.)*

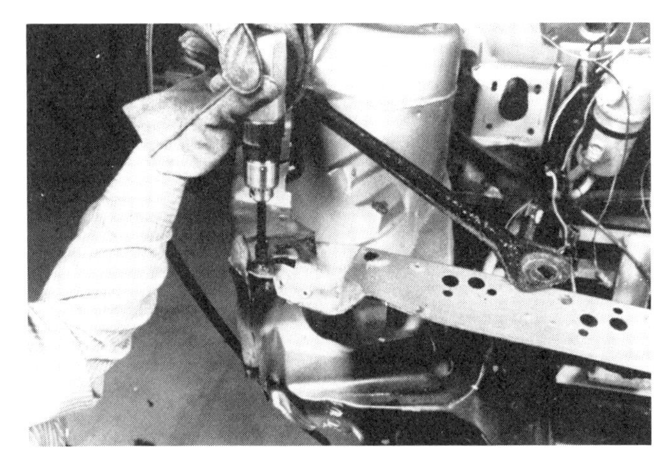

FIGURE 4-4 Removing spot welds with a spot weld cutter

TABLE 4-1: SEPARATION OF SPOT WELDS

Type			Application Method	Characteristics
Spot Cutter	Drill Type	Small	Places where the replacement panel is between other panels and welding cannot be done from the backside. Places where the replacement panel is on top and the weld is small.	The separation can be accomplished without damaging the bottom panel. Since the nugget is not left in the bottom panel, finishing is easy.
		Large	When the replacement panel is on top. When the panel is thick (places where nuggets are large) Places where the weld shape is destroyed.	
	Hole Saw Type		When the replacement panel is on top	Separation can be accomplished without damaging the bottom panel. Since only the circumference of the nugget is cut, it is necessary to remove the nugget remaining in the bottom panel after the panels are separated.
Drill			When the replacement panel is on bottom When the replacement panel is between and welding can be done from the backside (Select a drill diameter that is appropriate for the panel thickness and the weld diameter.)	Lower cost Recently, a labor saving spot weld removing tool has been developed that is easy to use and has a built-in attaching clamp.

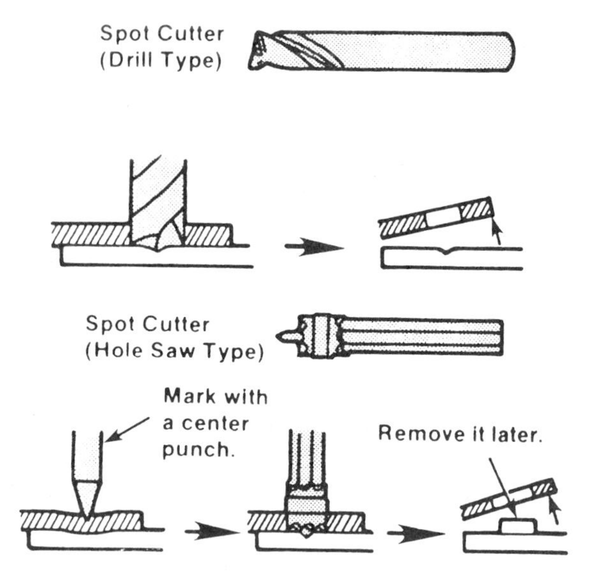

FIGURE 4-5 Spot weld cutters: (A) drill type and (B) hole saw type *(Courtesy of Toyota Motor Corp.)*

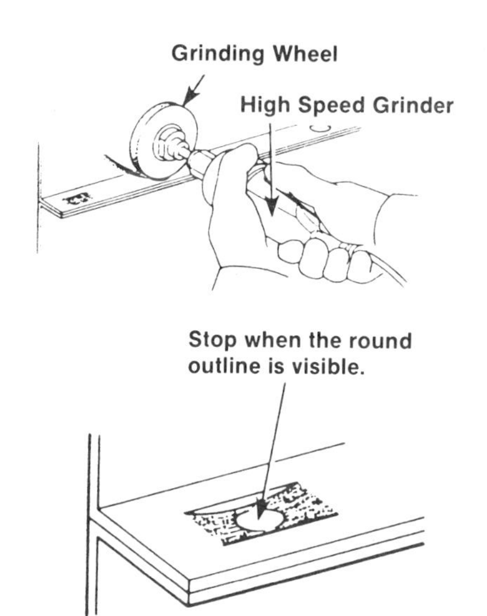

FIGURE 4-8 Removing spot welds with a grinder *(Courtesy of Toyota Motor Corp.)*

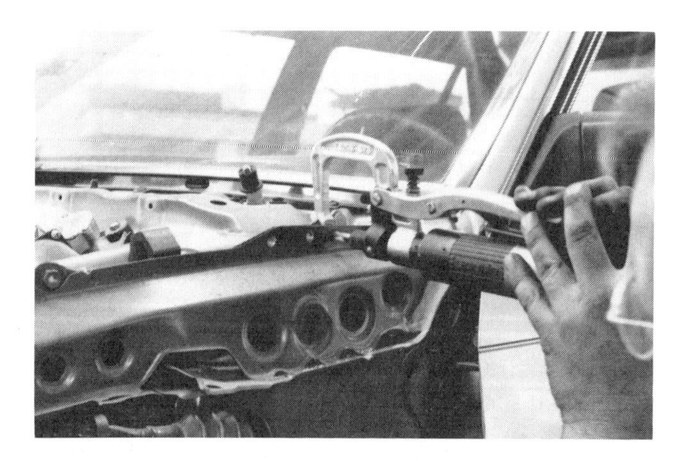

FIGURE 4-6 A spot weld removing tool

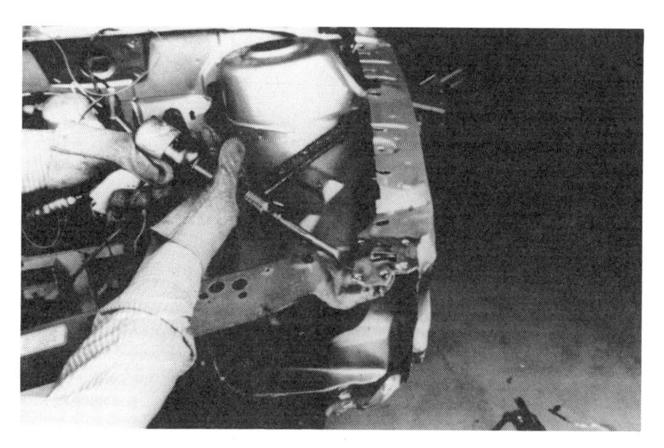

FIGURE 4-9 Separating panels with an air chisel

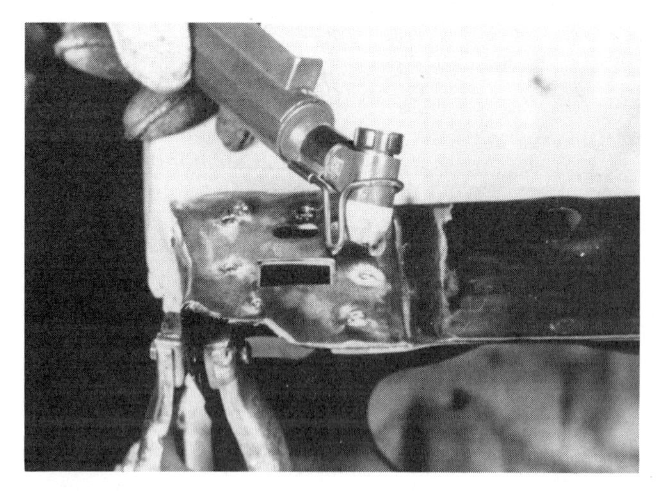

FIGURE 4-7 Removing spot welds with a plasma torch

panel and be sure to cut out the plugs precisely to avoid creating an excessively large hole.

Drilling out the numerous spot welds in a panel can be tedious. To make the job easier, use a spot removing drill with an integral clamping mechanism (Figure 4-6). Hand pressure forces the special rounded bit into the weld.

The removal of spot welds using a plasma torch (Figure 4-7) is much faster. The plasma torch works somewhat like an acetylene torch; removing spot welds blows a hole in all the thicknesses of metal at the same time. Obviously, the use of a plasma torch does not preserve the integrity of the underlying panels.

A high-speed grinding wheel can also be used to separate spot-welded panels (Figure 4-8). Use this technique only when the weld is not accessible with a drill, where the replacement panel is on top, or where a plug weld (from a previous repair) is too large to be drilled out.

After the spot welds have been drilled out, blown out, or ground down, drive an air chisel between the panels to separate them (Figure 4-9). Be careful not to cut or bend the undamaged panel.

Separating Continuous Welds

In some vehicles, panels are joined by continuous MIG welding. Since the welding bead is long, use a grinding wheel or high-speed grinder to separate the panels. As shown in Figures 4-10 and 4-11, cut through the weld without cutting into or through the panels, holding the grinding wheel at a 45-degree angle to the lap joint. After grinding through the weld, use a hammer and chisel to separate the panels.

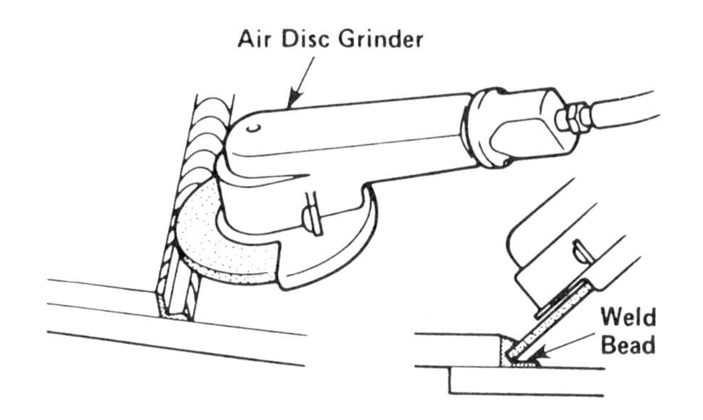

FIGURE 4-10 Removing a continuous weld with a disc grinder *(Courtesy of Toyota Motor Corp.)*

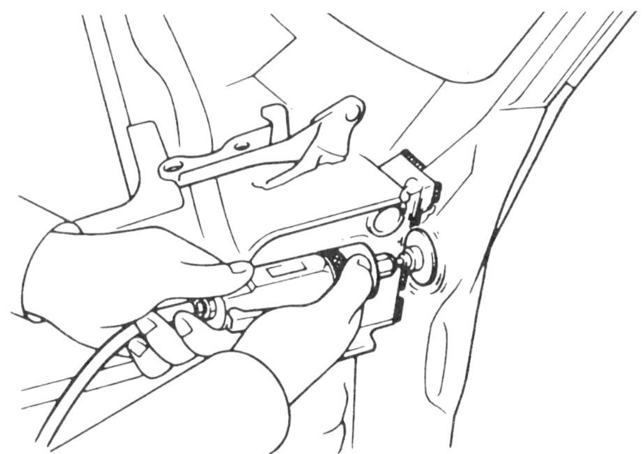

FIGURE 4-11 Using a high-speed grinder to remove a continuous weld *(Courtesy of Toyota Motor Corp.)*

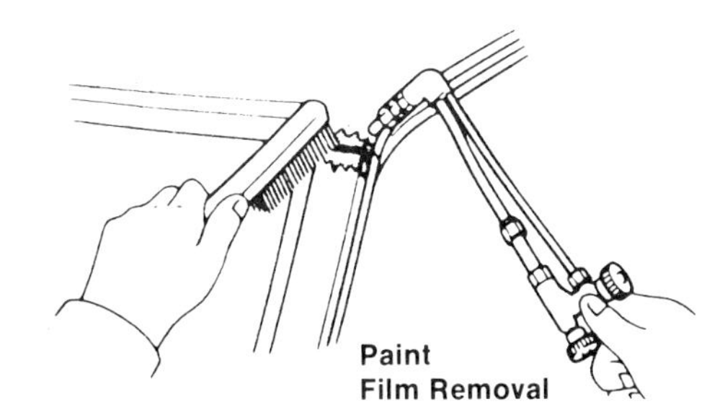

FIGURE 4-12 Removing paint from a brazed area *(Courtesy of Toyota Motor Corp.)*

Separating Brazed Areas

Brazing is used at the ends of outer panels or at the joints of the roof and body pillars to improve the finish quality and to seal the body. Generally, separation of brazed areas is done by melting the metal with an oxyacetylene or propane torch. However, in areas where arc brazing is used, the fusion temperature of the metal is higher than with ordinary brazing, and melting the metal would result in damaging the panels underneath. Therefore, areas that are arc brazed are normally separated by grinding. Ordinary brazing can be distinguished from arc brazing by the color of the metal. Ordinary brazed areas are the color of brass, while arc-brazed areas are a reddish copper color.

First, soften the paint with an oxyacetylene torch and remove it with a wire brush or scraper (Figure 4-12). Next, heat the brazing metal until it starts to melt and puddle, then quickly brush it off. Be careful not to overheat the surrounding sheet metal. Drive a chisel or screwdriver between the panels and separate them (Figure 4-13). Keep the panels separated until the brazing metal cools and hardens.

SHOP TALK

It is easier to separate brazed areas after all the other welded parts have been separated.

If after removing the paint, the brazed joint is determined to be arc brazing, use a high-speed grinder and grinding wheel to cut through it (Figure 4-14). If replacing the top panel, do not cut through the panel underneath. After grinding through the brazing, separate the lapped panels by using a chisel and a hammer.

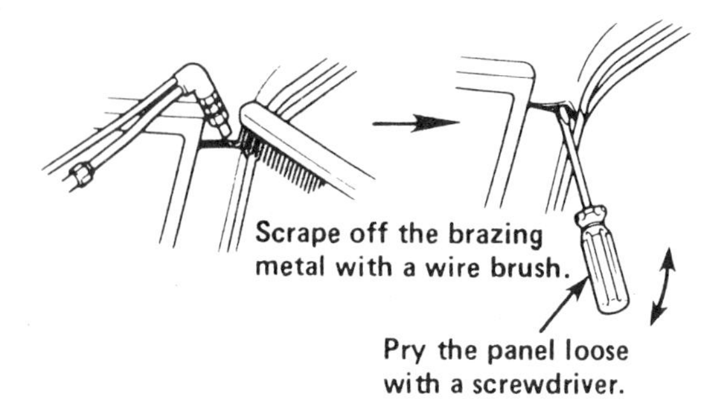

FIGURE 4-13 Separating brazed joints *(Courtesy of Toyota Motor Corp.)*

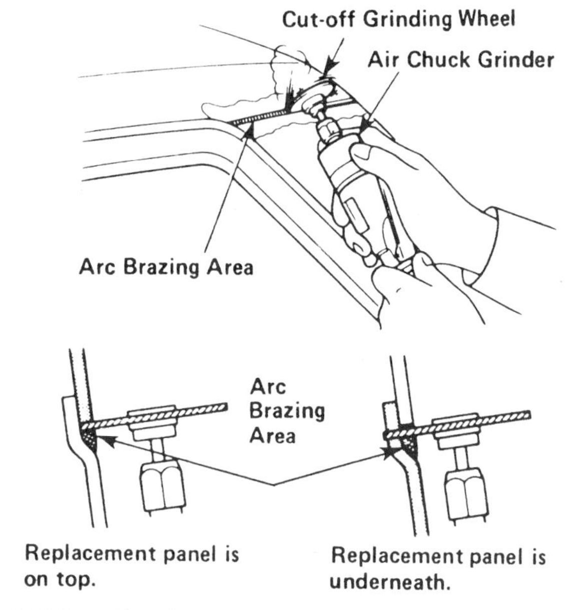

FIGURE 4-14 Separating panels connected by arc brazing *(Courtesy of Toyota Motor Corp.)*

INSTALLING NEW PANELS

Welding replacement panels requires thorough preparation and careful alignment. The following procedure is typical of many panel replacement operations. Always refer to the appropriate body repair manual provided by the manufacturer for the type and placement of welds.

VEHICLE PREPARATION

After removal of the damaged panels, prepare the vehicle for installation of the new panels. To do this, follow these steps:

1. Grind off the welding marks from the spot welding areas (Figure 4-15). Use a wire brush to remove dirt, rust, paint, sealers, zinc coatings, and so on from the joint surfaces. Do not grind the flanges of structural panels. Grinding will remove metal, thinning the section and weakening the joint.
2. On parts that will be spot welded during installation, remove paint and undercoating from the back sides of the panel joining surfaces.

FIGURE 4-15 Removing paint and rust from weld joints

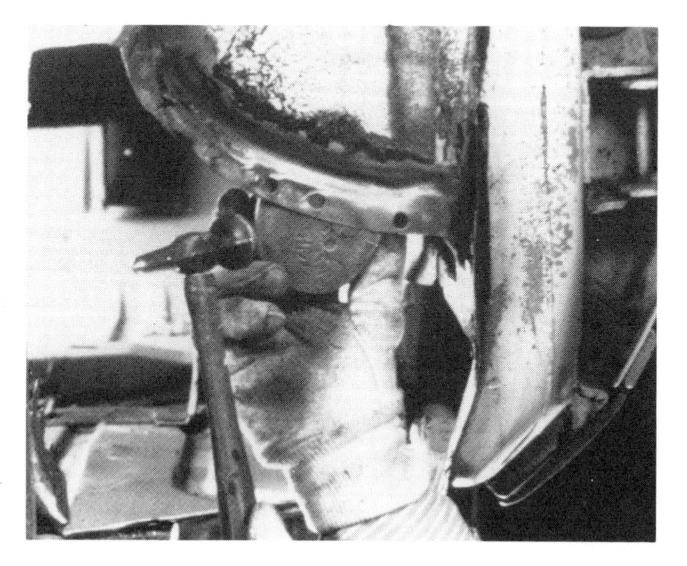

FIGURE 4-16 Smoothing a panel flange with a hammer and dolly

3. Smooth the dents and bumps in the mating flanges with a hammer and dolly (Figure 4-16).
4. Apply weld-through primer to areas where the base metal is exposed after the paint film and rust have been removed from the joining surfaces. It is very important to apply the primer to joining surfaces or to areas where painting cannot be done in later processes.

REPLACEMENT PANEL PREPARATION

Since all new parts are coated with a primer, the primer and zinc coating must be removed from the mating flanges to allow the welding current to flow properly during spot welding. Also, drill holes for plug welds where spot welding is not possible. Use plug hole diameters that correspond to the thickness of the panels. To prepare the new panel for welding, follow these steps:

1. Use a disc sander to remove the paint from both sides of the spot welding area. Do not grind into the steel panel and do not heat the panel so that it turns blue or begins to warp.

WARNING: Wherever possible, grind so that sparks fly down and away. Always wear proper eye protection when grinding.

---

2. Apply weld-through primer (as an antirust treatment) to the welding surfaces where the paint film was removed. Apply the weld-through carefully so that it does not ooze out from the joining surfaces. If it does ooze out, it will have a detrimental effect on painting, necessitating extra work. So remove any excess with a solvent-soaked rag.
3. Make holes for plug welding with a punch or drill. Always refer to the specific body repair manual for the required number of plug welding holes. Duplicate the number and location of factory spot welds. Be sure to make plug welding holes the proper diameter. If the size of the holes is too large or too small for the thickness of the panel, either the metal will melt through or the weld will be inadequate (Table 4-2). When drilling high-strength steel, use a variable speed drill and lower RPM's, and strive for uniform cutting. If the metal chips begin to change color to gold, purple, or blue, the RPM's should be reduced. Use high-speed or carbide drill bits for drilling high-strength steel.

POSITIONING NEW PANELS

Aligning new parts with the existing body is a very important step in repairing vehicles with major damage. Improperly aligned panels will affect both the appearance and driveability of the repaired vehicle. Basically, there are two methods of positioning body panels. One way is to use dimension measuring instruments to determine the installation position and the other is to determine the position by the relationship between the new part and the surrounding panels. The dimensional accuracy of structural panels making up the engine compartment or underbody parts such as the fender aprons or the front side members and rear side members has a direct effect on wheel alignment and driving characteristics. Therefore, when replacing structural panels in unibody vehicles, use the dimensional measurement positioning method because it is more accurate. There is also a definite relationship between the fit of the new and old parts and the finish appearance, so whether structural or cosmetic panels are being replaced, the emphasis is on proper fit. Of course, it is desirable to use both methods together and, therefore, assure the accuracy and the finish necessary for a high-quality vehicle repair.

CAUTION: All measurements must be accurate before finish welding structural panels in position, regardless of the measurement system used for replacing structural panels. Because there are no shims on unibodies, no adjustments can be made to the outer panels. Therefore, each panel must be precisely positioned before welding.

TABLE 4-2: PLUG HOLE DIAMETER FOR RESPECTIVE WELDED PORTION

Welded Area Panel Thickness	Plug Hole Diameter
Less than 1/32"	3/16"-1/4"
1/32"-3/32"	1/4"-13/32"
More than 3/32"	13/32" or more

Positioning by Dimensional Measurement Methods

When using a dedicated bench (Figure 4-17) or a mechanical universal (Figure 4-18) system of measurement, the vehicle must be properly positioned on the bench before the new panel can be correctly aligned. This will usually have already been done in order to pull and straighten damaged panels that do not require replacing. All straightening must be done before replacing any panels. Otherwise, proper alignment of the new panels will be impossible.

FIGURE 4-17 Aligning a new panel on a dedicated bench fixture

FIGURE 4-18 Aligning a new panel on a universal bench

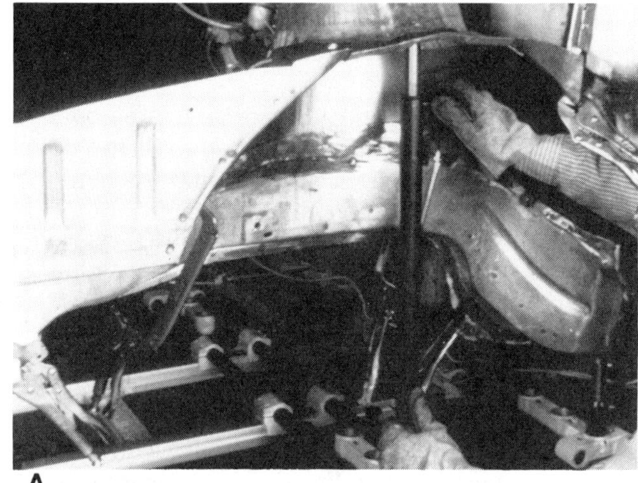

A

B

FIGURE 4-19 Attaching the new part: (A) clamping in place and (B) welding

So, when using a bench system for panel replacement at a factory joint, place the fixtures or gauges on the bench in their correct locations and tie them to the bench. Then, place the new panel in position on the fixtures and see how it lines up with the undamaged panels on the car. Make any necessary adjustments, clamp the panel in place, and weld it (Figure 4-19). Measure it again, then plug weld the panel into place. Grind the welds for appearance, if necessary.

Positioning by Visual Inspection

Nonstructural outer panels can sometimes be visually aligned with adjacent panels without the precise measurements necessary when replacing structural panels. The emphasis here is on appearance. The panel is carefully aligned with adjacent body parts and secured with spot welds.

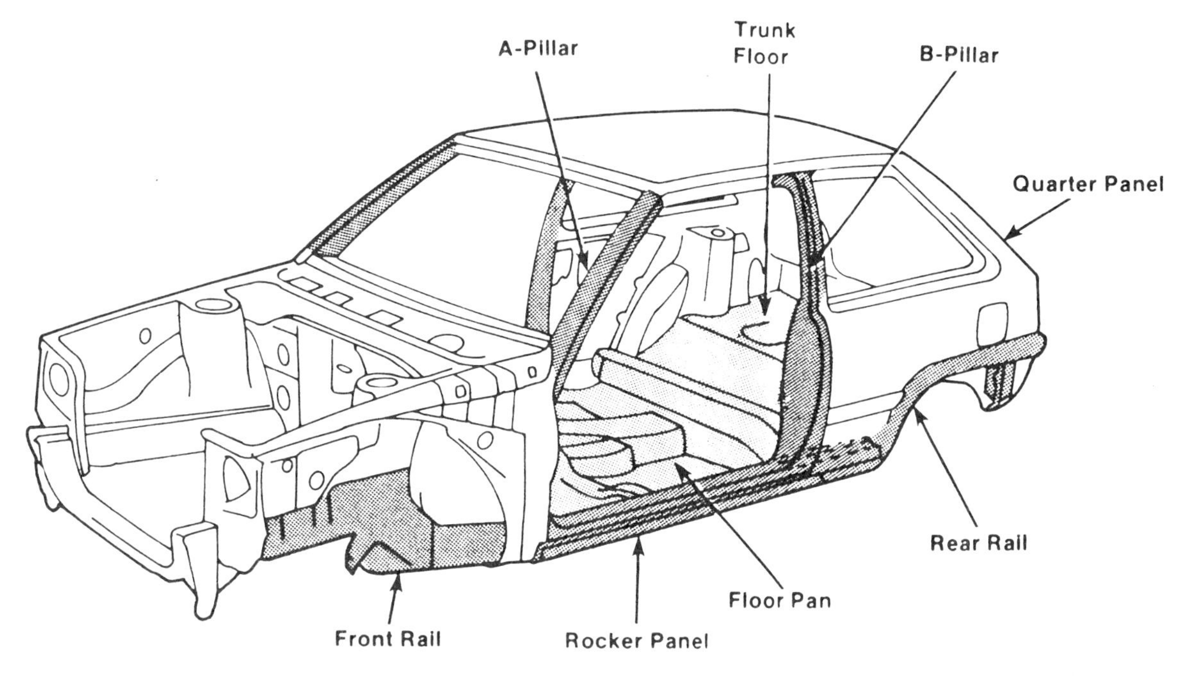

FIGURE 4-20 Sectioning areas

SECTIONING

Sectioning, as previously stated, is a practical and efficient alternative to replacing parts at factory seams. The following procedures are considered proper techniques for unibody structural components. Various manufacturers and independent testing agencies have examined these procedures and found that they provide structurally adequate repairs for the specified areas. However, be sure to follow the carmaker's recommendations on the subject of sectioning.

Major auto manufacturers are continually investigating the sectioning of structural components and are publishing repair procedures as they are developed and verified. Always give careful attention to detailed damage repair techniques where prescribed in technical manuals or notices published by the manufacturers.

This discussion of panel sectioning and replacement covers the repair of the following components (Figure 4-20): rocker panels, quarter panels, floor pan, front rails, rear rails, trunk floor, B-pillars, and A-pillars. These unibody structural components involve two basic types of construction design: closed sections, such as rocker panels, pillars, and body rails (Figure 4-21), and open surface or single layer overlap joint components, such as floor pans and trunk floors.

The closed type sections are the most critical, because they provide the principal strength in the unibody structure. They possess much greater strength per pound of material than other types of sections.

CRUSH ZONES

Certain structural components have crush zones, or buckling points, designed into them for absorbing the impact energy in a collision. This is particularly true of the front and rear rails because

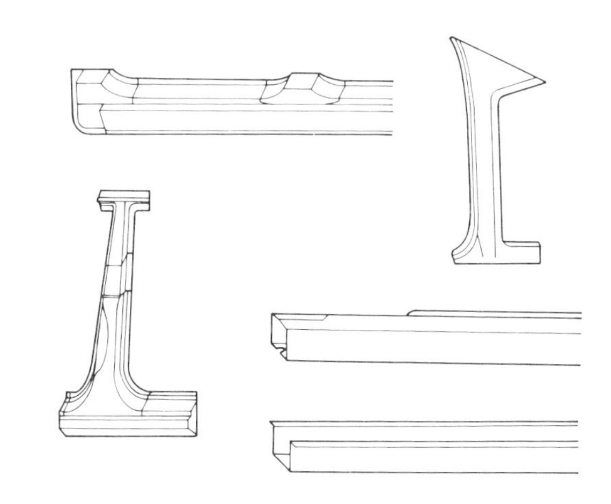

FIGURE 4-21 Closed unibody sections

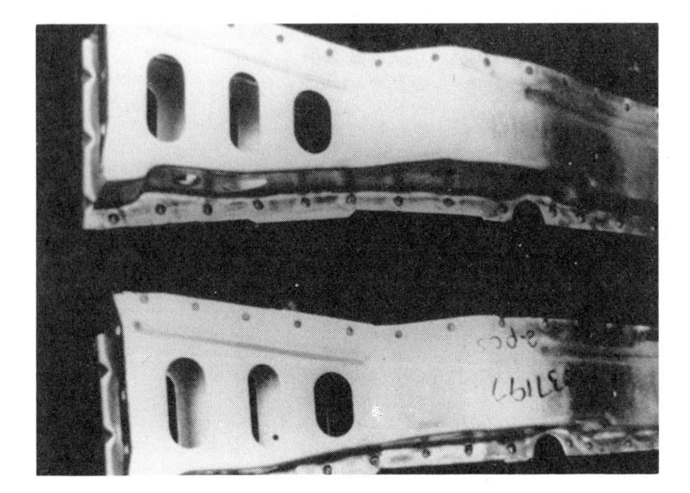

FIGURE 4-22 Typical crush zones

FIGURE 4-23 Crush zones protect suspension bearing areas

they take the brunt of the impact in most collisions. Crush zones are in all front and rear rails (Figure 4-22). Identify these crush zones by how they look. Some are in the form of convoluted or crinkled areas, some are in the form of dents or dimples, and others are in the form of holes or slots put in deliberately so the rail will collapse at these points. In virtually all cases, the crush zones are ahead of the front suspension and behind the rear suspension (Figure 4-23).

Avoid crush zones whenever possible. Sectioning procedures can change the designed collapsibility if improperly located. If a rail has suffered major damage, it will normally already be buckled in the crush zone, so the crush zone will usually be easy to locate. Where only moderate damage has occurred, be very careful. The hit might not have used up the entire crush zone, so be aware of other potential areas where designed-in collapse might occur.

OTHER CAUTION AREAS

There are other areas to stay away from when making cuts. Stay away from holes in the component. Do not cut through any inner reinforcements, meaning double layers in the metal. Stay away from anchor points, such as suspension anchor points, seat belt anchor points in the floor, and shoulder belt D-ring anchor points. For example, when sectioning a B-pillar, make an offset cut around the D-ring area to avoid disturbing the anchor reinforcement. Other areas to avoid are mounts for the engine or drivetrain and compound shapes.

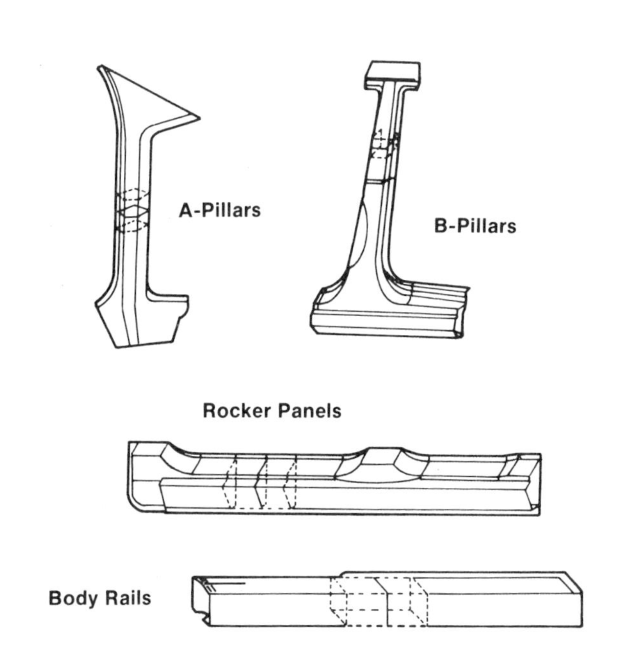

FIGURE 4-24 Butt joint with insert

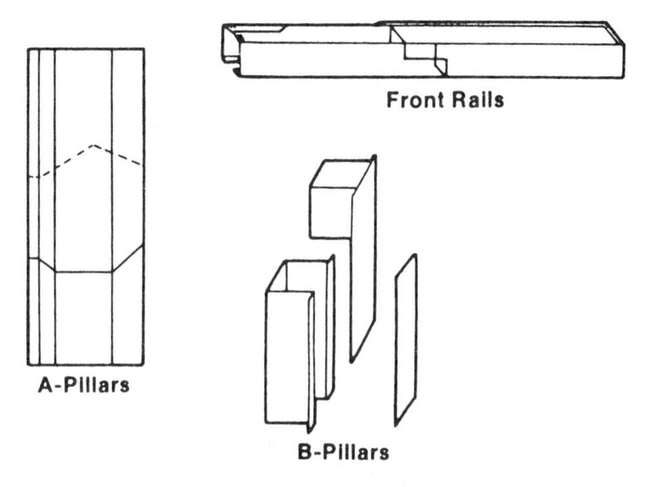

FIGURE 4-25 Offset butt joint without an insert

BASIC TYPES OF SECTIONING JOINTS

Correct structural sectioning procedures and techniques involve three basic types of joints, as well as certain variations and combinations of them. One is a butt joint with an insert (Figure 4-24) and is used mainly on closed sections, such as rocker panels, A-and B-pillars, and body rails. Inserts help make it easy to fit and align the joints correctly and help make the welding process easier.

Another basic type is an offset butt joint without an insert, also known as a staggered butt joint (Figure 4-25). This type is used on A- and B-pillars and front rails. The third type is an overlap joint (Figure 4-26). The overlap joint is used on rear rails, floor pans, trunk floors, and B-pillars.

The configuration and makeup of the component being sectioned might call for a combination of joint types. Sectioning a B-pillar, for instance, might require the use of an offset cut with a butt joint in the outside piece and a lap joint in the inside piece.

PREPARING TO SECTION

When the sectioning method is used to replace a damaged part, the replacement section may either be new or recycled. When preparing to section and replace any structural member with a recycled part, certain steps must be taken to ensure the quality of the repair. Specific instructions must be provided to the recycler as to the placement of the section and the method of sectioning to be used. Other important considerations are the welding techniques used and the cleanliness of the joint metal.

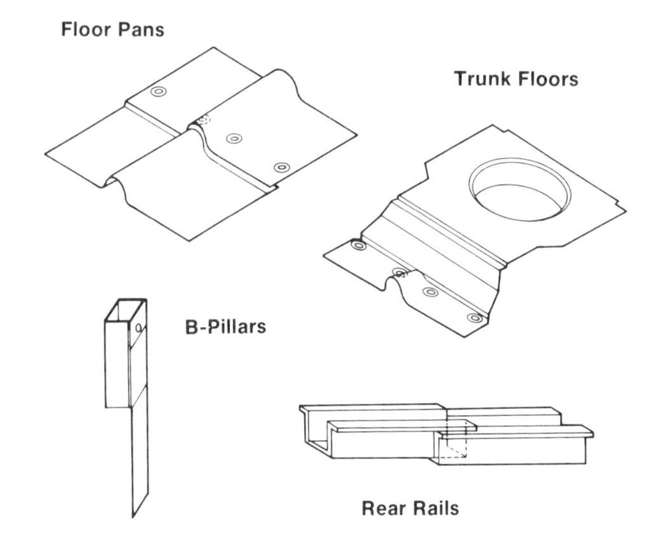

FIGURE 4-26 Overlap joints

FIGURE 4-27 Sectioning a recycled part

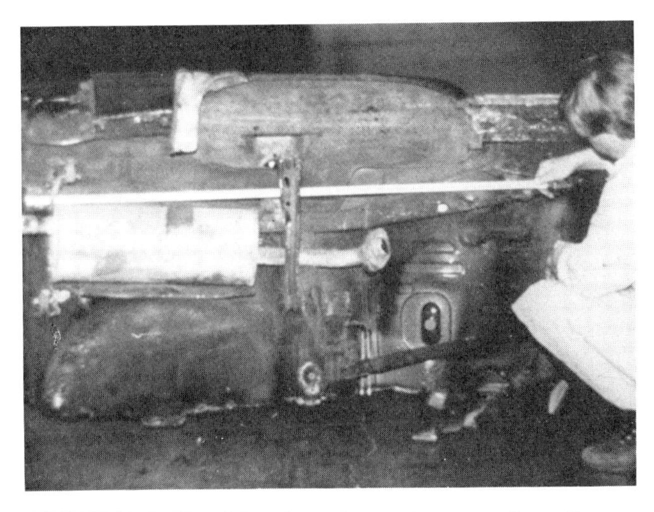

FIGURE 4-28 Check salvaged parts for dimensioned accuracy.

USING RECYCLED OR SALVAGED PARTS

When using recycled parts, tell the recycler exactly where to make the cuts. Preferably, have the required part removed with a metal saw (Figure 4-27). If the recycler uses a cutting torch, make sure that at least 2 inches of extra length are left on the part to insure that the heat dispersion from the cut does not invade the joint area. Instruct the recycler to make the cut so that reinforcing pieces welded inside the component are not cut through.

When a recycled or salvaged part is received, examine it for corrosion. If it has a lot of rust on it, do not use it. Ask for another one. Before installing a recycled part, check it for possible damage and make sure it is dimensionally accurate (Figure 4-28). Remember: Using quality material is a must to achieve a quality repair.

FIGURE 4-29 MIG welding is a must when sectioning.

MIG WELDING

Proper structural sectioning requires the use of MIG welding equipment and the appropriate techniques that have already been covered (Figure 4-29). When it comes to structural sectioning, use steel wire—a wire that meets or exceeds the requirements of American Welding Society standard AWS-EX-705-6 when using 75 percent Argon/25 percent CO_2 gas or other comparable argon-based gas mixtures.

JOINT PREPARATION

Careful preparation of the sectioned joint is another necessity for doing a proper job of structural sectioning. Before starting to weld, be sure to thoroughly clean the surfaces to be joined. The weld site must be completely free of any foreign material that might contaminate the weld. Otherwise a brittle, porous weld of poor integrity will be produced. Use a scraper and an oxyacetylene torch (Figure 4-30) to remove heavy undercoatings, rustproofing, tar, caulking and sealants, road dirt and oil, primers, and the like.

Do the finish cleaning with a wire brush and torch. Keep the metal at a very low heat—just enough to get the job done. Also, be sure to remove rustproofing, lead, plastic filler, and other contaminants from the inside of structural closed sections when preparing them for welding. A wire brush and a torch with a carburizing flame are generally the most effective and efficient cleaning tools. The same goes for removing paint and sealants that have been painted over; the wire brush and torch usually work best for this operation.

Make sure the surfaces to be welded are completely free of rust and scale. This is best removed by sanding or sandblasting until there is a clean metal welding surface. In some cases it is possible to do this with a power wire brush (Figure 4-31). In addition, attach the work cable clamp to a clean surface to have a trouble-free welding circuit (Figure 4-32).

Still another essential for achieving high-quality structural sectioning is that the MIG machine must be properly set up for the specific welding job. It must be precisely adjusted for the thickness of metal being welded and the type of weld being made.

To make sure that the MIG machine is correctly adjusted for the specific joint being welded, always do a test weld. Especially when welding a closed section where the backside of the weld cannot be checked, a test weld is the only way to ensure that the welding techniques and machine adjustments will restore the original strength, integrity, and alignment of the panel.

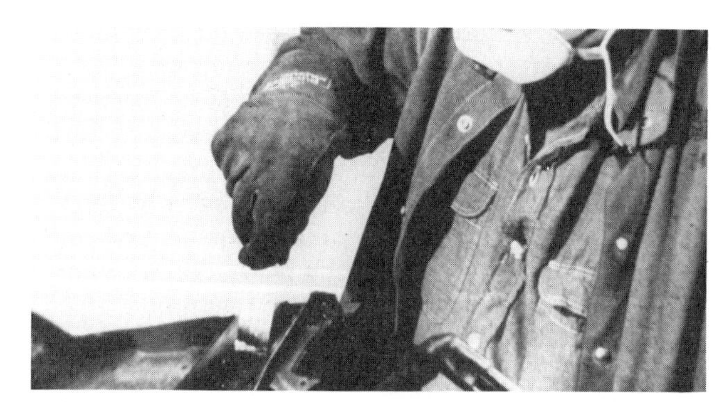

FIGURE 4-30 Removing coverings with a scraper and torch

FIGURE 4-31 Grinding rust from flanges

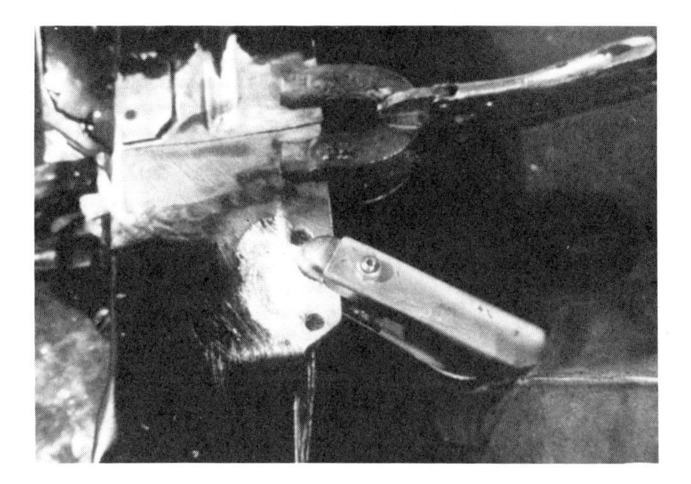

FIGURE 4-32 Clamp prior to cleaning metal.

FIGURE 4-33 A closed rail

To make sure complete penetration and full fusion is achieved, do test welds on sample pieces that duplicate the intended workpiece. In other words, make the same types of welds on the same type and configuration of joint and the same gauge metal. The ideal way to do this is to use pieces of excess material from the components being joined on the car: the scrap cut off to make the fit-up. While making the test welds, adjust the MIG machine to suit the given situation.

After completing the test welds, check them for strength.

SECTIONING BODY RAILS

Virtually all front and rear rails are closed sections, but the closures are of two distinct types. One is called a *closed section.* It comes from the factory or the recycler with all four sides intact. Sometimes it is referred to as a *box section* (Figure 4-33). The other type comes as an open hat channel and is closed on the fourth side by being joined to some other component in the body structure (Figure 4-34).

The butt joint with an insert is the procedure for repairing a closed section rail (Figure 4-35). Most rear rails, plus various makes of front rails, are of the hat channel type. Some of the hat channel closures are vertical, such as a front rail joined to a side apron. Some of them are horizontal, such as a rear rail joined to a trunk floor (Figure 4-34A).

In most cases when sectioning the open hat channel type of rail, the procedure should be a lap joint with plug welds in the overlap areas and a continuous lap weld along the edge of the overlap (Figure 4-36). Whenever sectioning front or rear rails, always remember that they contain crush zones that must be avoided when making cuts. Also remember to stay away from all other caution areas.

SECTIONING ROCKER PANELS

Before replacing a rocker panel, check the vehicle underbody for dimensional accuracy and make any repairs necessary for proper panel alignment. Once the panel is aligned, check all nearby seams for separation and cracking of the sealant. Any dam-

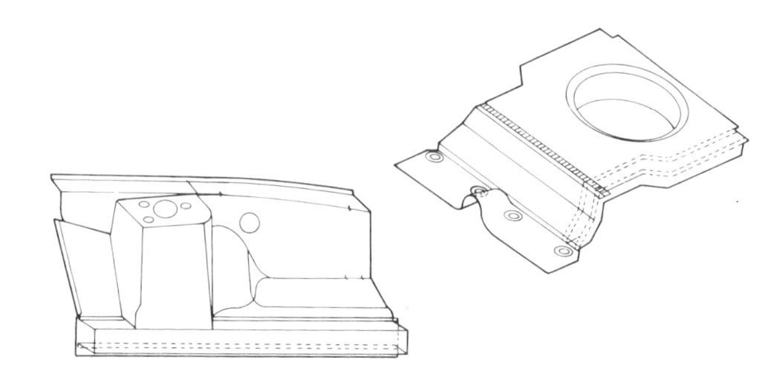

FIGURE 4-34 Typical hat channels: (A) front rail and (B) rear rail

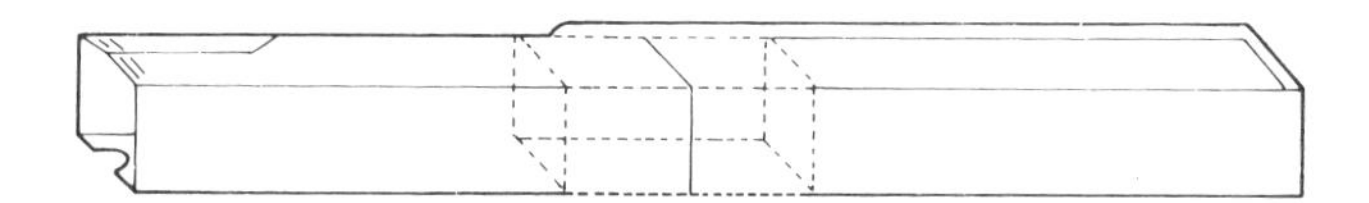

FIGURE 4-35 A rear rail with a butt joint and insert

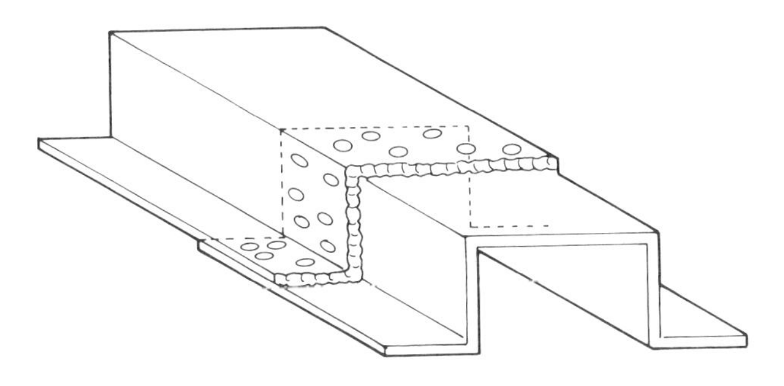

FIGURE 4-36 Joining open rails

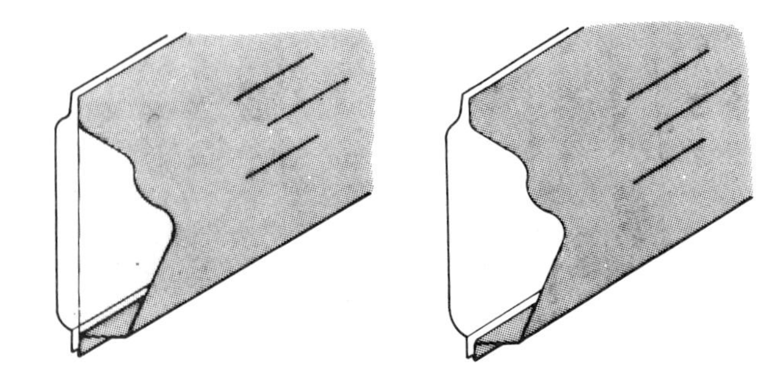

FIGURE 4-37 Rocker panel profiles

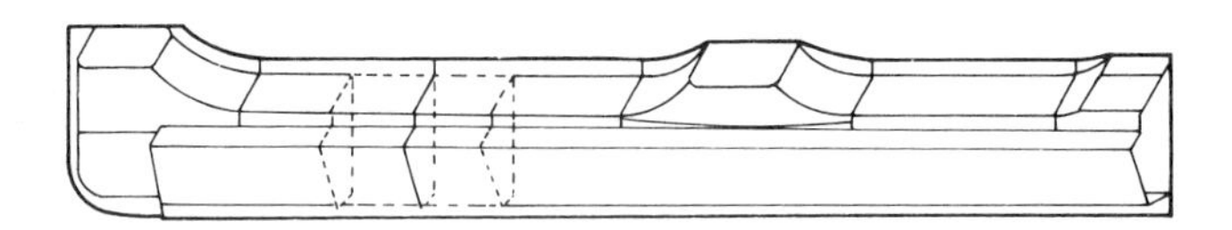

FIGURE 4-38 Joining the rocker panel section with a butt joint and insert

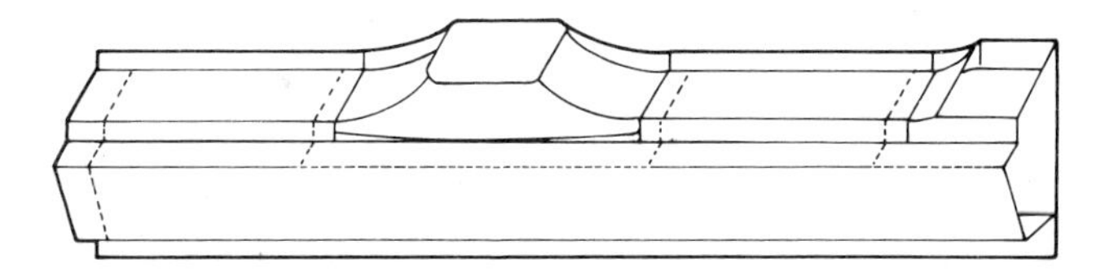

FIGURE 4-39 Replacing the outer rocker panel piece

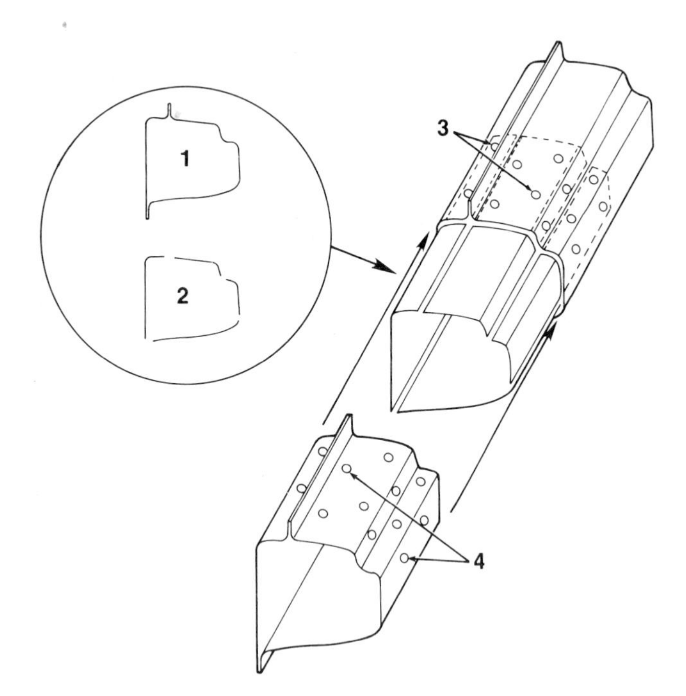

FIGURE 4-40 Cutting an insert to fit a rocker panel

aged seam must be repaired since it could create a corrosion problem.

Rocker panels come in two-piece or three-piece designs, depending on the make and model of the unibody (Figure 4-37). In both cases, the rocker panel might contain reinforcements, and the reinforcements can be intermittent or continuous. Depending on the nature of the damage, the rocker panel can be replaced with the B-pillar or without it.

To section and repair the rocker panel, a straight-cut butt joint with an insert can be used (Figure 4-38), or the outside piece of the rocker panel can be cut and the repair piece installed with overlap joints (Figure 4-39). Generally speaking, the butt joint with an insert is used when installing a recycled rocker panel with the B-pillar attached and when installing a recycled quarter panel.

To do a butt joint with an insert, cut straight across the panel. An insert is fashioned out of one or

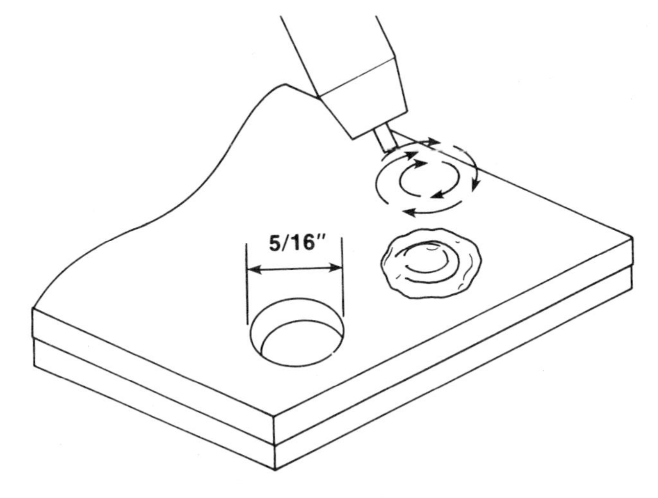

FIGURE 4-41 A circular motion of the gun is required in this 5/16-inch hole.

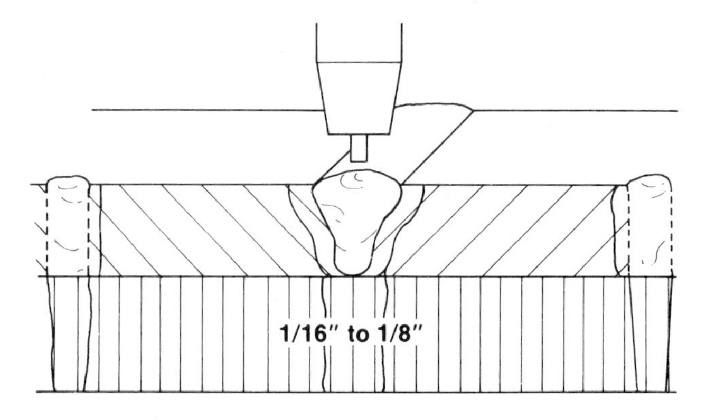

FIGURE 4-42 Butt weld gap

more pieces cut from the excess length on the repair panel or from the end of the damaged panel. The insert should be 6 to 12 inches long and should be cut lengthwise into two to four pieces, depending on the rocker panel configuration (Figure 4-40). Remove the pinch weld flange so that the insert will fit inside the rocker panel. With the insert in place, secure it with plug welds. For structural sectioning, 5/16-inch plug weld holes are needed to achieve an adequate nugget and acceptable weld strength. This 5/16-inch hole requires a circular motion of the gun to properly fuse the edge of the hole to the base metal (Figure 4-41).

When installing an insert in a closed section, whether it is a rocker panel, A- or B-pillar, or body rail, make sure the closing weld fully penetrates the insert. When closing the job with a butt weld, leave a gap that is wide enough to allow thorough penetration of the insert. The width of the gap depends on the thickness of the metal, but ideally it should not be less than 1/16 inch nor more than 1/8 inch (Figure 4-42). Be careful to remove the burrs from the cut edges before welding. Otherwise, the weld metal tends to travel around and under the burr. This can create a flawed weld, resulting in stress concentration that can cause cracks and weaken the joint (Figure 4-43).

FIGURE 4-43 Burrs in the joint will weaken the weld.

In general, use the overlap procedure on a rocker panel when installing only the outer rocker or a portion of it. Leave the inner piece intact and cut only the outer piece. One way to make an overlap joint is to make the cut in the front door opening and allow for an overlap when measuring. When making this cut, stay several inches away from the base of the B-pillar to avoid cutting any reinforcement underneath it (Figure 4-44).

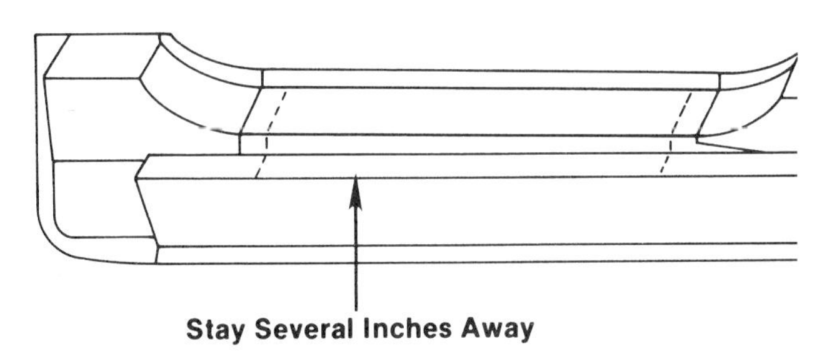

FIGURE 4-44 Overlapping the outer rocker panel section

1. Cut around the bases of the B- and C-pillars, leaving overlap areas around each (Figure 4-45).
2. Cut out the new outer rocker panel so that it overlaps around the bases of the pillars and around the original piece of the outer rocker still affixed to the car.
3. In the pinch weld flanges, use plug welds to replace the factory spot welds (Figure 4-46).
4. Plug weld the overlaps around the B- and C-pillars, using approximately the same spacing as in the pinch weld flanges (Figure 4-47).
5. Lap weld the edges with about a 30 percent intermittent seam; about 1/2 inch of weld in

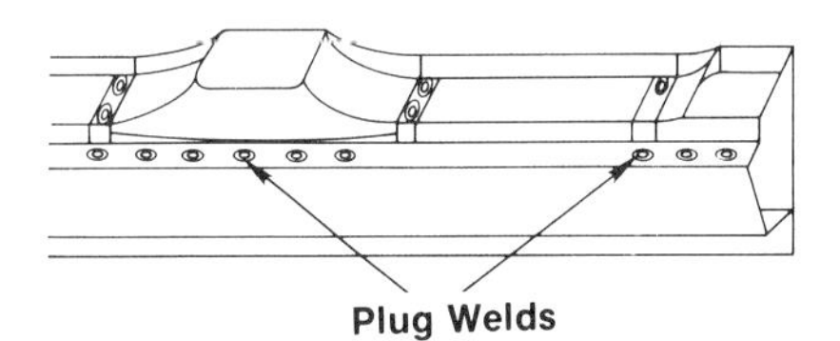

FIGURE 4-46 Plug weld flanges

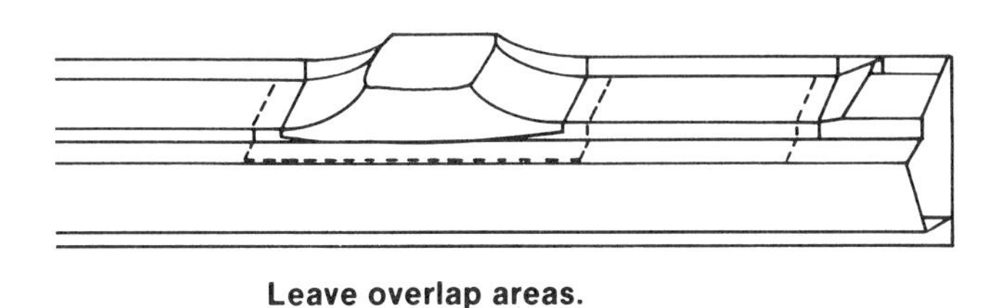

FIGURE 4-45 An overlap joint around a pillar

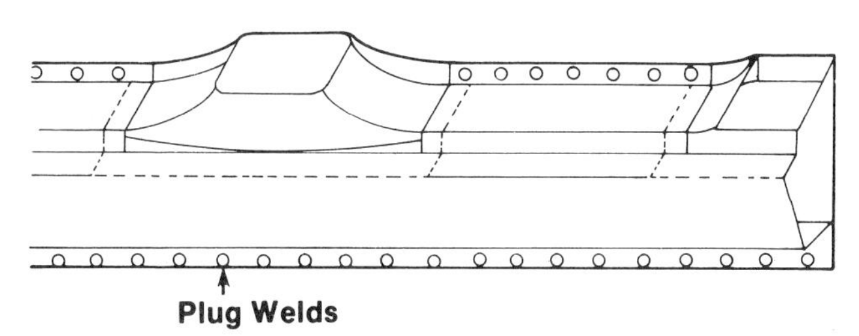

FIGURE 4-47 Plug weld overlaps

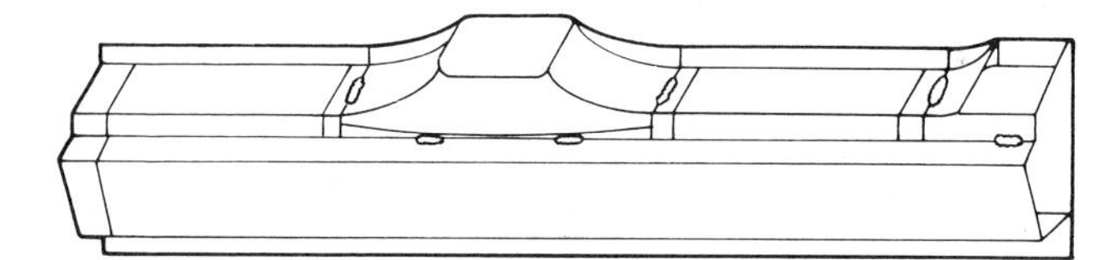

FIGURE 4-48 Intermittent lap weld seams

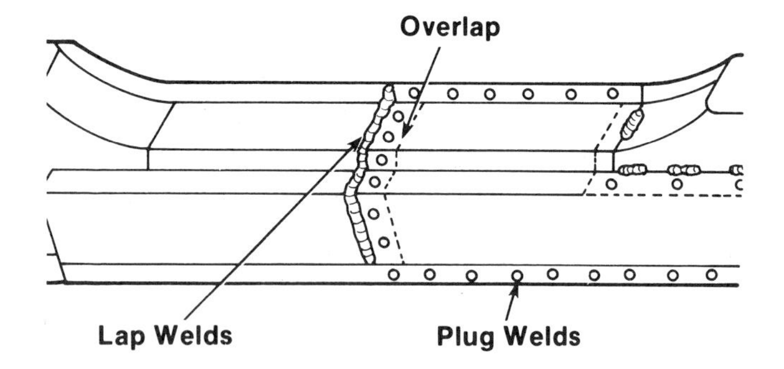

FIGURE 4-49 A plug weld and lap weld overlapping section

every 1-1/2 inches of overlap edge (Figure 4-48).

6. Put plug welds in the overlap area in the door opening, and lap weld around the edges to close the joint (Figure 4-49).

Of course, it might be necessary to reverse this procedure due to the nature of the hit. Make the overlap cut in the rear door opening and cut out and overlap around the bases of the A- and B-pillars. Use this same basic technique to replace the entire outer rocker. In this version, cut around the bases of all three pillars and overlap all three bases in the same way as before.

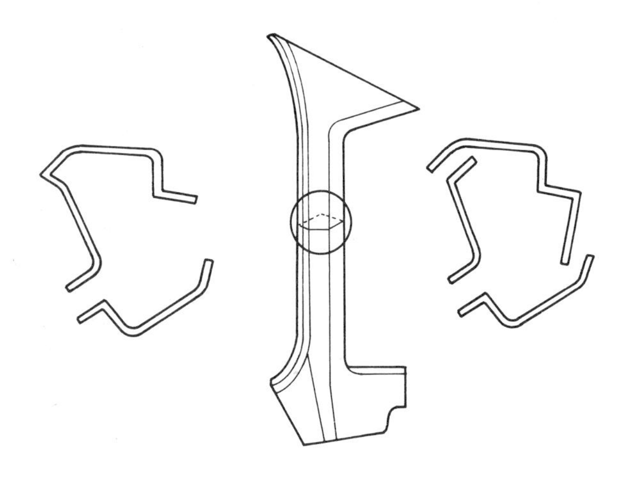

FIGURE 4-50 Two- and three-piece A-pillar sections

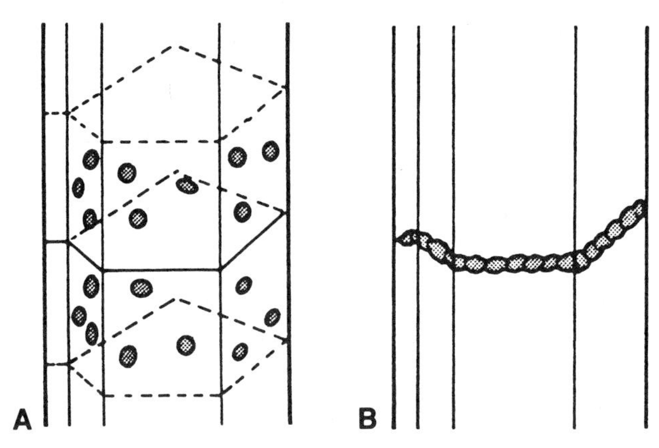

FIGURE 4-51 Welding the butt joint with an insert in an A-pillar using (A) plug welds and (B) butt welds

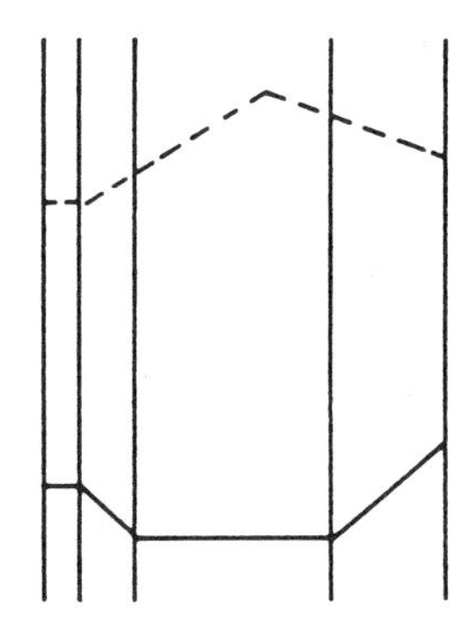

FIGURE 4-52 An offset butt joint

SECTIONING A-PILLARS

A-pillars can be either two-piece or three-piece components (Figure 4-50). They can be reinforced at the upper end or the lower end, or both. They are not likely to be reinforced in the middle. Therefore, A-pillars should be cut near the middle to avoid cutting through any reinforcing pieces. It is also the easiest place to work.

To section an A-pillar, use a straight-cut butt joint with an insert or an offset butt joint without an insert. The butt joint with an insert is made in the same manner as already described for the rocker panel. The A-pillar insert should be 4 to 6 inches in length. After cutting the insert lengthwise and removing any flanges, tap the pieces into place. Secure the insert with plug welds and close all around the pillar with a continuous butt weld (Figure 4-51).

To make the offset butt joint, cut the inner piece of the pillar at a different point than the other piece was cut, creating the offset (Figure 4-52). Whenever possible, try to make the cuts between the factory spot welds, so that it will not be difficult to drill them out, and make the cuts no closer to each other than 2

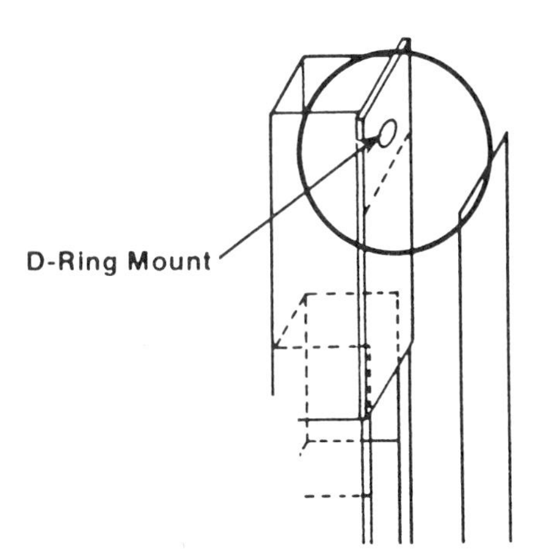

FIGURE 4-53 Two-piece B-pillar

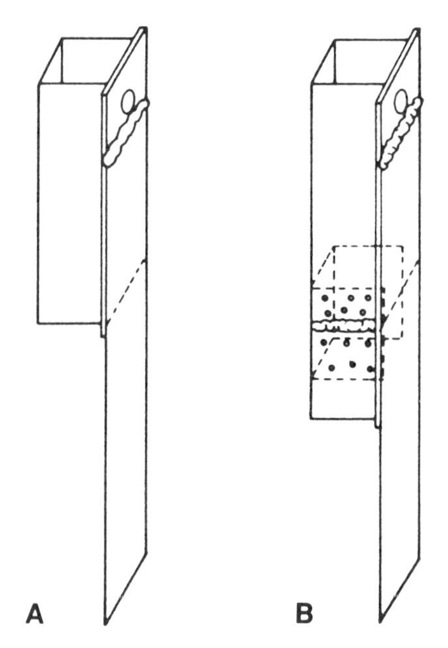

FIGURE 4-54 (A) Lap weld the inner panel, and (B) plug and butt weld the outer panel.

to 4 inches. Butt the sections together and continuous-weld them all around.

SECTIONING B-PILLARS

For sectioning B-pillars, two types of joints can be used: the butt joint with an insert and a combination offset cut and overlap. The butt joint with an insert is usually easier to align and fit up when the B-pillar is a simple two-piece cross section without a lot of internal reinforcing members. The insert provides additional strength (Figure 4-53).

Be sure to cut far enough below the D-ring mount to avoid cutting through the D-ring anchor reinforcement. The majority of B-pillars have them. In the case of the B-pillar, use a channel insert in only the outside piece of the pillar. The D-ring anchor reinforcement welded to the inside piece prevents the installation of an insert there.

Begin by overlapping the new inside piece on the existing one, rather than butting them together, and lap weld the edge (Figure 4-54A). Then, secure the insert with plug welds and close the joint with a continuous butt weld around the outer pillar (Figure 4-54B).

On occasion it is practical to obtain a recycled B-pillar and rocker panel assembly and replace them as a unit, because any time a B-pillar is hit so hard that it needs to be replaced, the rocker panel is almost invariably damaged, too. Install the upper end of the B-pillar with either of the two approved types of joints and make a butt joint with an insert in the rocker panel in the manner already shown. If the main damage is in the rear door opening, make the butt joint with an insert in the front door opening and install the other end of the rocker in its entirety. If the main damage is in the front door opening, reverse the procedure.

Generally speaking, the combination offset and overlap joint (Figure 4-55) is used more often when installing new parts and when working with separate inside and outside pieces.

1. Cut a butt joint in the outside piece above the level of the D-ring anchor reinforcement.
2. Make an overlap cut in the inside piece below the D-ring anchor reinforcement.
3. Install the inside piece with the new segment overlapping the existing segment.
4. Lap weld the edge (Figure 4-56A).
5. Put the outside pieces in place, make plug welds in the flanges, and close the section with a continuous weld at the butt joint (Figure 4-56B).

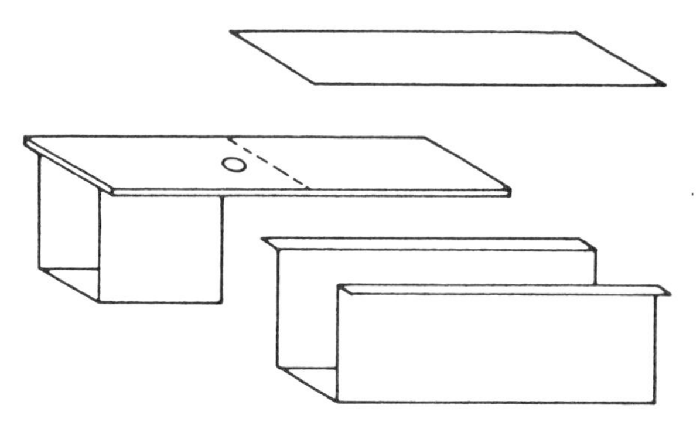

FIGURE 4-55 A combination offset and overlap joint

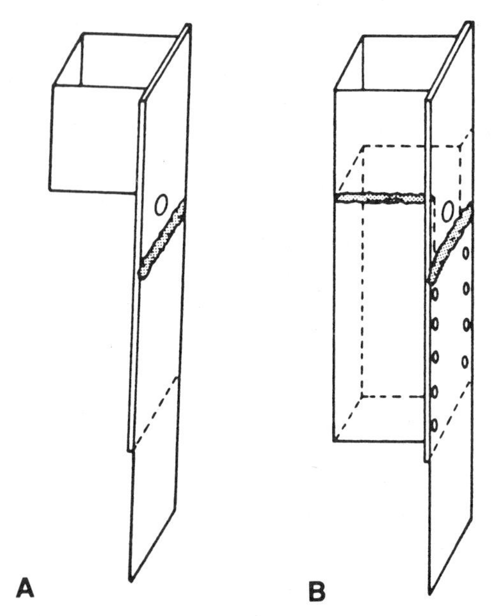

FIGURE 4-56 Creating the combination offset and overlap joint: (A) lap welding inside and (B) plug and lap welding outside

Usually, it is advantangeous to use the offset and overlap joint on a B-pillar with three or more pieces in its cross section. In fact, sometimes the offset and overlap procedure is mandatory, because it is actually not possible to install an insert.

SECTIONING FLOOR PANS

When sectioning a floor pan, do not cut through any reinforcements, such as seat belt anchorages. The rear section should overlap the front section by at least 50 mm, so the edge of the bottom piece, under the car, is always pointing toward the rear. This is so road splash, which moves from the front to the rear, streams past the bottom edge and does not strike it head-on (Figure 4-57).

1. Join all floor pan sections with an overlap.
2. Punch or drill plug weld holes in the top section.
3. If the mating surfaces are ungalvanized metal, apply a zinc-rich weld-through primer to them.
4. Fit the sections together and clamp them.
5. Plug weld the overlap (Figure 4-58).
6. On the bottom side, lap weld the underlapping edge with a continuous bead.
7. Caulk the forward edge of the top side with a flexible body caulk.
8. Cover the lap weld with a primer, a seam sealer, and a topcoat. The primer helps the sealer hold better, and the topcoat completes the protection. This assures that there will be no carbon monoxide intrusion through the joint into the passenger compartment.

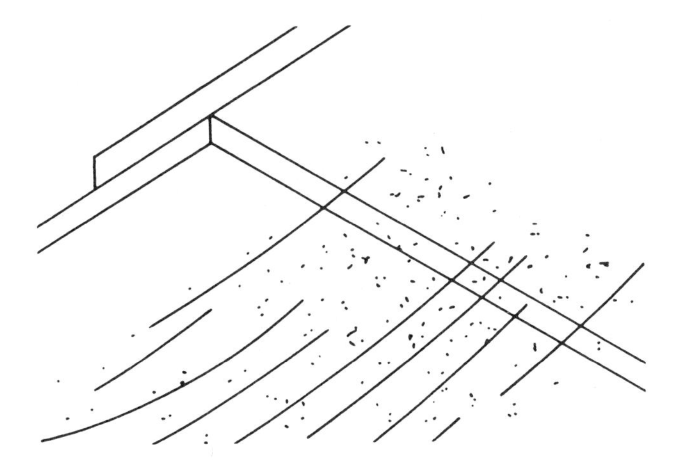

FIGURE 4-57 The rear section overlap shields the joint from wind.

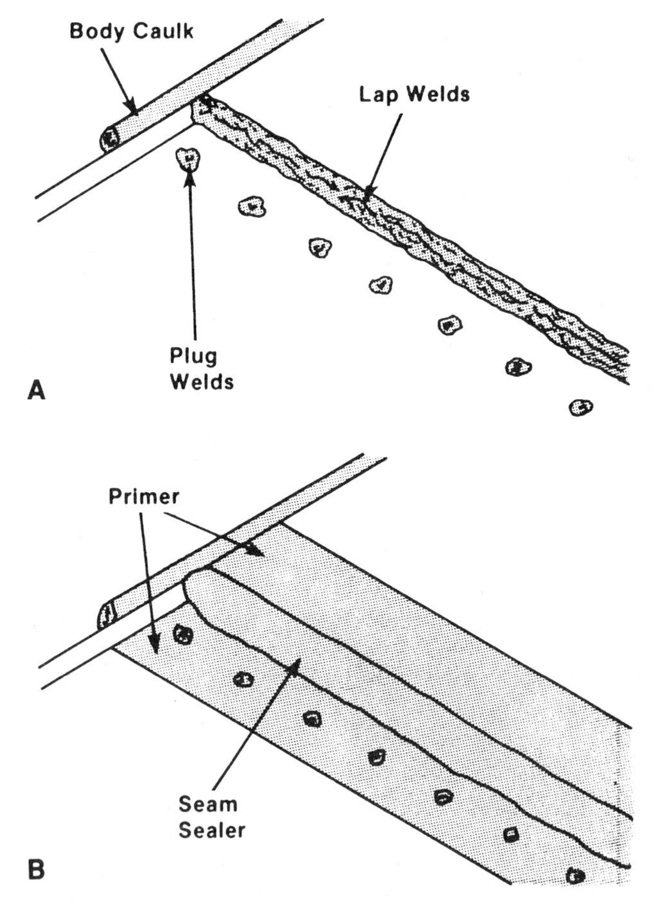

FIGURE 4-58 Plug weld from the top and lap weld, seal, and prime the bottom edge.

FIGURE 4-59 Sectioning a trunk floor above the cross member

SECTIONING TRUNK FLOORS

When a trunk floor is being sectioned, follow the basic procedures just described for the floor pan with some variations.

SHOP TALK

In a collision that necessitates sectioning of the trunk floor, the rear rail usually requires sectioning as well.

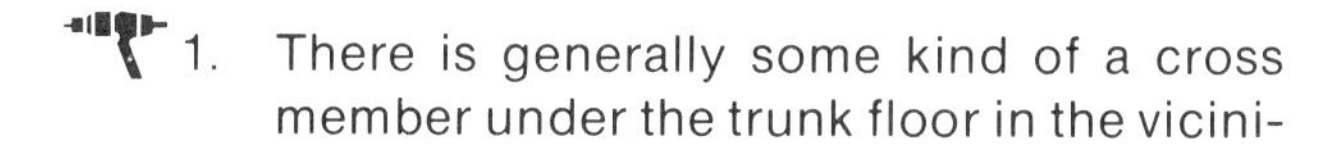

1. There is generally some kind of a cross member under the trunk floor in the vicinity of the rear suspension. Whenever possible (Figure 4-59) section the trunk floor above the cross member's rear flange. And section the rail just to the rear of the cross member.
2. Plug weld the trunk floor overlap joint to the cross member, again putting the plugs in from the topside, downward, as in a floor pan (Figure 4-60).
3. Caulk the topside, forward edge just like a floor pan seam (Figure 4-61).
4. A lap weld on the underlapping edge of the bottom side is not necessary because of the strength provided by the cross member. However, on cars where the trunk floor section is not above a cross member, the lower edge must be lap welded.

In both cases cover the bottom-side seam with a primer, a seam sealer, and a topcoat. With the trunk floor, sealing against carbon monoxide intrusion is critical because of the proximity of the tail pipe.

FULL BODY SECTIONING

To do a complete vehicle reconstruction by full body sectioning, the front portion of one vehicle is joined to the rear portion of another. This type of repair is generally done on a vehicle that has sustained severe rear end damage (Figure 4-62). In such a case, sectioning is more practical and cost effective in comparison to conventional methods of repairing severely damaged unitized automobiles. This method reduces the extensive time and labor of dismantling the vehicle and disturbs less of the corrosion protection.

Front full body sectioning is not recommended since the VIN would not match the registration and the power train warranty could be affected. Also, the matching of mileage, condition, and accessories would be difficult.

FIGURE 4-60 Plug welding the trunk floor

FIGURE 4-61 Caulking the inside seam

FIGURE 4-62 Severe rear-end damage

The techniques and procedures just described for sectioning a rocker panel, an A-pillar, and a floor pan are repeated in a full body section. By sectioning both A-pillars, both rocker panels, and the floor pan, a vehicle can be fully sectioned.

When the individual components are properly sectioned, aligned, and welded using the proper techniques and procedures, full body sectioning is a suitable and satisfactory procedure. Vehicles repaired by full body sectioning are completely crashworthy. This has been tested and proven over time. Keep in mind, however, that full body sectioning is not a frequently required procedure and full disclosure should be given to the car owner before repairs are started.

A discussion between the insurance representative, car owner, and repairer must be conducted, with the following points covered:

- All repair procedures, including alignment and welding, must be fully explained to the car owner.
- The recycled sections—both body and mechanical—must be of like kind and quality. Always verify that all VIN code identifications and EPA emission control requirements are met and that all suspension, braking, and steering components are in proper working order.
- Carefully inspect the front and rear sections for proper alignment before cutting. If either is out of alignment, proper fit and lineup of the section joints will be difficult, if not impossible, to achieve.

Figure 4-63 shows a popular compact that was fully body sectioned in a commercial repair shop. The undamaged front half of one car was joined to the undamaged rear half of another. Butt joints and inserts were used in the middle of the A-pillars (Figure 4-64) and in the two rocker panels (Figure 4-65), and an overlap joint was used in the floor pan (Figure 4-66). The rocker panel and floor pan cuts were made in the middle of the front door opening to avoid any brackets or reinforcements in the A- and B-pillars.

Remember that the floor pan might have reinforcements and brackets that need to be removed

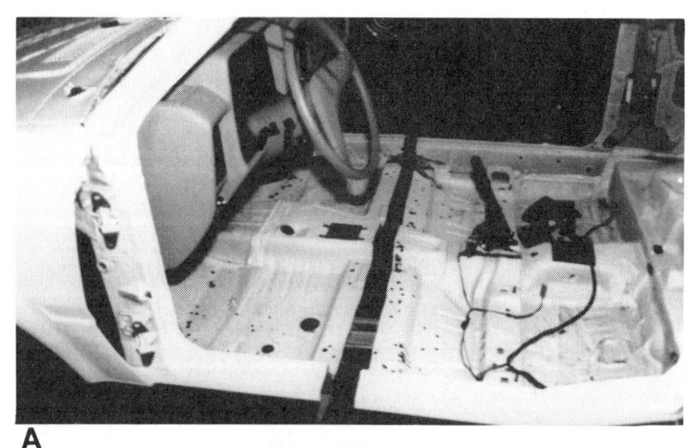

A

B

FIGURE 4-63 (A) Full body section and (B) replacement rear end

FIGURE 4-64 A welded A-pillar section

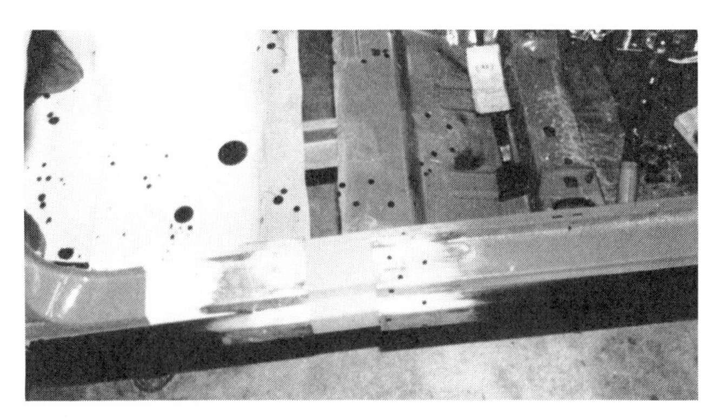

FIGURE 4-65 Rocker panel butt joint

before sectioning. Reinforcements can be left on the replacement rear half to aid in alignment. Proper corrosion protection must be restored when replacing brackets and reinforcements.

JOINING THE FULL BODY SECTIONS

After the front and rear sections have been trimmed to fit, drilled for plug welds, and primed with weld-through, follow these steps to join the sections:

1. Install the rocker and pillar inserts. Clamp them in place with sheet metal screws.
2. Place the A-pillar inserts in the upper or lower portion of the windshield pillar, depending on the angle and contour of the windshield.
3. Fit the two halves together by first joining the rocker panels and then the A-pillars. Clamp the rocker and pillar flanges to prevent the sections from pulling apart.
4. Check the windshield and door openings for proper dimension, using a tram gauge or a steel rule. If possible, install the doors and windshield to verify proper alignment.

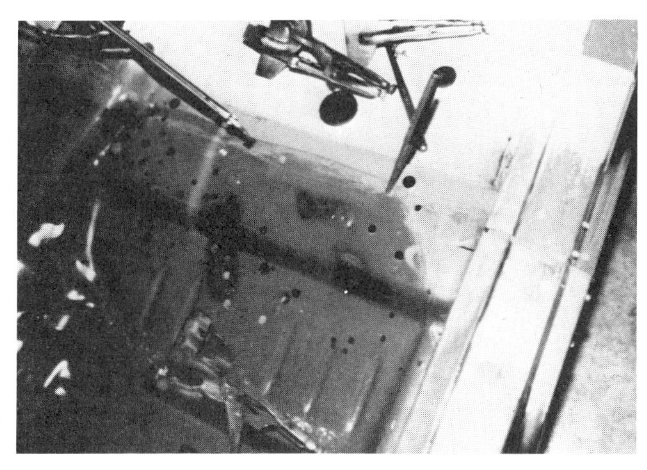

FIGURE 4-66 Floor pan overlap

5. When proper alignment is achieved, secure the overlapping areas with sheet metal screws to pull the seam areas together and hold the sections together during welding.
6. Using centerline gauges and a tram gauge, double-check the vehicle dimensions and section alignment before welding the sections together.
7. Weld the sections together using the techniques already described in this chapter for joining rocker panels, A-pillars, and the floor pan.

REVIEW QUESTIONS

1. Which of the following statements is incorrect?
   a. A structural member with only small cracks need not be replaced.
   b. High-strength steels are less ductile than mild steels.
   c. Parts that are kinked should be replaced.
   d. High-strength steels are prone to crack during pulling and straightening.

2. Replacement at factory seams sidesteps which of the following?
   a. extensive dismantling
   b. destruction of numerous factory spot welds
   c. destruction of factory corrosion protection
   d. all of the above
   e. none of the above

3. To remove a spot-welded panel, spot welds can be ______________.
   a. drilled out
   b. blown out with a plasma torch
   c. ground out with a high-speed grinding wheel
   d. all of the above
   e. none of the above

4. How should replacement parts be aligned with the existing unibody?
   a. using dimension measuring instruments
   b. determining the installation position by the relationship between the new part and the surrounding panels
   c. both a and b
   d. neither a nor b

5. Which type of unibody structural component is most critical?
   a. closed sections
   b. open surface components
   c. single layer overlap joint components
   d. none of the above

6. Which of the following joints is used to repair floor pans and trunk floors?
   a. butt joint with an insert
   b. offset butt joint
   c. overlap joint
   d. all of the above

7. What is the preferred tool for cutting a part that is being recycled for sectioning?
   a. metal saw
   b. propane torch
   c. plasma torch
   d. both a and c

8. What is the best method for removing rust-proofing and other contaminants from the inside of structural closed sections?
   a. sanding
   b. sandblasting
   c. wire brush and carburizing flame
   d. both b and c

9. How many pieces in a rocker panel design?
   a. one
   b. two
   c. three
   d. both b and c

10. Welder A uses a butt joint with an insert for sectioning a B-pillar. Welder B uses a combination offset cut and overlap. Who is right?
    a. Welder A
    b. Welder B
    c. Both A and B
    d. Neither A nor B

11. When sectioning a floor pan, Welder A overlaps the front section with the rear section. Welder B overlaps the rear section with the front section. Who is right?
    a. Welder A
    b. Welder B
    c. Both A and B
    d. Neither A nor B

12. Full body sectioning is recommended to repair extensive damage to the ______________.
    a. front end
    b. side
    c. rear end
    d. all of the above

13. When performing a full body section, which of the following should be joined first?
    a. rocker panels
    b. A-pillars
    c. B-pillars
    d. floor pan

14. Where should A-pillars be cut?
    a. lower end
    b. upper end
    c. middle
    d. both a and b

15. Proper sectional welding requires the use of ______________.
    a. TIG welding equipment
    b. oxyacetylene welding equipment
    c. MIG welding equipment
    d. none of the above

CHAPTER FIVE

TIG AND OXYACETYLENE WELDING; PLASMA ARC CUTTING

Objectives

After reading this chapter, you should be able to

- Explain the principles of TIG welding, oxyacetylene welding, and plasma arc cutting.
- Describe the differences between direct current straight polarity, direct current reverse polarity, and alternating current.
- Explain the importance of electrode size, tip condition, and position in the gun for achieving quality welds.
- Name the basic differences involved in TIG welding mild steel, stainless steel, and aluminum.
- List the equipment that makes up a typical oxyacetylene welding outfit.
- Describe how an oxyacetylene flame is adjusted.
- List the factors that determine the quality of a plasma arc cut.

Although both oxyacetylene and TIG have been viable and trusted welding methods for many years, neither one has a great influence in today's auto body shop. As has been seen, modern car metals are better suited to MIG and resistance spot welding; the result has been a marked decline in the use of oxyacetylene and TIG. However, since there will undoubtedly be instances in which these welding methods will still occasionally come into use, they are discussed in this chapter. And although plasma welding is seldom used nowadays in the body shop, plasma arc cutting is fast becoming the first choice among body technicians for cutting automotive metals. Thus, though none of the three procedures discussed in this chapter match the significance of MIG or resistance spot welding, it is important that their theory and operation are understood by the auto body technician.

TIG WELDING

Tungsten inert gas, or TIG, welding is of limited use in auto body repair applications. However, it is valuable for general auto repair purposes, such as repairing cracks in aluminum cylinder heads, reconstructing combustion chambers, and general maintenance welding of other automotive components. MIG welders deposit weld beads at an average of 25 inches per minute; TIG welding is much slower, with weld speeds ranging between 5 to 10 inches per minute. However, this slower speed gives much more control, and the end result is the best looking weld obtainable.

Like MIG welding, the TIG process uses an inert gas such as argon or helium to surround the weld area and prevent oxygen and nitrogen in the atmosphere from contaminating the weld. But instead of having a wire feed welding electrode like MIG units, TIG machines use a tungsten electrode with a very high melting point (about 6900 degrees Fahrenheit) to strike an arc between the welding gun and the work (Figure 5-1).

Since the tungsten electrode has such a high melting point, it is not consumed during the welding process, so a filler rod must be used when welding thicker materials. Because the torch is held in one hand and the filler rod is held in the other (Figure 5-2), TIG welding is similar in some ways to oxyacetylene welding.

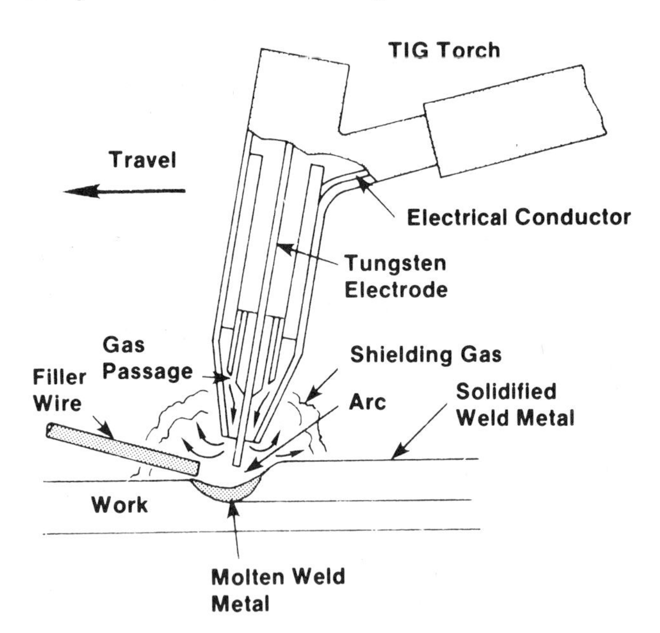

FIGURE 5-1 Principles of TIG welding

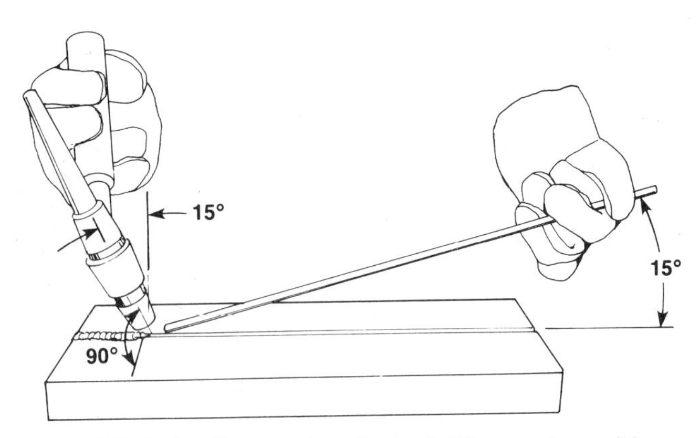

FIGURE 5-2 Proper torch and filler rod position for manual TIG welding

TIG POWER SUPPLY

Three major components make up a TIG welding machine: the power supply, the welding gun, and the gas cylinder with flowmeter (Figure 5-3). Most TIG units have a sophisticated power supply system that can supply current to the electrode as direct current straight polarity, direct current reverse polarity, and alternating current (Figure 5-4). The choice of current depends on the type of material being welded, the desired weld shape, and the penetration of the weld bead. Generally speaking, the two direct currents are used for steel, cast iron, and other ferrous metals; alternating current is primarily for nonferrous metals such as aluminum.

With direct current straight polarity, the electron flow at the arc is from the electrode to the

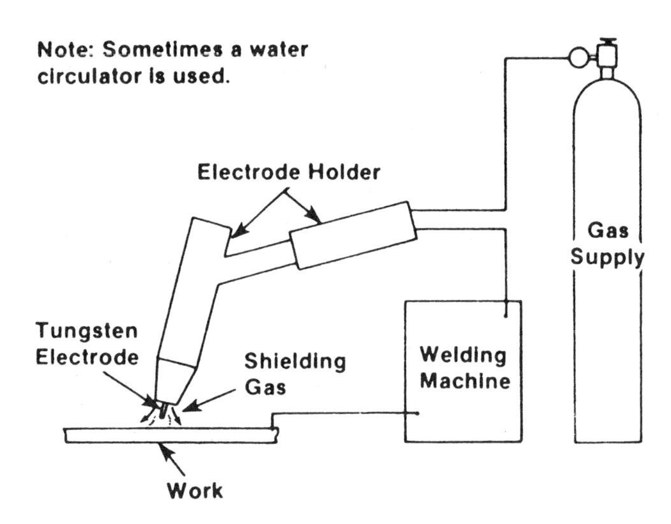

FIGURE 5-3 Schematic of a manual TIG welding outfit

FIGURE 5-4 Typical TIG welder

workpiece. Because the electrons strike the weld at a high rate of speed, heat buildup occurs; this results in a narrow weld with a high degree of penetration. The one notable disadvantage to direct current straight polarity is that it does not provide cleaning action on the workpiece.

Cleaning action does take place with direct current reverse polarity. Since the electron flow is reversed, heat builds up in the electrode instead of the weld. To handle this added heat, a larger diameter electrode is required. Usually an electrode for reverse polarity is approximately four times the diameter of electrodes used on straight polarity. The weld bead tends to be wide with a minimal amount of penetration.

With alternating current, the features of both direct current modes are found to a degree. Cleaning occurs on each straight polarity half cycle and heat is developed at the workpiece. The area cleaned on the straight polarity half cycle remains clean during the reverse polarity half cycle while it is shielded with an inert gas. The newest TIG power supplies include a solid-state control board that makes it possible to adjust the alternating current so as to favor either the straight or reverse polarity half cycle. Generally speaking, when maximum cleaning is desired, reverse polarity should be favored; when maximum heat is desired, straight polarity should be favored. Figure 5-5 illustrates the differences between the three current types for TIG welding.

TIG WELDING GUN

The TIG welding gun holds the electrode with a collet that screws into the body of the gun (Figure 5-6). A ceramic cup blows the shielding gas around the weld area. The cups are available in a variety of sizes and flow rates to provide optimum shielding for different types of welds. Some guns also have a screened gas nozzle to eliminate troublesome turbulence, which can interfere with the shielding gases' ability to surround the weld. Many TIG welders have a trigger on the gun to control current and gas flow; others utilize a foot control.

The gun can be either water or air cooled. High-production or high-amperage guns are usually water cooled, while lighter duty guns for low amperage applications can be air cooled. The water-cooled type is designed so that water is circulated through it and the power cable. The power cable is contained inside a hose, and the water returning from the gun flows around the power cable, thus cooling it. In this way, the power cable can be relatively small, making the entire cable assembly light and easily maneuverable.

When using a water-cooled TIG welding gun, a lack of cooling water or no cooling water at all will create a heat buildup that can cause the polyethylene sheath to melt or possibly burn the power cable in two. The gun manufacturer's specifications will designate the required amount of cooling water. A safety device known as a fuse assembly can be installed in the power cable; this assembly contains a fuse link, which is also cooled by the water. If there is no cooling water circulating, the fuse link will melt in two and prevent damage to other more expensive components. The fuse link is easily replaced. When this happens, the water flow is maintained and welding can continue.

Air-cooled torches are popular for lower amperage applications. They require no additional cooling other than the surrounding air. The power

Current Type	DC	DC	AC (balanced)
Electrode Polarity	Straight	Reverse	
Electron and Ion Flow Penetration Characteristics	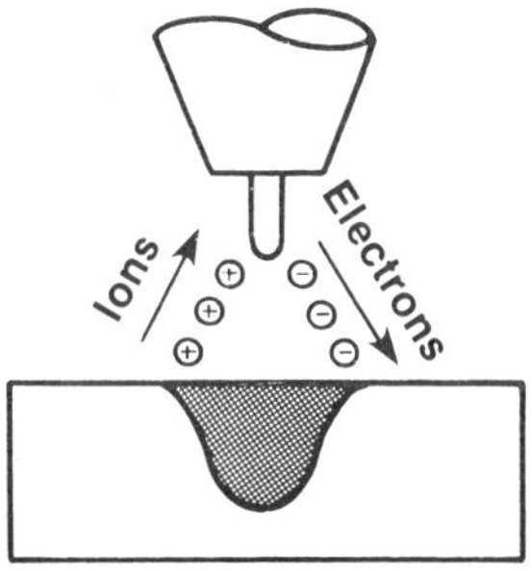 Ions / Electrons	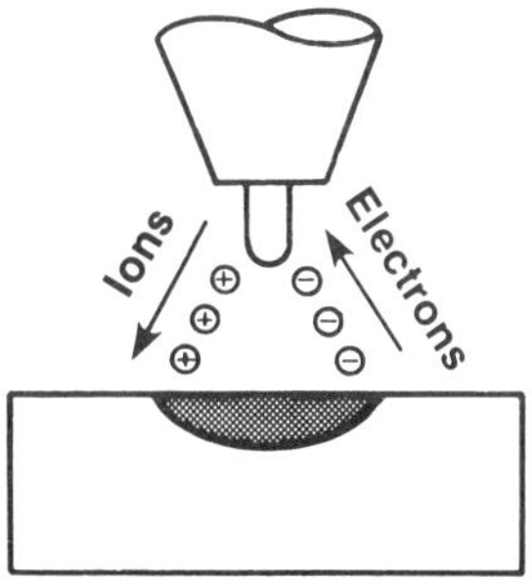 Ions / Electrons	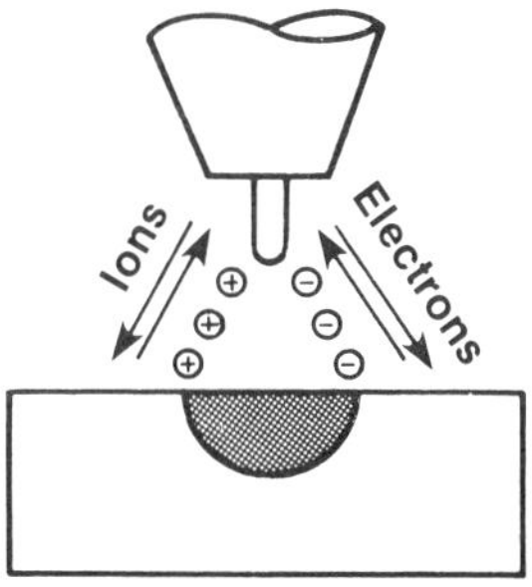 Ions / Electrons
Oxide Cleaning Action	No	Yes	Yes—once every half cycle
Heat Balance in the Arc (Approx.)	70% at work end 30% at electrode end	30% at work end 70% at electrode end	50% at work end 50% at electrode end
Penetration	Deep; narrow	Shallow; wide	Medium
Electrode Capacity	Excellent e.g., 1/8 in. (3.18 mm); 400A	Poor e.g., 1/4 in. (6.35 mm); 120A	Good e.g., 1/8 in. (3.18 mm); 225A

FIGURE 5-5 Characteristics of current types for TIG welding

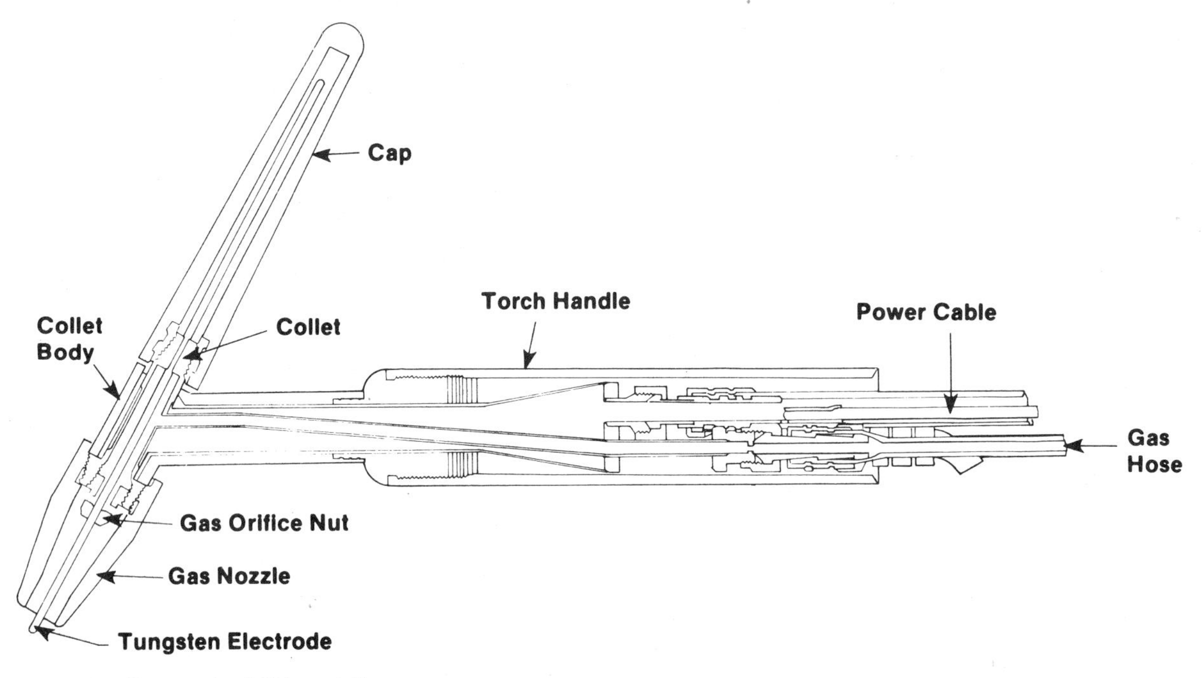

FIGURE 5-6 Parts of a TIG welding gun

cable, which must be heavier than the cable in a water-cooled torch, is wound around the gas-carrying hose or is located inside the hose.

SHIELDING GASES

Although almost any gas could be used to shield the arc in the TIG welding process, argon and helium are the best choices because of their availability, economy, and efficiency (Figure 5-7). A higher gas flow is required with helium than with argon, but helium permits faster welding speeds and greater penetration. Differentials in rates of gas flow are often almost equalized in welding thicker sections. Deeper penetration is obtained with helium, because the arc in a helium atmosphere is hotter than in argon. Gas selection is based on the metals to be welded.

Many operators prefer argon for manual TIG welding, because the arc has greater stability. Additives to argon increase the arc temperature and provide some of the advantages of using a helium shielding gas. Generally, when 10 percent or more helium is added to argon, arc penetration increases significantly. Hydrogen is also sometimes added in amounts up to 25 percent. Hydrogen increases the arc voltage and produces more heat than the pure inert gas. A higher arc voltage is desirable when welding thick materials and metals that have high heat conductivity; it is also an advantage in high-speed automatic welding. Hydrogen, however, cannot be added and used with all metals. The addition of hydrogen can damage some metals and alloys, including all aluminums, coppers, and magnesium-base alloys. It can be used successfully on certain stainless steels and nickel alloys.

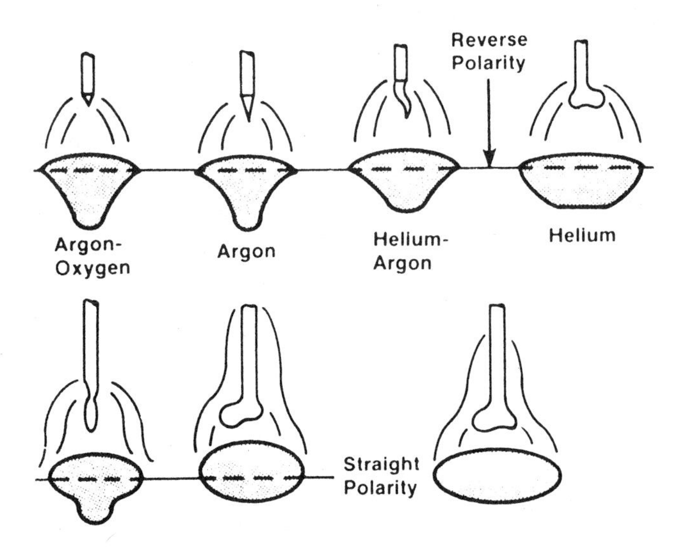

FIGURE 5-7 Electrode tip shape, weld bead contour, and penetration patterns for argon and helium shielding gases

Purity of the inert gas is very important; only welding-grade gases should be used. Furthermore, contamination of the gas must be prevented during its passage from the cylinder to the gun. Moisture, oil residues or fumes, and dust are common contaminants to guard against. Hoses previously used with acetylene, compressed air, or water are not suitable for TIG welding.

Inert gas for TIG welding is available in pressurized cylinders. For consistent results, a pressure-reducing regulator flowmeter of the proper type for the gas being used should be included in the system. Shutoff can be manual (by hanging the torch on a hook arm that actuates a shutoff valve) but automatic control is preferred. This is accomplished by connecting a time-delay solenoid valve in the gas line. Production TIG welding equipment has automatic gas and water-control valves built into the welding unit. For example, a gas flow solenoid can be installed that permits the shielding gas to flow while the current is flowing. When the welding begins, the solenoid is electrically energized, allowing the gas to flow. When welding is stopped, the gas should continue to flow for a short time to protect the electrode from contamination while it cools. Some machines are equipped with a postpurge timer that permits adjustment for different postflow lengths of time.

ELECTRODES

The size, tip condition, and position in the gun of the tungsten electrode are important factors that determine the quality of TIG welding. Electrode size is chosen so that the tip is maintained at a temperature near its melting point when welding is done in any given alternating current range. The maximum electrode extension beyond the end of the gas cup should be about equal to the diameter of the cup for butt welding, and slightly greater for fillet welding. However, extension of the electrode somewhat less than these maximums is always preferable, because it helps improve weld shielding.

Electrodes for alternating current TIG welding are available in sizes ranging from 0.020 inch to 1/4 inch (0.5 to 6.35 mm). Selecting the right size electrode for each job is important in preventing poor welds and electrode damage caused by using too high or too low a current for the size electrode used. When electrode size is matched with the correct current, the tip will become a molten hemisphere. There are four types of nonconsumable tungsten electrodes: pure tungsten, 1 percent thorium, 2 percent thorium, and half of 1 percent zirconium added (Figure 5-8). The addition of thorium and zirconium makes the tungsten alloy more able to emit electrons when hot. It also provides for increased current-carrying capacity of the electrode and provides for a more stable arc and better arc starting.

When installing a new tungsten electrode in the gun, the color tip should be at the back end of the gun so that it is not destroyed by the arc. The collet should be checked to see that it is the proper size for

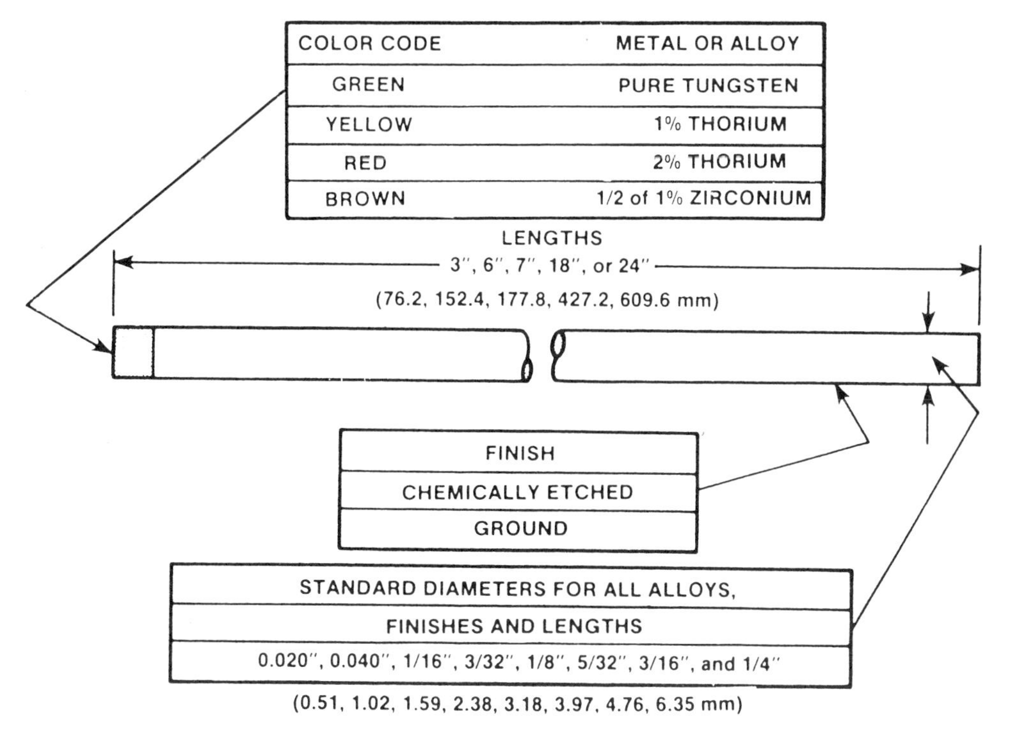

FIGURE 5-8 Tungsten electrodes available for TIG welding

the electrode being used. The entire assembly must be tight to the tungsten and the collet tight to the gun so that the heat of the arc is transmitted to the gun body where it can be carried away. Pure tungsten electrodes are the least expensive and should be used for the less critical operations or for general purpose work on different metals. The 1 percent thoriated tungsten provides for easier arc starting, gives a more stable arc, and can be operated at slightly higher temperatures. The 2 percent thoriated tungsten is even better for arc starting, is more stable, and has a higher current-carrying capacity. Zirconated tungsten electrodes also provide for longer life and more stable operation.

As mentioned earlier, electrode size must be selected carefully and must be related to the type of current, the amount of current, and the particular work surface. Of course, welder preference may also enter into the size selection. However, experience is also important; for example, if the welder is using a pure tungsten electrode and it tends to become overheated or appears to have a wet surface, the current is too high for the size of the electrode. When the tungsten has this wet surface appearance it becomes more susceptible to picking up contamination from the base metal; a larger electrode of the same type should be selected. Too much current or an electrode that is too small will cause excessive tungsten erosion, and tungsten particles might become deposited in the weld metal.

Electrode Spitting

Electrode spitting occurs when particles of the electrode are ejected across the arc; this condition simply cannot be tolerated for long before it affects the weld. One result of spitting is that the electrode particles contaminate the weld metal and reduce the mechanical properties of the weld. Another is unnecessary and wasteful consumption of the electrode. Also, the time required to reset the electrode tip and eventually to replace it adds unnecessarily to job time.

In alternating current TIG welding, the transfer of electrode particles into the weld pool can be caused by partial rectification of the welding current. The use of batteries in the circuit as a means of reducing partial rectification is a possible solution. In all three current modes of TIG welding, the transfer of electrode particles can be caused by using too small an electrode for the welding current. This overheats the molten hemisphere at the electrode tip, causing it to drop off.

Spitting of electrode particles across the arc can also occur if too large an electrode is used. In this case, the path of the current or focal point of the arc wanders over the tip of the electrode and particles of tungsten fly off. It is easy to detect whether the tungsten particles are flying off or whether the tip is becoming overheated and dropping off.

Electrode Care

An electrode should have a clean, silvery appearance at all times. A dirty, rough electrode usually means that the inert gas was shut off before the electrode cooled, that there was air leakage in the gas-supply system of the torch, or that the electrode tip was contaminated by touching metal. Remember that the tungsten electrode should never be allowed to touch molten metal.

A dirty tungsten electrode can sometimes be cleaned satisfactorily with a fine emery cloth. The contamination can also be removed by using a high current while welding on scrap material until the tip has a sizable molten section, then quickly twisting the torch to flip off a drop of molten metal. If severely contaminated, the electrode should be replaced or the tip broken off and dressed.

Grinding of electrodes is a common practice for removing contamination. Using correct grinding techniques is important; poor techniques can cause problems. Grinding should be done on a fine grit, hard abrasive wheel. Tungsten is harder than most grinding wheels, therefore it must be chipped away rather than cut away. The grinding marks should run lengthwise with the point. If the grinding is done on a coarse stone and the grinding marks are concentric with the electrode, a series of ridges will form on the surface of the ground area. There is a possibility of the small ridges melting off and floating across the arc. If the stone used for grinding is not clean, contaminating particles can be lodged in the grinding crevices and dislodge during welding, ending up in the deposit. A grinding wheel used on tungsten electrodes should be used for no other material.

Pointing of electrodes is a much discussed subject. There are many theories and opinions on the degree of the point; again, the application has a bearing on the exact configuration. Some general recommendations follow to serve as a guide to pointing of electrodes.

A common practice in pointing electrodes is to grind the taper for a length of 2 to 2-1/2 electrode diameters (Figure 5-9). Using this rule for a 1/8-inch electrode, the ground surface would be 1/4 to 5/16

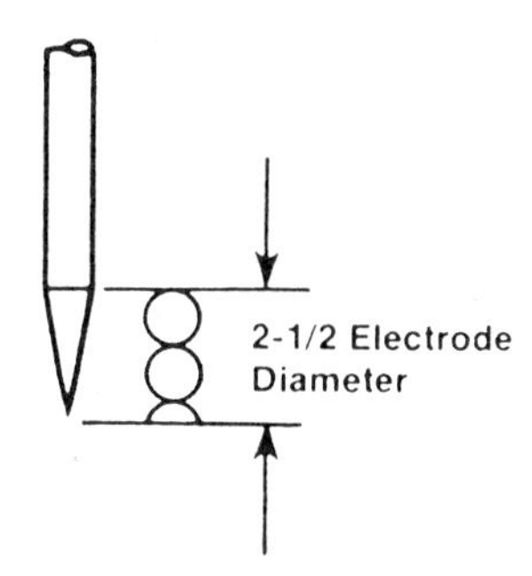

FIGURE 5-9 Sharpened electrode

inch long. In some applications, a sharp needle point is preferred to aid low amperage starting. Needle pointed electrodes are also usually preferred also on very thin metals in the range of 0.005 to 0.040 inch. In other applications a slightly blunted end is preferred because the extreme point can be melted off and end up in the deposit.

Figure 5-10 shows several 1/8-inch tungsten electrodes. Notice the following different tip configurations:

- Electrode A has a ball end. This pure tungsten was used with alternating current on aluminum. The end is uniform in shape and has a bright and shiny appearance.
- Electrode B is ground to a taper and was used with direct current straight polarity.
- Electrode C was used with alternating current on aluminum. Note that this electrode has several small ball-shaped projections rather than a round complete ball end like Electrode A.
- Electrode D is a pure tungsten that was used with alternating current on aluminum. This electrode was subjected to a current above the rated capacity. Notice that the ball started to droop to one side. It became very molten during operation; continuing to operate the welder would have caused the molten end to drop into the weld puddle.
- Electrode E was tapered to a point and used on direct current straight polarity. Notice the ball tip characteristic of pure tungsten. Pointing of pure tungsten is not recommended because the extreme point will always melt when the arc is established. The electrode in this illustration melted back; however, often the point will melt and drop into the weld puddle.
- Electrode F was severely contaminated by touching the filler rod to the molten metal. In this case the contaminated area must be broken off and the electrode reshaped.
- Electrode G did not have sufficient gas postflow. Notice the black surface which is oxidized because the atmosphere contacted the

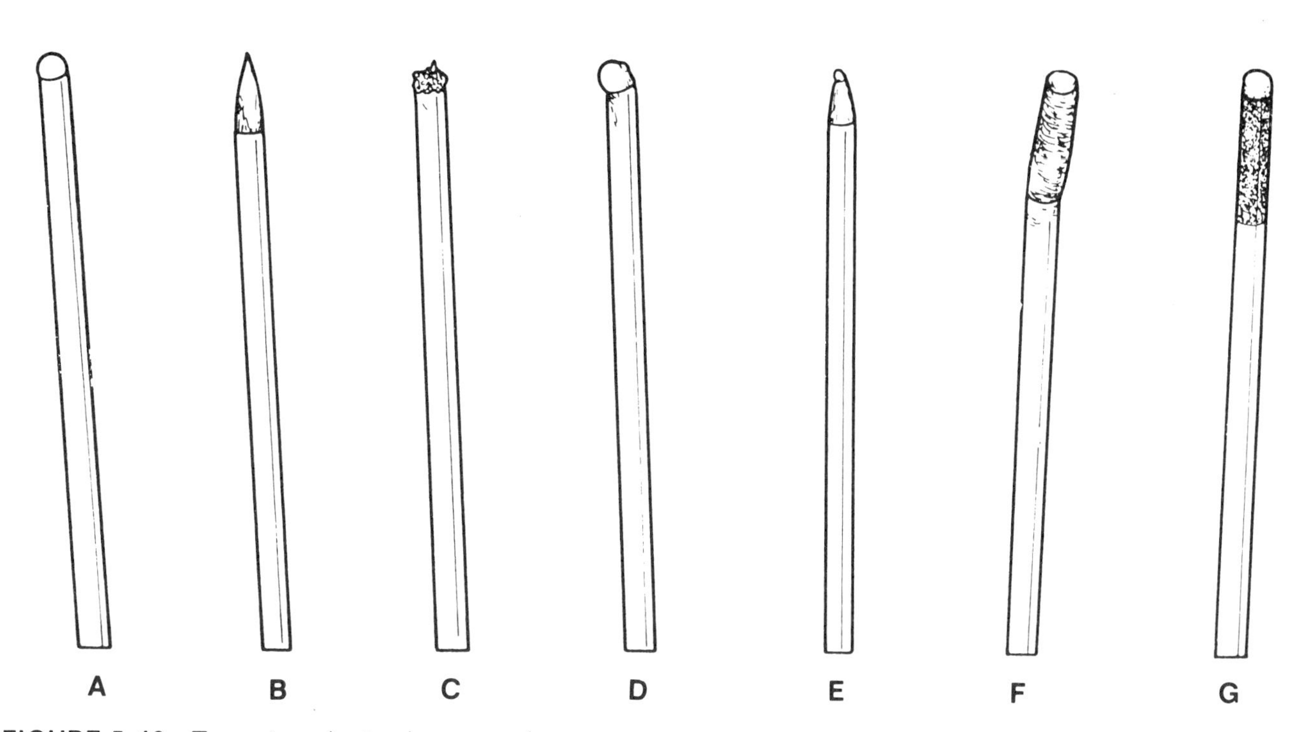

FIGURE 5-10 Tungsten electrode comparison

electrode before it cooled sufficiently. If this electrode were used, the oxidized surface would flake off and drop into the weld puddle. Postflow time should be increased so the appearance is like electrode A after welding.

Filler Metal

Filler metal is not used when welding extremely thin metals; however, for most applications it is added to the puddle to reinforce the joint. The composition of the filler metal should always match that of the base metal. The size of the filler metal rod depends on the thickness of the base metal. Filler metal is normally added to the puddle manually, but an automatic feed is sometimes used.

TIG WELDING TECHNIQUES

Good TIG welds start with good setups. The electrode must be adjusted the correct distance from the ceramic cup (Figure 5-11), the right filler rod must be used, the current must be set, and the proper gas at the proper flow rate must be selected. All this information can be found in the welding charts in the manufacturer's instruction manual.

How to start the arc depends on the type of current being used. Normally, with direct current, the electrode must momentarily touch the plate before the arc will start. However, some direct current units have a momentary high-frequency generator, so that starting the arc manually is not necessary. With these machines as well as alternating current welders, there is no need to touch the electrode to the metal—the arc will automatically start when the electrode is about 1/8 inch from the material.

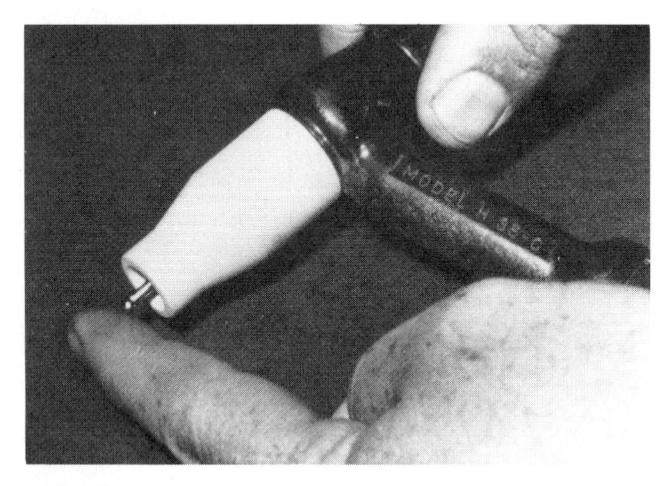

FIGURE 5-11 Adjust the electrode so it extends beyond the end of the gas cup or nozzle.

Once the arc is started, the gun should be held at about a 75-degree angle to the welding surface so that the weld will trail the gun. Use a small circular motion to create a molten puddle of metal. Once the puddle develops, hold the gun steady. Then begin moving the gun slightly along the weld so that the puddle follows the electrode. That is all that is needed when it comes to welding thin-gauge materials (Figure 5-12).

Adding the filler rod when welding thicker materials is more difficult. Begin by starting the initial puddle in the same way. Then, with the filler rod held at a 15-degree angle opposite to the gun, move the gun to the rear of the puddle and add filler rod by quickly touching the rod to the leading edge of the puddle. Remove the rod and move the gun back up

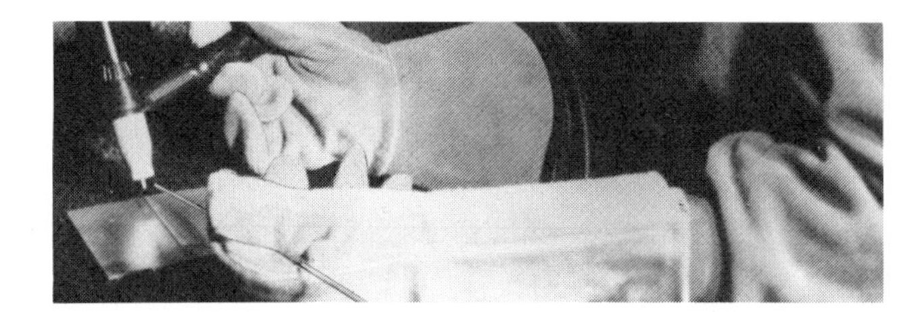

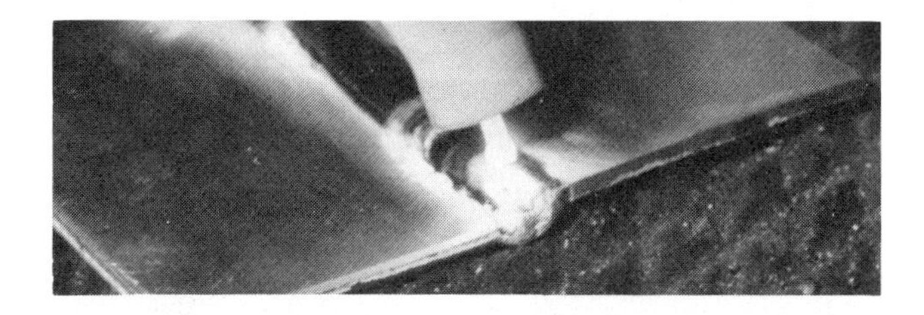

FIGURE 5-12 Steps in tack welding with a TIG welder

FIGURE 5-13 Example of a good TIG weld

to the leading edge of the puddle to start the process again.

Figure 5-13 shows what a good TIG weld looks like. Figure 5-14 illustrates some of the control factors of TIG welding.

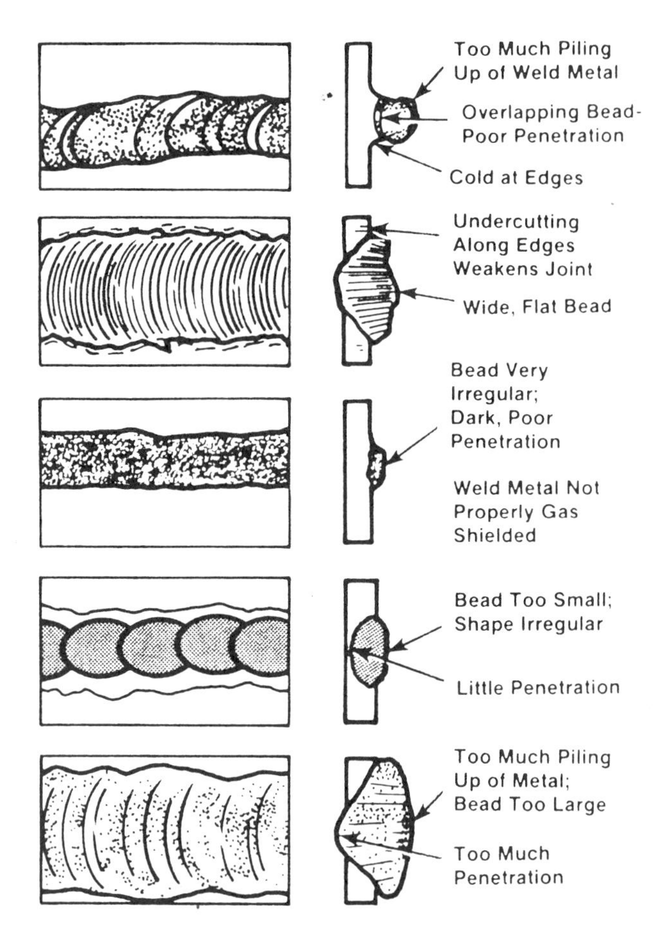

FIGURE 5-14 TIG welding quality control factors

TIG Welding Stainless Steel

Stainless steel is a term commonly used to refer to chromium alloyed and chromium-nickel alloyed steels. The latter are readily TIG weldable, provided the filler metal used is compatible with the base metal. The heat conductivity of these steels is about 50 percent that of mild steels; therefore, the heat is concentrated in the weld area rather than being dissipated throughout the work. However, thermal expansion is usually about 50 percent greater than mild steel, which increases the tendency for distortion on thin sections.

The weld area should be thoroughly cleaned. Protective paper or plastic coatings can be applied to many stainless steel sheets; otherwise, foreign material might cause carburetion of the surface, which will lessen the corrosion-resisting properties. Any wire brushing should be done with stainless steel wire brushes to prevent iron pickup on the surfaces. As with other welding procedures, clean, dry filler metal should be used, and proper precautions taken to prevent contamination during welding. Table 5-1 provides a guide for manual TIG welding of stainless steel using direct current straight polarity. Keep in mind that particular job conditions might affect actual amperages, flow rates, and materials used.

Heat input can be critical. On many applications it is best to keep the heat as low as possible in order to prevent a metallurgical change known as carbide precipitation from taking place. Some of the corrosion resistance properties are lost in the weld and adjacent areas when heated above the temperature where carbide precipitation occurs (800 to 1400 degrees Fahrenheit). Keeping heat input to a minimum is important in this situation. The longer the work is at the 800 to 1200 degrees Fahrenheit range, the greater the carbide precipitation. Rapid cooling through this range will help keep carbide precipitation at a minimum.

TIG Welding Mild Steel

Low carbon steels, commonly referred to as *mild steels*, are readily welded by the TIG process. This group of steels is available in many different alloys and types. The familiar structural shapes—

TABLE 5-1: TIG WELDING STAINLESS STEEL

Metal Thickness (Inches)	Joint Type	Tungsten Electrode Diameter (Inches)	Filler Rod Diameter (If Required) (Inches)	Amperage	Gas	
					Type	Flow-CFH
1/16	Butt	1/16	1/16	40-60	Argon	15
	Lap	1/16	1/16	50-70	Argon	15
	Corner	1/16	1/16	40-60	Argon	15
	Fillet	1/16	1/16	50-70	Argon	15
1/8	Butt	3/32	3/32	65-85	Argon	15
	Lap	3/32	3/32	90-110	Argon	15
	Corner	3/32	3/32	65-85	Argon	15
	Fillet	3/32	3/32	90-110	Argon	15
3/16	Butt	3/32	1/8	100-125	Argon	20
	Lap	3/32	1/8	125-150	Argon	20
	Corner	3/32	1/8	100-125	Argon	20
	Fillet	3/32	1/8	125-150	Argon	20
1/4	Butt	1/8	5/32	135-160	Argon	20
	Lap	1/8	5/32	160-180	Argon	20
	Corner	1/8	5/32	135-160	Argon	20
	Fillet	1/8	5/32	160-180	Argon	20

plates and hot rolled sheet metal—are usually of a semi-killed steel; this means the steel has been partially deoxidized during manufacture but still contains oxygen. This oxygen can cause problems when TIG welding. It will appear in the form of bubbles in the weld puddle, and possibly in the finished weld. Killed steel has had more of its oxygen removed and therefore presents less of a problem while welding.

A filler wire, with silicon and manganese added as deoxidizers, is necessary to TIG weld mild steel. Lower grade filler rods used for oxyacetylene welding are not suitable for making high-quality TIG welds. Direct current, straight polarity is recommended, along with high-frequency starts. A touch start can be used if it does not pose a contamination problem. A 2 percent thorium tungsten electrode is recommended; a point or taper on the electrode can be used.

The work surface should be mechanically cleaned prior to welding. Rust, paint, oil, grease, and any other surface contaminants should be removed. For best results, any varnish-type coating should also be removed before welding. Table 5-2 can be used as a guide for manual TIG welding of mild steel using direct current straight polarity.

TIG Welding Aluminum

Aluminum is very susceptible to contaminants, which can cause considerable problems during the welding process. Aluminum has a natural surface oxide that must be removed. Although the electron flow from the workpiece lifts away any loose oxides, keep in mind that this action should not be relied on to do all the cleaning (Figure 5-15).

Mechanical cleaning methods are needed to remove heavy oxide, paint, grease, oil, or any other materials that will hinder proper fusion. Mechanical cleaning can be done with abrasive wheels or wire brushes; special abrasive wheels are available for aluminum and stainless steel wire brushes work well. The important point is that the abrasive wheels and wire brushes should be used only on the material being cleaned. If a wire brush were used on rusty steel and then on aluminum, it could carry contaminants from one piece to another. The vigorous brushing can imbed the contaminants into the aluminum. The same is true of the abrasive wheel and equipment used to cut and form aluminum.

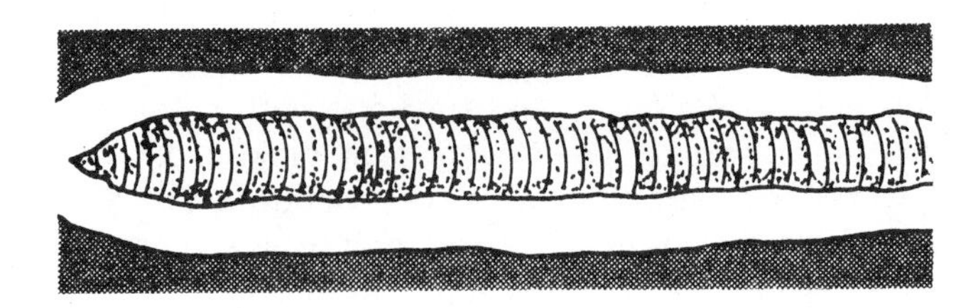

FIGURE 5-15 The electron flow removes some of the oxides from aluminun, as shown by the light areas.

TABLE 5-2: TIG WELDING MILD STEEL						
Metal Thickness (Inches)	Joint Type	Tungsten Electrode Diameter (Inches)	Filler Rod Diameter (If Required) (Inches)	Amperage	Gas	
					Type	Flow-CFH
1/16	Butt	1/16	1/16	60-70	Argon	15
	Lap	1/16	1/16	70-90	Argon	15
	Corner	1/16	1/16	60-70	Argon	15
	Fillet	1/16	1/16	70-90	Argon	15
1/8	Butt	1/16-3/32	3/32	80-100	Argon	15
	Lap	1/16-3/32	3/32	90-115	Argon	15
	Corner	1/16-3/32	3/32	80-100	Argon	15
	Fillet	1/16-3/32	3/32	90-115	Argon	15
3/16	Butt	3/32	1/8	115-135	Argon	20
	Lap	3/32	1/8	140-165	Argon	20
	Corner	3/32	1/8	115-135	Argon	20
	Fillet	3/32	1/8	140-170	Argon	20
1/4	Butt	1/8	5/32	160-175	Argon	20
	Lap	1/8	5/32	170-200	Argon	20
	Corner	1/8	5/32	160-175	Argon	20
	Fillet	1/8	5/32	175-210	Argon	20

Another problem sometimes encountered occurs when only the side of the joint being welded is cleaned, thus allowing contamination from the backside or between butting edges to be drawn into the arc area. It is recommended that both sides of the joint be cleaned. Another frequent source of contamination is the filler metal. Aluminum filler wire and rod oxidizes just like the base metal; if it is severe enough, the rod must be cleaned prior to use. And remember that the operator can transfer contaminants from dirty welding gloves onto the filler rod and, consequently, into the weld area. The information contained in Table 5-3 will serve as a guide to manual TIG welding of aluminum using high-frequency alternating current.

Repairing Aluminum Cylinder Heads

Aluminum cylinder heads have become increasingly popular in recent years, so the ability to repair them through TIG welding is very significant. Though most are found on four-cylinder engines, a limited number of V-6 and V-8 engines also feature aluminum heads. Aluminum presents its own unique problems; probably the most troublesome is its thermal expansion characteristics.

Aluminum expands and contracts almost twice as much as cast iron in response to temperature changes, and this creates a number of problems. When an aluminum head is mated to an iron block (which most are), the difference in thermal expansion between the head and block creates a great deal of scrubbing stress on the head gasket. Unless the gasket is engineered to take such punishment, the result can be leakage and premature gasket failure.

Increased thermal expansion and stress also lead to cracking. The most crack-prone areas in the head are usually the areas around the valve seats. The interference fit of the seat in the head, combined with different rates of expansion between the seat and head, high combustion temperatures, and the constant pounding of the valve against the seat, often cause cracking between the intake and exhaust seats or just under the exhaust seat. If the seats are not machined to very close tolerances and installed properly, cracking is virtually guaranteed.

The difference in thermal expansion between aluminum and iron creates a lot of stress throughout the head. The head wants to expand in all directions at once as it heats up, but the head bolts keep it from going sideways or lengthwise. The only place to go is up, so the head has a tendency to bow up in the middle.

This is why scuffed cam bores are often found in many aluminum heads with overhead cams. Add a bad thermostat, clogged water passages, a leaky radiator, or any other problem in the cooling system that causes the engine to overheat, and it is the perfect combination for severe warpage and cracking. This often results in combustion leaks, head gasket failure, timing belt or chain breakage, and, in some cases, even cam breakage.

TABLE 5-3: MIG WELDING ALUMINUM

Metal Thickness (Inches)	Joint Type	Tungsten Electrode Diameter (Inches)	Filler Rod Diameter (If Required) (Inches)	Amperage	Gas	
					Type	Flow-CFH
1/16	Butt	1/16	1/16	60-85	Argon	15
	Lap	1/16	1/16	70-90	Argon	15
	Corner	1/16	1/16	60-85	Argon	15
	Fillet	1/16	1/16	75-100	Argon	15
1/8	Butt	3/32-1/8	3/32	125-150	Argon	20
	Lap	3/32-1/8	3/32	130-160	Argon	20
	Corner	3/32-1/8	3/32	120-140	Argon	20
	Fillet	3/32-1/8	3/32	130-160	Argon	20
3/16	Butt	1/8-5/32	1/8	180-225	Argon	20
	Lap	1/8-5/32	1/8	190-240	Argon	20
	Corner	1/8-5/32	1/8	180-225	Argon	20
	Fillet	1/8-5/32	1/8	190-240	Argon	20
1/4	Butt	5/32-3/16	3/16	240-280	Argon	25
	Lap	5/32-3/16	3/16	250-320	Argon	25
	Corner	5/32-3/16	3/16	240-280	Argon	25
	Fillet	5/32-3/16	3/16	250-320	Argon	25

Diagnosing Cracks. Aluminum is nonmagnetic, so magnetic crack detection equipment cannot be used to find cracks. Penetrating dyes work well on aluminum, as does pressure testing. Dye is good for pinpointing surface cracks in combustion chambers or on the face of the head but is more difficult to use when it comes to locating cracks in valve ports or internal areas. Pressure testing, on the other hand, is very good at finding hidden cracks in the water jackets, but it cannot identify cracks elsewhere, such as around head bolt bosses or valve guides. It also requires the use of a test fixture and sealing plugs. It is probably best to use both test methods to check for cracks. To use a dye, all dirt and oil should first be cleaned from the head. The dye can then be sprayed on and allowed to dry. Wipe off the excess dye dust, then spray on the developer. Any cracks should appear as dark lines against the light metal.

When pressure testing a head, all water jacket openings are sealed on the test fixture. Once sealed, the head is then pressurized with 25 to 60 psi of air. Spraying soapy water on the head will reveal any cracks by the appearance of bubbles.

Welding Cracks. As discussed earlier, aluminum oxides are harmful to the welding process. Luckily, the back and forth heating action of TIG welding serves to "cook off" the oxides and prevent weld contamination. It is very important to clean the weld area thoroughly with a stainless steel wire brush. As for the choice of electrodes, it is generally thought that tungsten thorium (1 percent or 2 percent) works best with aluminum.

When repairing a crack in the head, it is crucial that the full extent of the damage be identified so the crack can be completely ground out. Extend the grinding a short distance beyond the visible ends of the crack to make sure all damage has been eliminated. The area can then be cleaned up by bead blasting or brushing prior to welding.

Preheating the head to 200 degrees Fahrenheit with a torch or putting the head in a baking oven helps cook out all oil or grease that might contaminate the weld. Preheating also helps eliminate the formation of stress cracks during the cooling-off period. It is a good idea to do all welding before attempting to straighten a warped cylinder head because welding can increase warpage. On the other hand, some technicians believe that it is better to straighten the head first to see if it can be salvaged before spending a lot of time trying to weld cracks.

It is important not to touch the metal with the electrode; doing so can contaminate the electrode. Using the high-frequency alternating current setting, the electrode should be held about an eighth of an inch above the work surface while welding. As soon as the weld is completed, allow a long cool-down period to reduce the possibility of warping or cracking. This can be accomplished by putting the head in an oven or by wrapping it to slow the heat loss.

Before milling the head to finish the repair, a light coating of wax or silicone spray can help prevent the aluminum filings from sticking to the cutter bits. Material removal should be limited to 0.001 inch per pass. After the machine work on the repair area is done, the head should once again be pressure checked to make sure the original crack has been sealed and no new cracks have opened up. This might seem like an unnecessary step, but it can prevent comebacks and is worth the effort.

REBUILDING ALUMINUM CYLINDER HEADS

TIG welding is an important part of the aluminum cylinder head rebuilding process. To see its importance, one need look no further than the popular Ford Escort. From 1981 to 1985, the Escort featured a 1.6-liter engine; a 1.9 liter was introduced in 1985 and is still in use today. Both cylinder heads use identical valve seats, keepers, and in some cases, valves, valve springs, and rocker arms.

The most common problems on these cylinder heads are cracks and warpage. To check for warpage, position a straightedge over the cylinder head sealing surface. Try to slide a thin feeler gauge strip under any part of the straightedge; if it goes under, the head is warped (Figure 5-16). The guides and valve stems wear surprisingly well, making extensive guide work unnecessary for the most part. The 1981, 1982, and some 1983 1.6-liter engines were not free-wheeling designs. Valve bending and guide damage are common on these engines whenever they jump time or the timing belt breaks. These engines can be identified before teardown by the decal they carry advising replacement of the timing belt at 60,000 miles.

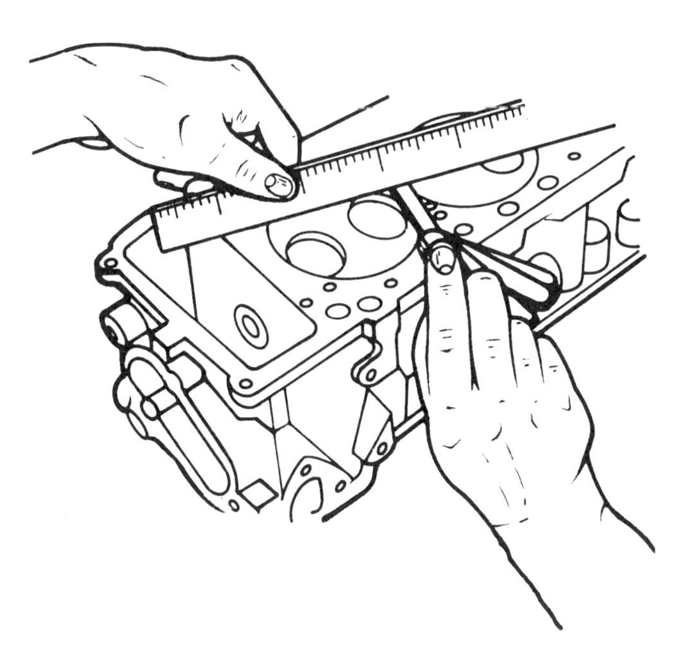

FIGURE 5-16 Using a straightedge and feeler gauge to check for cylinder warpage

Most of the equipment needed to repair Escort cylinder heads is available in any well-equipped shop, including a baking oven, a TIG welder with a water-cooled tip, and pressure test equipment. Keep in mind that none of the required equipment is unique; it can be used to rebuild all aluminum cylinder heads, not just Escorts.

A TIG welder is recommended for this job as opposed to a wire feed because it can be difficult reaching some spots on the Escort head with the wire feed tip; there is also the additional flexibility of being able to arc weld with the TIG welder. A water-cooled tip makes it possible to weld for longer periods of time without overheating the tip, thus producing more consistent welds.

Teardown of these heads is fairly straightforward. Watch for the square fulcrum washer found under the rocker arms. Do not force any heavily varnished lifters out of the bore. Wait until the camshaft is removed and push them through to the inside of the head. Mark the lifters carefully for replacement if they are to be reused. Check any sleeved lifter bores for proper alignment and press fit. The oil hole in the sleeve must match up to the oil gallery.

After diagnosing the cracks using the procedures described earlier, pull any seat rings that are affected by cracks. A quick and easy way to remove Escort rings is by using an air-powered or high-speed electric cutoff tool to cut through both rings and the head at the same time. Since the head is already cracked, this will not cause any undue damage. The rings can now be simply popped out from underneath with a chisel.

The cracks must be gouged out from end-to-end, and the areas that will be welded should be glass beaded, including the surrounding water jacket. Additives, rust, and other contaminates will make the welding difficult, so it is important that they are removed. It is not necessary to cut out each crack to its full depth. Leave about 1/16 inch to allow the weld to sag a bit and not drop through. Use a lubricant when gouging out the cracks to keep the cutter from clogging. Be sure to clean any traces of lubricant and all the glass bead residue from the areas to be welded to avoid contaminating the weld. Hot, soapy water and a stainless steel brush work nicely, but do not use the brush for anything other than aluminum.

Begin the welding process by heating the cylinder head in an oven. This preheating should be done at 350 to 400 degrees for approximately 80 minutes. If an isolated area away from drafts and air cross flow is not available, weld in an open-front oven lined with firebrick to help retain the heat. At a minimum, shield the head from anything that will cause premature cooling.

Argon gas set at 22 to 25 cubic feet per hour is used to shield the tip. Use a #5 ceramic cup with an ID of about 5/16 inch. The smaller cup seems to work better than a larger #8 or #9 cup. The tip material should be 1 to 2 percent Thoriated Tungsten. Use a 3/32-inch 4043 aluminum rod with a silicon alloy. It flows better and "wets" out better than a 5356. The 4043 has excellent bonding characteristics and machines very well. A larger diameter rod might be required in areas needing lots of fill.

The heat setting will vary according to the rod size and technique. A good starting point is to set the machine on full amperage and depress the foot or hand control until the tip will actually melt the aluminum. The rod should be flowed into the cracked area, melting out the crack entirely if it was not completely ground out previously. Failure to eliminate the crack completely will lead to future cracking in the same spot. Cracks in the seat ring area are welded at the crack only.

It is not necessary to fill in the entire counterbore, but plan on cutting for oversize seat rings to insure proper press fit if using this method. If welding a crack that runs through the floor of the exhaust port and across the seat ring and combustion chamber, be sure to weld both the floor of the port and the combustion chamber. Do not try to catch both sides from the chamber side. Any crack that goes through the oil feed hole to one of the intake lifters is easily repaired by grooving out the entire crack to below the feed hole and filling the groove with weld.

Redrill the hole with a 3/16-inch drill by coming in through the lifter bore itself. Do not let the weld sag too much on either side of the gallery; this will create interference with the camshaft and it will be difficult trying to grind it out of that area. In the areas adjacent to the lifter gallery, less is actually better. These areas are not important structural points—the point is just to close up the crack.

Normal time spent welding minimal damage from cracks is about 15 minutes, but more extensive cracks might require 35 to 40 minutes. Do not let the head cool too much. It is important to maintain a constant temperature throughout the welding process. Reheat it after 20 to 25 minutes if the welding is not being done in an oven environment. If even heat is not maintained throughout the cylinder head, cracks will form in areas that have cooled excessively.

When the welding is completed, the head must be placed back in the oven to post heat even if the welding was done in an open-front oven environment. Put it back in immediately and post heat for 30 minutes at 350 degrees. Let it cool slowly over a minimum 5- to 6-hour period; 8 to 10 hours is best. Slow cooling is an absolute must regardless of which method is used during the welding process to keep the head hot. The head should be cool enough to handle before beginning to work on it.

Once the head has cooled it is ready for rough machining of the welded areas. Recut the seat ring counterbores using a lift spring under the cutter. This will reduce the chances of breaking the cutter as the excess filler material is cut out. Allow for a minimum of 0.006-inch crush on the seat rings. Smooth out the combustion chamber, the ports, the topside, and anywhere else welding was done. The head should be pressure tested at this point, before the seat rings are installed. Waiting until the seat rings are installed can mask potential leaks. When this technique has been perfected, a less than 5 percent leak rate can be expected, though initially the rate will undoubtedly be higher.

If pressure testing reveals that the cylinder head is porous from exposure to heat or from chemicals that have affected the vacuum, commercial sealers are available. These gels are spread over the casting's external surfaces and dry within 3 minutes. However, be forewarned that such products are not intended for sealing cracks. Conduct further pressure tests after the sealer application to make sure the problem has been eliminated. It is important to keep in mind that the remainder of the rebuilding procedure hinges on the cylinder head being repaired properly. If *all* of the cracks have not been sufficiently repaired, there is no reason to proceed further. Thus, the welding essentially lays the cornerstone for a successful rebuilding job.

OXYACETYLENE WELDING

Oxyacetylene is a form of fusion welding in which oxygen and acetylene are used in combination. The process is simple: the two are mixed in a chamber, ignited at the tip of the torch, and used to melt and join the welding rod and base metal together (Figure 5-17). However, with the advances in recent years in other welding processes (particularly

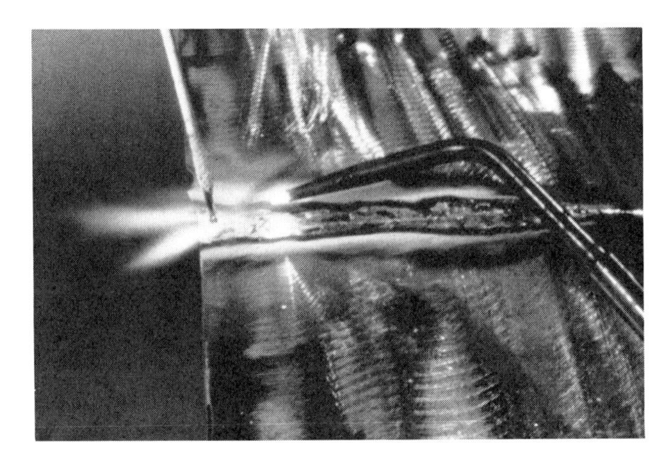

FIGURE 5-17 Oxyacetylene welder in action

MIG and TIG), the use of oxyacetylene has declined greatly. These processes are faster, cleaner, and cause less distortion than oxyacetylene. In addition, because it is difficult to concentrate the heat in one place, the heat affects the surrounding areas and reduces the strength of the steel panels. For this reason, automakers do not recommend the use of oxyacetylene in repairs of damaged vehicles. But although oxyacetylene is in disfavor with most manufacturers—with good reason—it has some use in the body shop. The oxyacetylene flame is still used in some heat shrinking operations, brazing and soldering, surface cleaning, and cutting of nonstructural parts. Oxyacetylene should not be used to cut structural parts of any vehicle unless special care is taken.

So, despite its obvious drawbacks, it is not time to give up on oxyacetylene welding yet. Because of its portability, it can go outside into the storage area to cut off damaged quarter panels. It can be moved both inside and outside the shop with ease, an obvious advantage when compared to plasma cutting equipment. Until plasma gouging becomes practical for small plasma cutters, the oxyacetylene torch is the best tool for removing rusted bolts. And cleaning caulk from a seam with oxyacetylene is still faster than any other method.

OXYACETYLENE EQUIPMENT

A typical oxyacetylene welding outfit (Figure 5-18) consists of the following:

- Steel tanks (cylinders) filled with oxygen and acetylene.
- Regulators reduce the pressure coming from the tanks to the desired level and maintain a constant flow rate. Each regulator has a high-pressure and a low-pressure gauge.

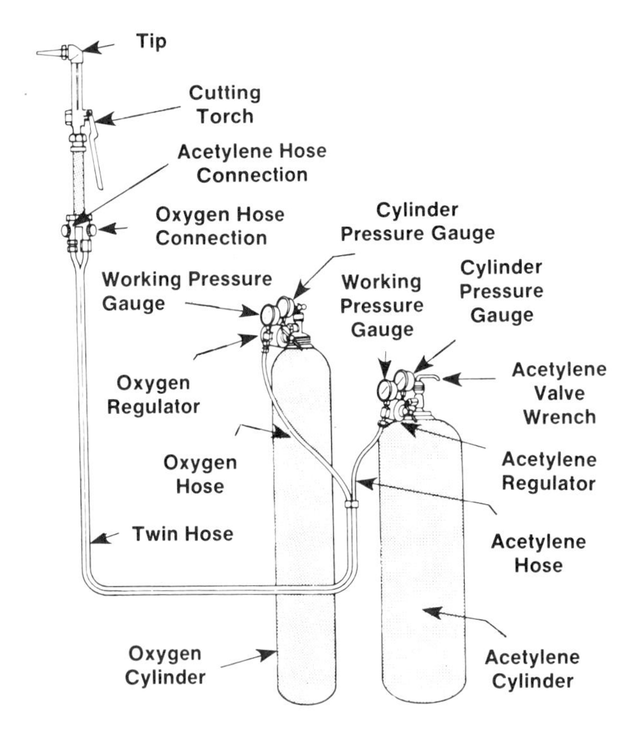

FIGURE 5-18 Typical oxyacetylene welding outfit

—Oxygen pressure: 15 to 100 psi
—Acetylene: 3 to 12 psi

- Hoses from the regulators and cylinders connect the oxygen and acetylene to the torch.
- The torch mixes the oxygen and acetylene from the tanks in the proper proportions and produces a heating flame capable of melting steel. There are two main types of torches: welding and cutting.

WARNING: The acetylene line pressure should never exceed 12 psi. Free acetylene has a tendency to dissociate at pressures above 12 psi and could cause an explosion.

To round out the equipment, safety gear should be worn as described in Chapter 1. All welding should be done with a number 4, 5, or 6 tinted filter shade. A spark lighter (Figure 5-19) is a necessity for oxyacetylene welding.

Torches

The oxyacetylene hand torch can be either a part of a combination welding and cutting torch set or a cutting torch only (Figure 5-20). The combination welding/cutting torch offers more flexibility be-

FIGURE 5-19 Typical spark lighter

cause a cutting head, welding tip, or heating tip can be attached quickly to the torch body. Combination torch sets are often used in schools, automotive repair shops, auto body shops, and small welding shops where flexibility is needed. A cut made with either type of torch has the same quality; however, the straight cutting torches are usually longer and have larger gas flow passages than the combination torches. The added length of the straight cutting torch helps keep the operator farther away from the heat and sparks and allows thicker material to be cut. This torch can be used at an extremely low acetylene pressure and has an injector nozzle. The important operational controls are shown in Figure 5-21.

Oxygen is mixed with the fuel gas to form a high-temperature preheating flame; the two gases must be completely mixed before they leave the tip and create the flame. There are two methods of mixing the gases: a mixing chamber and an injector chamber. The mixing chamber can be located in the torch body or in the tip, as shown in Figure 5-22. Torches that use a mixing chamber are known as equal-pressure torches because the gases must enter the mixing chamber under the same pressure. The mixing chamber is larger than both the gas inlet and the gas outlet. This larger size causes turbulence in the gases, resulting in their mixing thoroughly.

Injector torches work well with both equal gas pressures or low fuel gas pressures (Figure 5-23). The injector allows the oxygen to draw the fuel gas into the chamber even if the fuel gas pressure is low. The injector works by passing the oxygen through a venturi, which creates a low-pressure area that pulls the gases in and mixes them together. An injector-type torch must be used if a low-pressure acetylene generator or low-pressure natural gas is used as the fuel gas supply.

The cutting head can hold the cutting tip at a right angle to the torch body, or it can be held at a slight angle. Torches with the tip slightly angled are easier to use when cutting flat. Torches with a right-angle tip are easier to use when cutting uneven material shapes. Both types of torches can be used for any type of material being cut, but it is important to keep the cut square and accurate. The cutting lever may pivot from the front or back end of the torch body (Figure 5-24). Personal preference will determine which one is used.

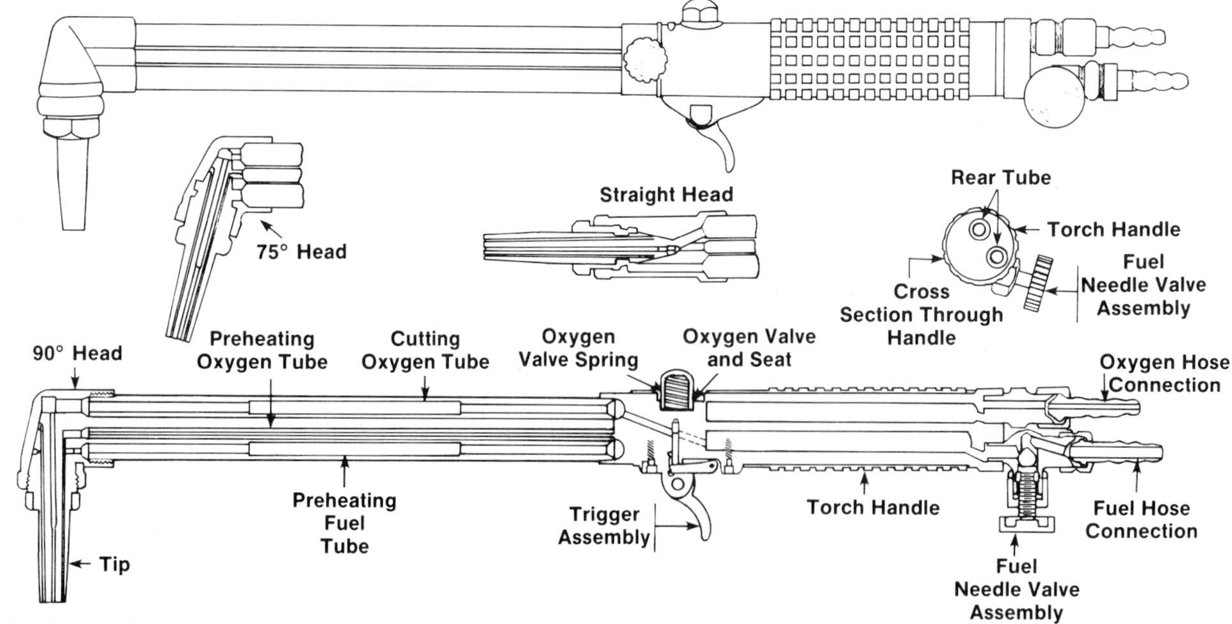

FIGURE 5-20 Parts of a typical cutting torch

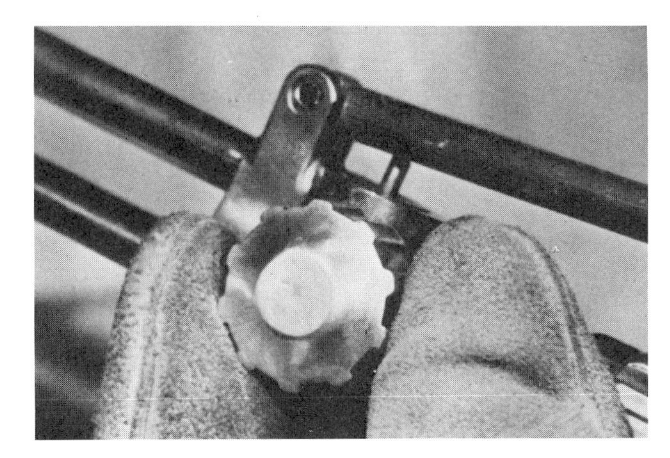

FIGURE 5-21 It is important to be familiar with the cutting torch adjustments.

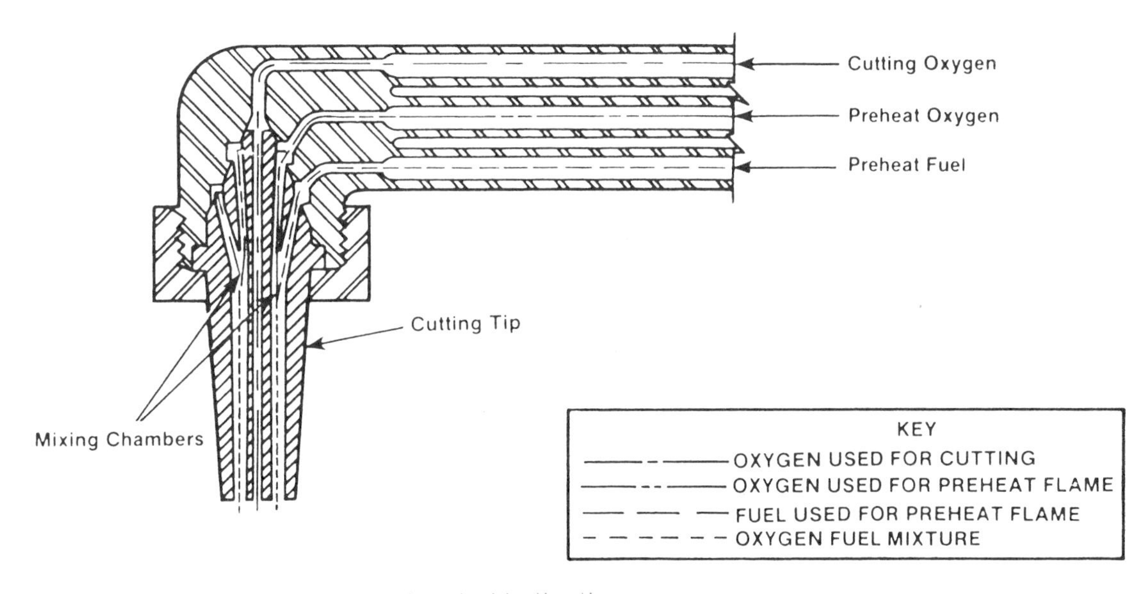

FIGURE 5-22 The mixing chamber is located in the tip.

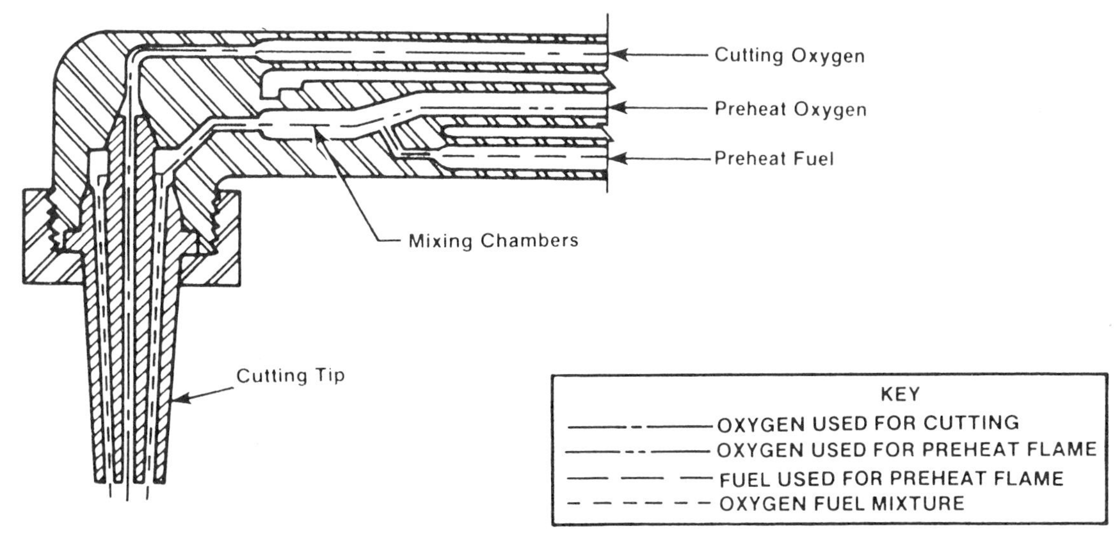

FIGURE 5-23 Injector mixing torch

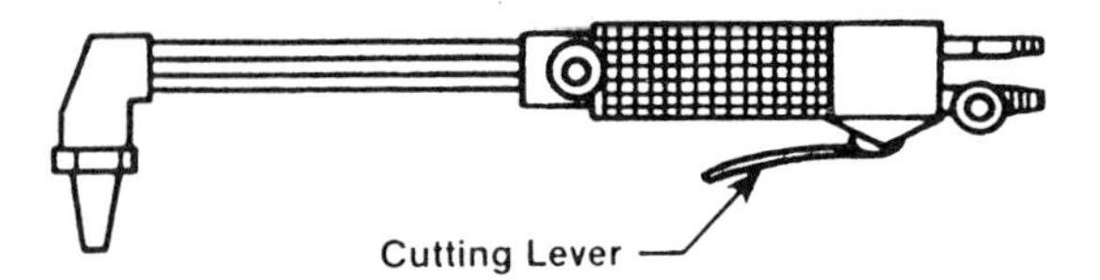

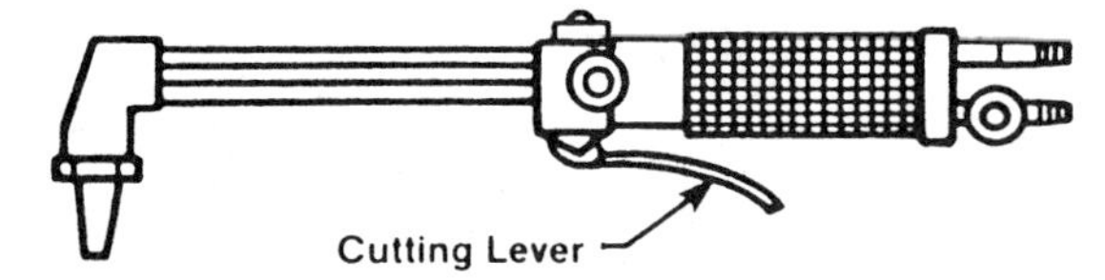

FIGURE 5-24 The cutting lever may be located on the front or back of the torch body.

A machine cutting torch (Figure 5-25), sometimes referred to as a blowpipe, operates similarly to a hand-cutting torch. The machine cutting torch might require two oxygen regulators, one for the preheat oxygen and the other for the cutting oxygen stream. The addition of a separate cutting oxygen supply allows the flame to be more accurately adjusted. It also allows the pressures to be changed during a cut without disturbing the other parts of the flame.

Cutting Tips

Most oxyacetylene cutting tips are made of copper alloy, and some are chrome plated to prolong their service life. Tip designs are variable, depending on their intended use. Those for straight cutting are either standard or high speed (Figure 5-26). The high-speed cutting tip is designed to allow a higher cutting oxygen pressure, which allows the torch to travel faster. High-speed tips are also available for different types of fuel gases.

The amount of preheat flame required to make a perfect cut is determined by the type of gas used and by the material's thickness, shape, and surface condition. Materials that are thick, round, or have surfaces covered with rust, paint, oil, and so on, require more preheat flame. The differences in the type or number of preheat holes also help determine the type of gas to be used in the tip. Acetylene is used in tips having from one to six preheat holes; MPS gases are used in tips having eight preheat holes and in two-piece tips that are not recessed; and propane or natural gas is used in two-piece tips that are deeply recessed.

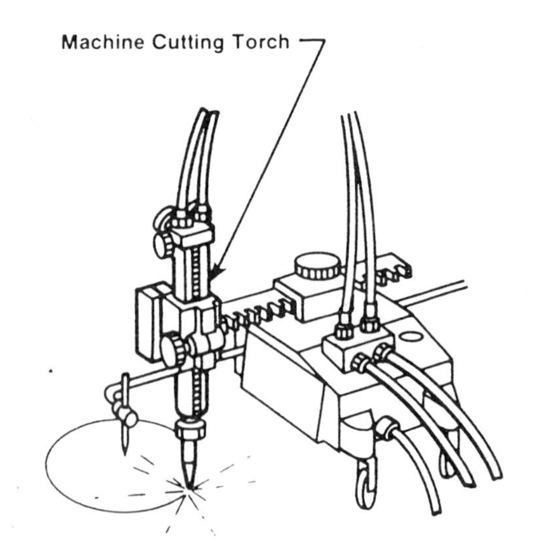

FIGURE 5-25 Portable flame-cutting machine

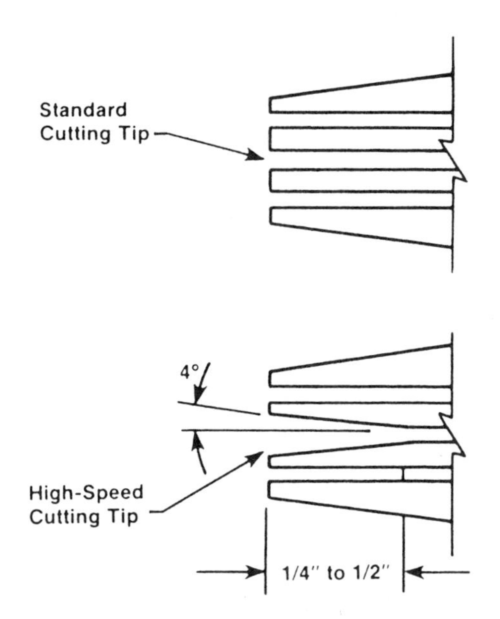

FIGURE 5-26 Comparing standard and high-speed cutting tips

Some cutting tips have metal-to-metal seals. When they are installed in the torch head, a wrench must be used to tighten the nut. Other cutting tips have fiber-packing gaskets to seal the tip to the torch (Figure 5-27). If a wrench is used to tighten the nut for this type of tip, the tip seat might be damaged. The torch owner's manual should be checked to learn the best way to tighten various tips.

When removing a cutting tip, if it is stuck in the torch head, tap the back of the head with a plastic hammer (Figure 5-28). Tapping on the side of the tip can damage the seat. To check the assembled torch tip for a good seal, place a thumb over the end of the tip, turn on the oxygen valve, and spray the tip with a leak-detecting solution.

If the cutting tip seat or the torch head seat is damaged, they can be repaired by using a reamer

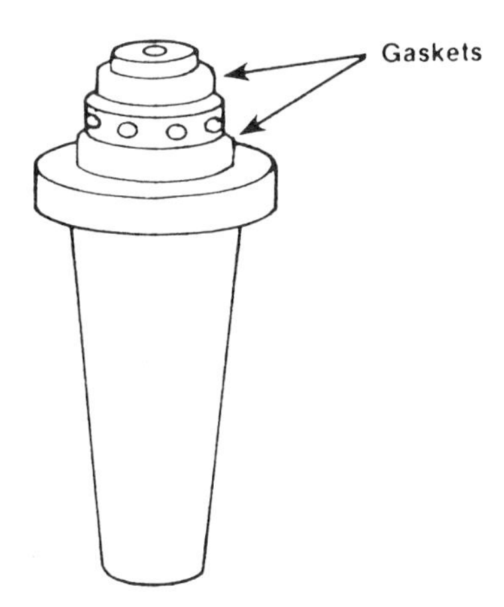

FIGURE 5-27 Some cutting tips use gaskets to make a tight seal.

designed for the specific torch tip and head. New fiber packings are available for tips with packings. Keep in mind that the leak-checking test should be repeated to be sure the new seal is good.

FLAMES AND ADJUSTMENT

When acetylene and oxygen are mixed and burned in the air, the condition of the flame varies depending on the volume of oxygen and acetylene used.

There are three types of flames:

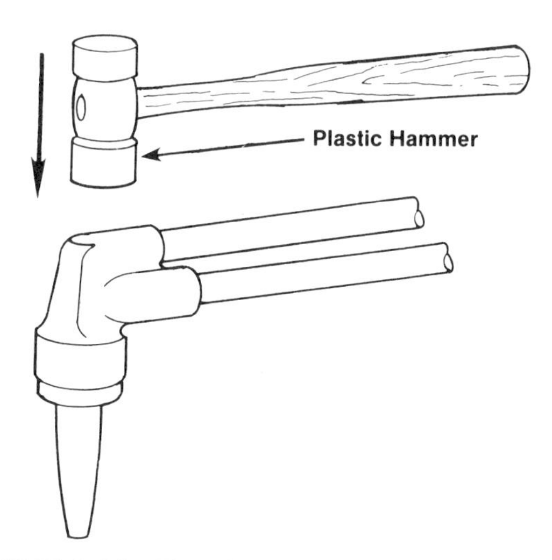

FIGURE 5-28 Tap the back of the torch head to remove a tip that is stuck.

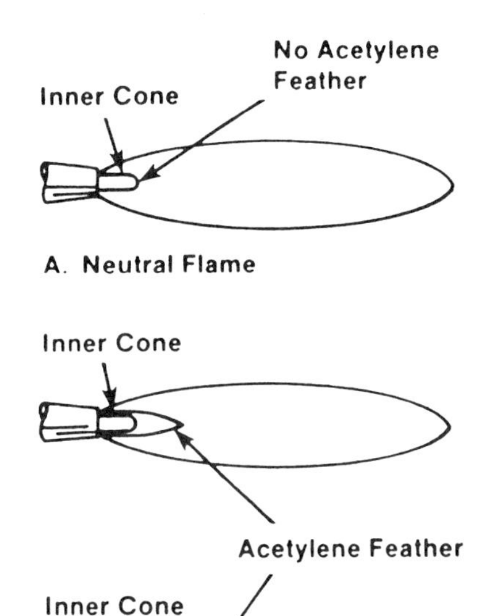

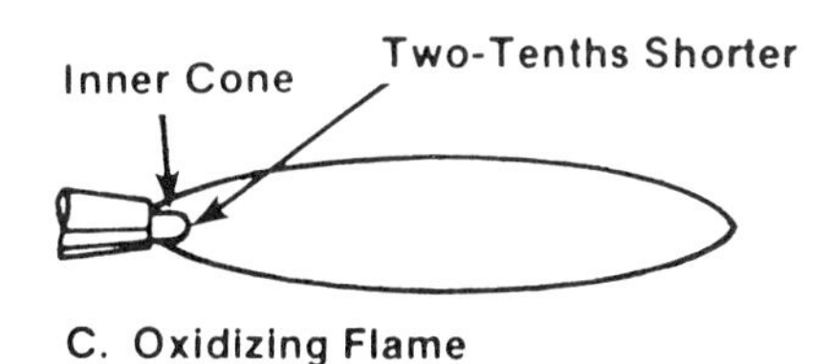

FIGURE 5-29 Types of cutting flames

1. *Neutral Flame.* The standard flame is said to be a neutral flame. It is produced by mixing acetylene and oxygen in a 1 to 1 ratio by volume. As shown in Figure 5-29A, a neutral flame has a brilliant white core surrounded by a clear blue outer flame. Its temperature is approximately 5900 degrees Fahrenheit.
2. *Carburizing Flame.* This flame, also called a *surplus* or *reduction flame,* is obtained by mixing slightly more acetylene than oxygen. Figure 5-29B shows that this flame differs from the neutral flame in that it has three parts. The core and the outer flames are the same as the neutral flame, but between them there is a light-colored acetylene cone enveloping the core. The length of the acetylene cone varies according to the amount of surplus acetylene in the gas mixture. For a double surplus flame, the oxygen-acetylene mixing ratio is about 1 to 1.4 (by volume). A carburizing flame is used for welding aluminum, nickel, and other alloys. Its temperature is approximately 6100 degrees Fahrenheit.
3. *Oxidizing Flame.* The oxidizing flame (Figure 5-29C) is obtained by mixing slightly

more oxygen than acetylene. It resembles the neutral flame in appearance, but the acetylene core is shorter and its color is a little more violet, while the outer flame is shorter and fuzzy at the end. Ordinarily, this flame oxidizes melted metal, so it is not used in the welding of mild steel but is used in the welding of brass and bronze. The temperature of an oxidizing flame is approximately 6300 degrees Fahrenheit.

Welding Torch Flame Adjustment

When using an oxyacetylene welding torch, proceed as follows:

1. Attach the appropriate tip to the end of the torch. Use the standard tip for sheet metal, keeping in mind that each torch manufacturer has a different system for measuring the size of the tip orifice.
2. Set the oxygen and acetylene regulators at the proper pressure: the oxygen at 8 to 25 psi and the acetylene at 3 to 8 psi.
3. Open the acetylene valve about half a turn, and ignite the gas. Continue to open the valve until the black smoke disappears and a reddish yellow flame appears. Slowly open the oxygen valve until a blue flame with a yellowish white cone appears. Further open the oxygen valve until the center cone becomes sharp and well defined. This is a neutral flame and is used for welding mild steel.

If acetylene is added to the flame or oxygen is removed from the flame, a carburizing flame will result. If oxygen is added to the flame or acetylene is removed from the flame, an oxidizing flame will result.

Gas Cutting Torch Flame Adjustment

The oxyacetylene cutting torch is sometimes used in collision repair shops to rough cut damaged panels. Gas cutting torch flame adjustment and cutting procedures are as follows:

1. Adjust the oxygen and acetylene valves for a preheating neutral flame.
2. Open the preheating oxygen valve slowly until an oxidizing flame appears. This will make it difficult for melted metal to remain on the surface of the cut panel, thus allowing for clean edges.
3. To cut a thick panel, heat a portion of the base metal until it is red hot. Just before it melts, open the high-pressure oxygen valve and cut the panel. Advance the torch forward while making sure the panel is melting and being cut apart. This method is widely used for thin panels (when there are several pieces overlapped together) or for a side member (even when there is an internal reinforcement).
4. To cut a thin panel, heat a small spot on the base metal until it is red hot. Just before it melts, open the high-pressure oxygen valve and incline the torch to cut the panel. When cutting thin material, incline the tip of the torch so that the cut will be clean and fast; this will prevent panel warpage.

Soft and Harsh Flames

After learning to adjust the flame so that the proportions of oxygen and acetylene are correct, it is important to learn how to obtain a soft flame. This is a flame produced when the gases flow to the welding tip at a comparatively low speed. If the gases flow to the welding tip at high speed under too much pressure, they produce a harsh flame that is easily recognized because it is noisy. A harsh flame destroys the weld puddle and causes the metal to splash around the edges of the puddle. It is very difficult to get the metal parts to fuse properly with a flame of this kind.

If there is any fluctuation in the flow of gases from the regulators, the mixture will change regardless of other conditions; hence, a good welder watches the flame constantly and makes any necessary adjustments to keep it neutral and soft. If a popping noise is heard, insufficient gas is reaching the tip. The answer is to immediately deliver a little more oxygen and acetylene by opening both needle vales slightly more.

Obstructed Tips. Should the flame adjustment and operating pressures be correct and a soft flame still cannot be produced, a dirty or obstructed welding tip is probably the reason. An obstructed welding tip does not permit the gas mixture to flow evenly. It restricts the source of heat required to melt the metal, thus creating welding difficulties.

If the orifice in the tip becomes clogged, it should be cleaned with the proper size "tip cleaner." Do not force a cleaner that is too large into the orifice. This can cause enough damage to require discarding the tip, or it might cause the cleaner to break off and thus render the tip useless. Likewise, one should not use a cleaner that is too small. This will cause the orifice to become out-of-round, which will, in turn, make proper flame adjustment impossible. The tip is made of soft copper, so excessive cleaning should be avoided.

If using a tip with a small orifice, be careful not to use excessive oxygen pressure. If this excess pressure cannot escape through the tip, it could back up into the acetylene side and pressurize it to dangerous levels.

CUTTING HSS FOR SALVAGE PURPOSES

If the use of an oxyacetylene torch is necessary when cutting HSS sheet metal components for salvage purposes, or for cutting a body structure for a "front/rear clip," factory engineers advise the following approach:

- Cut the metal structure at least 2 inches away from the desired cutline. Sheet metal within the heat-affected area will lose strength when subjected to the high heat levels of the torch.
- After cutting, use a grinding wheel disc, an air chisel, or a metal saw to make the final cut at the originally intended dimension line. HSS damage will then be "cut out" of the salvaged part.

Oxyacetylene equipment should not be used on HSS components for welding; vehicle engineers stress this point. There is just too much heat buildup that can reduce the structural strength. However, in some instances an oxyacetylene torch can be used to heat HSS components or parts, provided the critical 1200 degrees Fahrenheit temperature is not exceeded. (Check the manufacturer's shop manual on this point; some say 1000 degrees Fahrenheit is the critical temperature.)

High-strength steels should only be exposed to high temperatures from an oxyacetylene torch for a very short period of time. Three minutes is the recommended maximum time span for exposing HSS to a 1400 degrees Fahrenheit temperature. High-temperature exposure causes discoloration as shown in Figure 5-30.

In order to determine and control temperatures of high-strength steel parts and components being "heat worked" with oxyacetylene equipment, it is necessary to use a temperature indicating crayon (Figure 5-31). The metal should be marked closely adjacent to the area being worked with a crayon rated no more than 1400 degrees Fahrenheit. Using such a crayon will indicate to the welder whether or not an excessive amount of heat is being applied. In this way, metal temperatures can be controlled within safe levels and HSS damage easily prevented.

FIGURE 5-30 Discoloration of heat-affected HSS steel

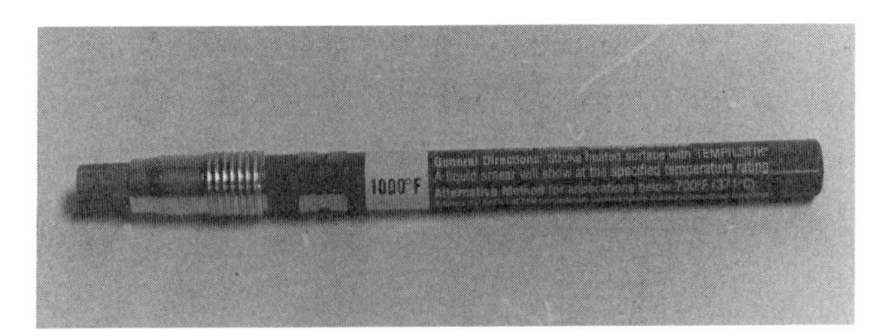

FIGURE 5-31 Typical temperature indicating crayon

CLEANING WITH A TORCH

It is important before starting any welding operation that the surfaces to be joined are thoroughly clean. The weld site must be completely free of any foreign material that might contaminate the weld. If this is not done, the finished weld is quite likely to be brittle, porous, and of poor integrity. To remove heavy undercoating, rustproofing, tars, caulking and sealants, road dirt and primers, and the like, first use a scraper and an oxyacetylene torch. Do a thorough cleaning with a wire brush and the torch, using a carburizing flame (Figure 5-32). In any event, keep the torch at a very low, controlled heat—just enough heat to get the job done.

FLAME ABNORMALITIES

When changes occur during the welding operation—for example, overheating of the flame outlet,

FIGURE 5-32 Cleaning metal with an oxyacetylene torch and a wire brush

adhesion of spatter, or fluctuations in the gas adjustment pressure—the result will be variations in the flame. Therefore the operator must always be aware of the condition of the flame. Flame abnormalities, their causes, and remedies are described in Table 5-4. Becoming familiar with the common trouble symptoms and how to correct them can save a great deal of time and frustration later, especially when in the middle of a welding procedure.

SHUTTING DOWN THE EQUIPMENT

Always turn off the torch when not in use; never lay down or hang a lighted torch. Standard practice is to turn off the torch when it is not being held in the hand. Do this by first closing the torch acetylene needle valve, then the torch oxygen needle valve.

TABLE 5-4: FLAME ABNORMALITIES AND REMEDIES

Symptom	Cause	Remedy
Flame fluctuations	Moisture in the gas; condensation in hose. Insufficient acetylene supply.	Remove moisture from the hose. Adjust acetylene pressure and have the tank refilled.
Explosive sound while lighting the torch	Oxygen or acetylene pressure is incorrect. Removal of mixed-in gases are incomplete. Tip orifice is too enlarged. Tip orifice is dirty.	Adjust the pressure. Remove the air from inside the torch. Replace the tip. lean the orifice in the tip.
Flame cut off	Oxygen pressure is too high. Flame outlet is clogged.	Adjust the oxygen pressure. Clean the tip.
Popping noises during operation	Tip is overheated. Tip is clogged. Gas pressure adjustment is incorrect. Metal deposited on the tip.	Cool the flame outlet (while letting a little oxygen flow). Clean the tip. Adjust the gas pressure. Clean the tip.
Reversed oxygen flow (Oxygen is flowing into the path of the acetylene.)	Tip is clogged. Oxygen pressure is too high. Torch is defective. (Tip or valve is loose.) There is contact with the tip and the deposit metal.	Clean the tip. Adjust the oxygen pressure. Repair or replace the torch. Clean the orifice.
Backfire (There is a whistling noise and the torch handle grip gets hot. Flame is sucked into the torch.)	Tip is clogged or dirty. Oxygen pressure is too low. Tip is overheated. Tip orifice is enlarged or deformed. A spark from the base metal enters the torch, causing an ignition of gas inside the torch. Amount of acetylene flowing through the torch is too low.	Clean the tip. Adjust the oxygen pressure. Cool the tip with water (letting a little oxygen flow). Replace the tip. Immediately shut off both torch valves. Let the torch cool down; then relight it. Readjust the flow rate.

Shutting the torch fuel gas valve first offers less chance of unburned acetylene escaping in the work area; that is, shutting off the flow of fuel gas will immediately extinguish the flame. If the oxygen is shut off first, the acetylene will continue to burn, throwing off a great deal of smoke and soot. These instructions are equally valid for both low-pressure and medium-pressure type torches.

When work is stopped for a half hour or less, gas pressure should be released from the hose lines using the following procedure:

1. Release the oxygen pressure adjusting screw on the oxygen regulator.
2. Open the torch oxygen valve.
3. Close the torch oxygen valve (Figure 5-33).

Repeat the same procedure with the acetylene system.

When stopping work for longer than a half hour, make it a habit to relieve both oxygen and acetylene systems of all pressure. This is accomplished by the following:

1. Close the oxygen and acetylene cylinder valves.
2. Open both torch valves and allow the oxygen and acetylene gauge hands to return to zero.
3. Turn out the oxygen and acetylene regulator pressure adjusting handles until they rotate freely.
4. Close both torch valves.

If fuel gas is supplied through a back pressure valve, simply close the service valve which supplies it.

Never move, transport by vehicle, or store any cylinder with a regulator attached. However, a cylinder in an immediate work area with a regulator attached may be moved from one work location to another as long as it is secured to a suitable cylinder cart and the cylinder valve is tightly closed. To disconnect a pressure-reducing regulator, use the following procedure:

1. Close the cylinder valve.
2. Open the torch valve to release all pressure from the hose and regulator.
3. Turn out the pressure-adjusting screw.
4. Close the torch valve.
5. Uncouple the regulator.

Closing the cylinder valve and then opening the torch valve relieves all pressure downstream of the cylinder valve, making it safe to disconnect the regulator from the cylinder. After the regulator gauge readings have reached zero, the pressure-adjusting screw should be released. This must always be done before the valve is opened on the new cylinder. The fuel gas and oxygen pressure should not be relieved simultaneously, and care should be taken that the release of fuel gas does not create a fire hazard.

When regulators are to be out of service for several weeks or longer, it is good practice to turn the pressure adjusting screw just enough to move the regulator valve off its seat. This aids in lengthening the life of the valve seat. Before putting the regulator into service again, the pressure adjusting screw should be turned counterclockwise until spring tension is released. To complete the shutdown, remove the hoses from the regulators, coil them (avoid kinking the hoses), and suspend them on a suitable holder or hanger. The regulators (if removed from the cylinders) and the torch should be carefully stored. The protective caps can be replaced on the cylinder.

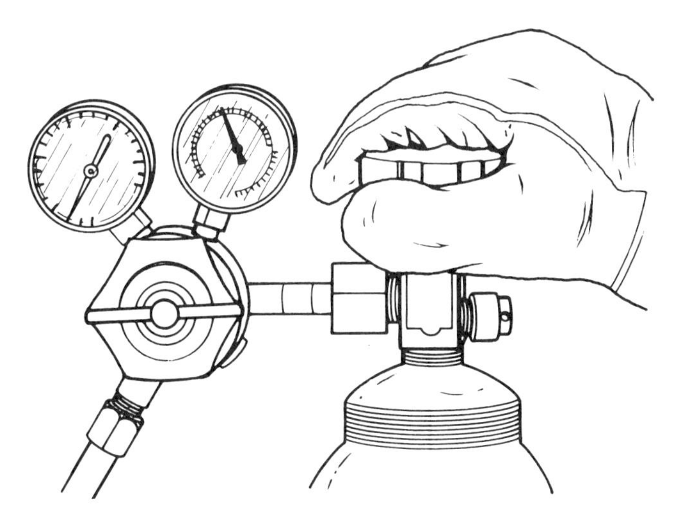

FIGURE 5-33 Closing the torch oxygen valve

PLASMA ARC CUTTING

As mentioned earlier, plasma arc cutting is replacing oxyacetylene cutting as the best way to cut modern car metals. It cuts mangled metal effectively and quickly (Figure 5-34), but will not destroy the properties of the base metal. This is important to today's metal shop, which sees more unibody cars with high-strength steel or high-strength alloy steel components. The old method of flame cutting just does not work that well anymore. The high heat, fast travel speed, low-heat input qualities, coupled with the fact that plasma cutting will cut rusted, painted, or coated metal with little difficulty, make it an ideal process for the auto body repair field.

FIGURE 5-34 Plasma arc is ideal for cutting mangled metal.

Basically, plasma cutting is an extension of the TIG process. It utilizes an electrically ionized gas that is capable of rapidly transferring its heat to the part being cut. To accomplish this, the electric current must pass through compressed air or an inert gas, usually nitrogen because it is a reasonably priced gas. The nitrogen gas (the atmosphere contains 78 percent nitrogen) is ionized, shaped, and then forced through an orifice. The nitrogen or compressed air is superheated by an electrode inside the torch and expands tremendously. The gas itself comes out of the torch at a temperature of up to 2000 degrees Fahrenheit. The gas or air, because it is so hot and moves so fast, heats up the metal and blows it away.

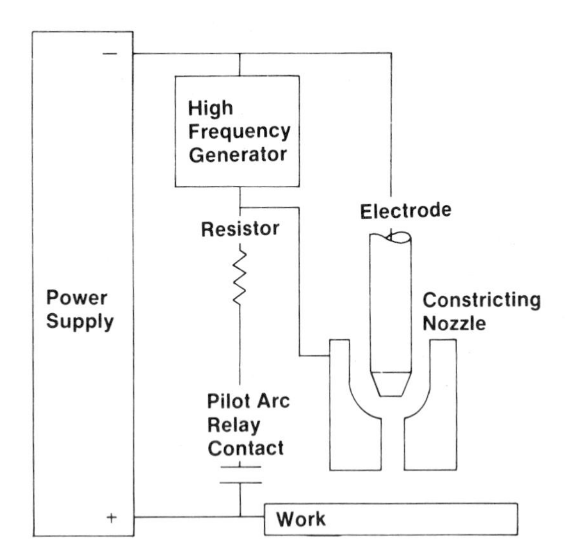

FIGURE 5-35 Basic plasma arc cutting circuitry

There are two circuits in plasma cutting (Figure 5-35). The pilot arc starts the ionization process; high frequency transfers from the electrode, which is negative, to the torch, which is positive. When the torch is brought down to the work, the ionized gas or air flows to the grounded workpiece, starting the cutting action.

The secondary gas (or shielded gas) is also drawn from the compressed air. The shielding gas performs exactly the same function as in MIG welding; it protects the cut from outside contaminants while the cutting is in progress.

CAUTION: Because of the presence of 78 percent nitrogen in the atmosphere, it is very important to maintain the volume and psi ratings recommended by the machine's manufacturer. Also keep in mind that moisture in the air will affect the quality of the cut somewhat.

ADVANTAGES AND LIMITATIONS

Early plasma arc cutting processes required the use of special gas mixtures and equipment. Today's solid-state machines and air-cooled, lightweight torches utilize compressed air and electricity to form the required arc (Figure 5-36). This has served to reduce costs dramatically and greatly expand the applications of plasma arc cutting. The cost effectiveness is achieved because of plasma arc's very high cutting speed (up to ten times that of oxyfuel cutting on 1-inch-thick steel) and its ease of use, which reduces operator training time. No preheating is required, and air plasma arc cutting is a continuous process: the time-consuming start-and-stop procedures typical of oxyfuel cutting are eliminated. These advantages offset the initially higher investment in plasma arc cutting equipment, permitting the equipment to pay for itself in a short time period.

The economic advantages of plasma arc versus oxyfuel cutting are clear in situations where long, continuous cuts must be made on several pieces. Plasma arc can be used for stack cutting, shape cutting, plate beveling, and piercing. In the case of thicknesses totaling 2-1/2 inches or more, the decision about whether to use plasma or oxyfuel should depend upon such factors as equipment cost, load factors, the material being cut, and specific application considerations.

Because of its power source requirements, plasma arc is less portable than oxyfuel and other

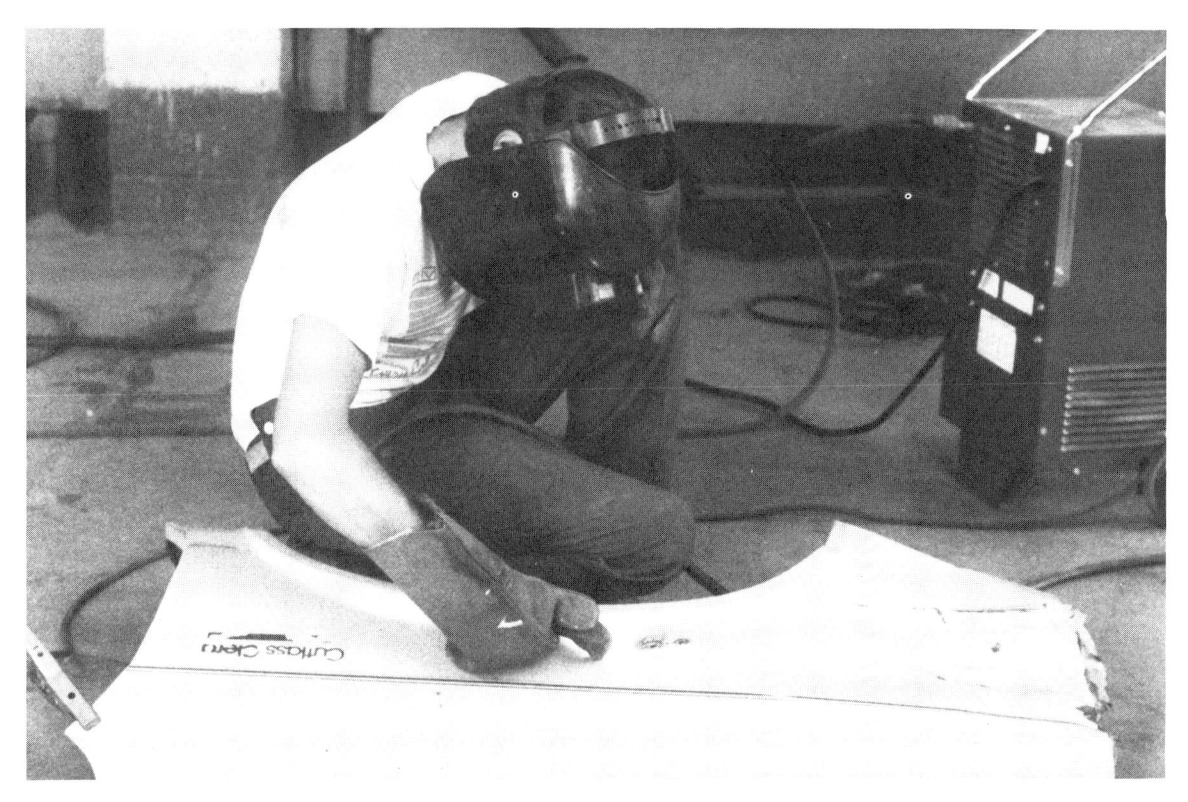

FIGURE 5-36 Making a quality plasma arc cut *(Courtesy of Century Mfg. Co.)*

welding processes. However, today's smaller and lighter power sources tend to make this a minor problem. A more important consideration might be the fact that it is difficult to make two straight up and down edges in one cut using plasma arc. Although the edges of the kerf will be quite smooth, only one edge will be exactly vertical; the other side will be slighty beveled, as much as 3 to 5 degrees. This bevel is caused by the swirling of the cutting gas as it leaves the nozzle. The swirling phenomenon takes a clockwise direction, meaning that the straight edge will be on the right side in the direction of the cut. Overall, the degree of plasma arc cutting quality will be determined by the following factors:

- *Travel Speed.* The thicker the material, the slower the speed. Naturally, this is exactly the opposite for thin material.
- *Stand Off.* This is the distance the torch is held above the workpiece. Generally speaking, the closer the nozzle is to the work, the better the cut.
- *Parts Wear.* The tip and electrode will erode with use. The greater the wear, the poorer the quality of the cut.
- *Air Quality.* Moist or oil-contaminated air will contribute to a poor quality cut.

A direct result of torch speed that is too slow or too fast is the tendency to produce dross; this is the resolidified, oxidized metal that adheres to the bottom edge of a cut. Stainless steel and aluminum are relatively dross-free, while nickel copper, nickel chromium iron, and copper alloys form dross quite readily.

BASIC OPERATING PRINCIPLES

The plasma arc cutting process operates on direct current straight polarity, electrode negative, with a constricted transferred arc. Its severing action is produced by melting a specific area with the heat of a constricted arc, then removing the molten material with a high-velocity jet of hot, ionized gas expelled from the nozzle orifice of the cutting torch. In any plasma cutting process, heated gas is transformed into a plasma gas, resulting in extremely high temperatures.

To establish the arc, a low current pilot arc is initiated by a high-voltage, high-frequency discharge between the electrode and the nozzle of the torch. When this pilot arc comes into contact with the electrically grounded part, it is automatically transferred to the workpiece, and a higher current arc is initiated. Once the plasma arc cutting current is established, the pilot arc circuit is opened and the unit begins to cut. After the cut is completed, the plasma cutting arc is removed from the workpiece, the electrical cutting circuit is opened, current stops flowing, and a postflow gas cools the torch components. This dissipates damaging heat from the torch, thus prolonging electrode and nozzle life.

Control consoles for plasma arc cutting can contain solenoid valves to turn the gases and cooling water on and off. They usually have flowmeters for the cutting gases and a water flow switch to stop the operation if the cooling water flow falls below a safe limit. Controls for high-power automatic plasma arc cutting can also contain programming features for upslope and downslope of current and orifice gas flow.

Gas Selection

Plasma cutting gas selection depends on the material being cut and the surface quality requirements. Most nonferrous metals are cut using nitrogen, nitrogen-hydrogen mixtures, or argon-hydrogen mixtures. Titanium and zirconium are cut with pure argon because of their susceptibility to embrittlement by reactive gases.

Carbon steels are cut using compressed air (80 percent nitrogen, 20 percent oxygen) or nitrogen. Nitrogen is used with the water injection method of plasma arc cutting. Some systems use nitrogen for the plasma forming gas, with oxygen injected into the plasma downstream of the electrode. This arrangement prolongs the life of the electrode by not exposing it to oxygen. For some nonferrous cutting with the dual-flow system, nitrogen is used as the plasma gas and carbon dioxide (CO_2) for shielding. For better quality cuts, argon-hydrogen plasma gas and nitrogen shielding are used.

Plasma Cutting Torches

The plasma cutting torch consists of an electrode holder that centers the electrode tip in the orifice in the constricting nozzle. The electrode and nozzle are water cooled to prolong their service lives. Plasma gas is injected into the torch around the electrode and exits through the nozzle orifice. Nozzles with various orifice diameters are available for each type of torch. Orifice diameter depends on the cutting current; the higher the current, the larger the diameter required. Nozzle design depends on the type of plasma arc cutting and the metal being cut.

Both single- and multiple-port nozzles may be used for plasma arc cutting. Multiple-port nozzles

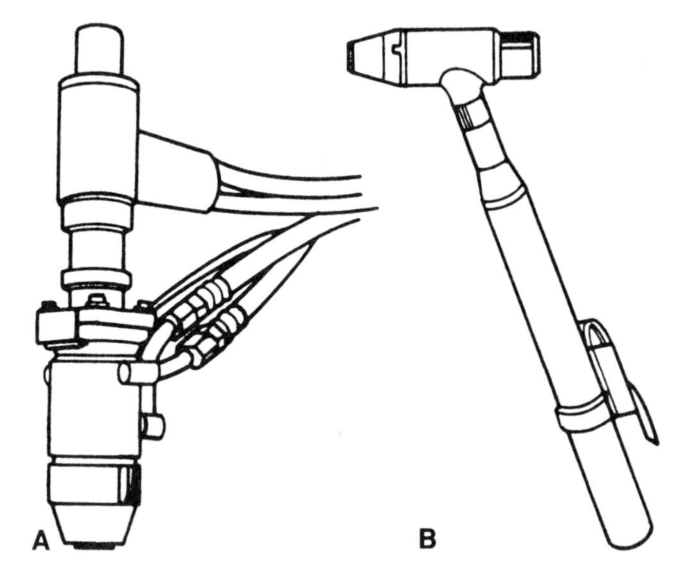

FIGURE 5-37 (A) A mechanized plasma arc cutting torch; and (B) a manual plasma arc cutting torch

have auxiliary gas ports arranged in a circle around the main orifice. All of the arc plasma passes through the main orifice with a high gas flow rate per unit area. Multiple-port nozzles produce better quality cuts than single-port nozzles at equivalent travel speeds; however, cut quality decreases as travel speed increases.

Torch designs for introducing a shielding gas or water around the plasma flame are available. Plasma arc cutting torches are similar to those used in TIG welding, both the manual and machine types. A typical mechanized plasma arc cutting torch is shown in Figure 5-37A; Figure 5-37B illustrates the manual type. Mechanized plasma arc cutting torches are mounted on shape-cutting machines similar to mechanized oxyfuel cutting equipment. The cutting can be controlled by photoelectric tracing, numerical control, or a computer.

OPERATING A PLASMA ARC CUTTER

To set up and operate a typical plasma arc cutter such as the one shown in Figure 5-38, use the following procedure. Keep in mind the importance of always using the manufacturer's recommendations as the best guide for working with the equipment. Also, be sure to provide local exhaust ventilation to control the metal fumes and potentially toxic gases within acceptable limits; this is especially critical when working in confined spaces.

1. Remove all grease, dirt, oil, and other foreign matter from the exterior of the nitrogen tank valve.

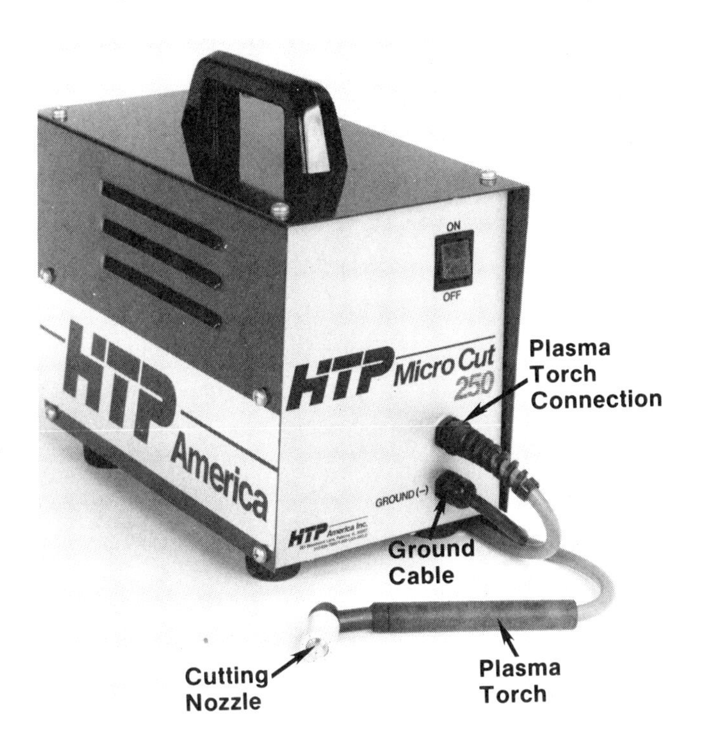

FIGURE 5-38 Typical plasma arc cutter *(Courtesy of HTP America Inc.)*

2. "Crack" the nitrogen valve, then close it immediately to remove any other foreign matter.
3. Connect the nitrogen regulator to the nitrogen tank.
4. Connect the nitrogen supply hose to both the nitrogen tank and the panel on the back of the welding machine.
5. Properly attach and adjust the torch and leads.
6. Inspect every connection cover and panel to insure that they are properly secured.
7. Fill the coolant reservoir to the cross wires.
8. Turn on the primary power. Let it run for a few minutes to allow the coolant reservoir system to rid itself of air.
9. Turn off the primary power and refill the coolant reservoir to the cross wires.
10. Attach the supply cable to the welding machine. The setup should look similar to the one shown in Figure 5-39.
11. Attach the ground clamp to the workpiece.
12. Follow the manufacturer's recommendations for the correct nitrogen pressure.

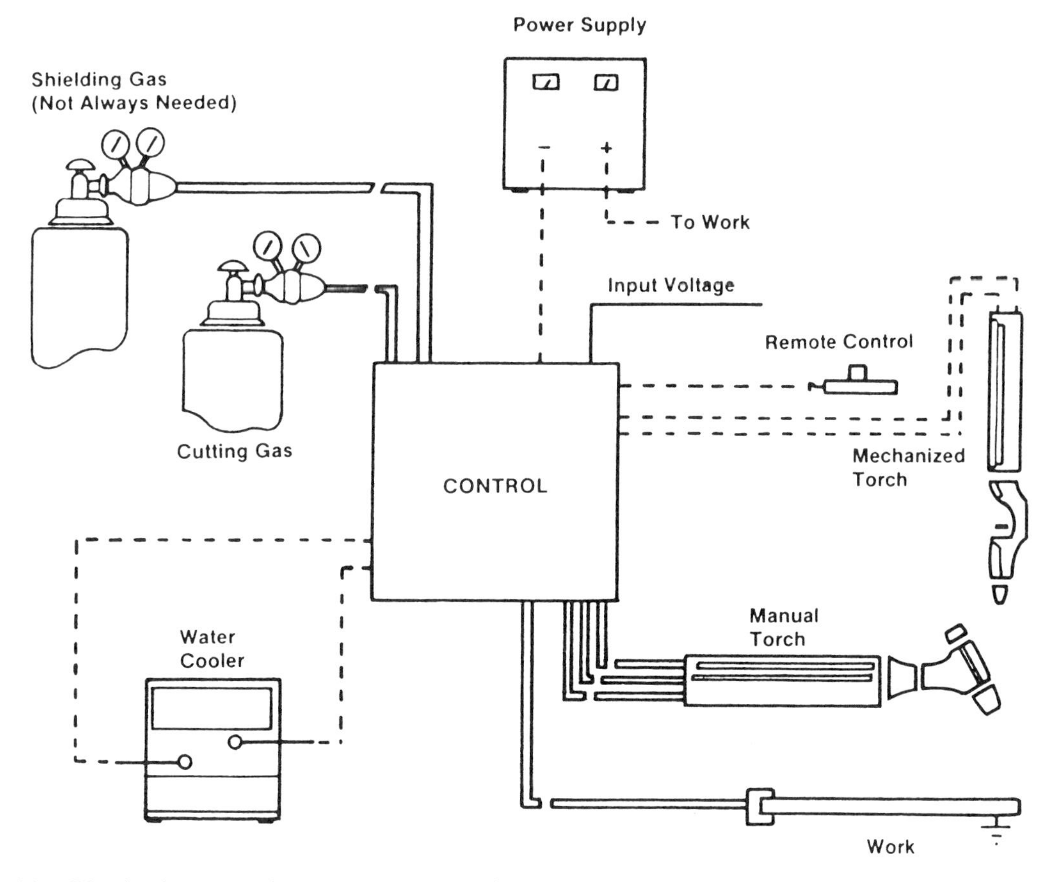

FIGURE 5-39 Block diagram of a plasma arc cutting setup

SHOP TALK

Flow rates vary with the orifice diameter from about 100 cfh for a 1/8-inch orifice to about 200 cfh for a 1/4-inch orifice. The gas used is almost exclusively nitrogen or argon, with a hydrogen addition of anywhere from zero to 30 percent. Manufacturer's recommendations must be consulted for each application.

13. Open the nitrogen supply valve on the tank.
14. Adjust the nitrogen pressure regulator until the desired gas pressure level is reached.
15. Purge the nitrogen for approximately 3 minutes to remove condensation and moisture.
16. Set the correct amperage and power levels as shown in Table 5-5.
17. The nozzle, much like the wire in a MIG welding gun, must come in contact with an electrically conductive part of the work before the plasma arc can start. This must be done to satisfy the work safety circuit. However, once the arc has started, the cutter will easily cut through painted surfaces.
18. Hold the torch so that the cutting nozzle is perpendicular to the work surface (Figure 5-40). Push the torch down. This will force the cutting nozzle down until it comes in contact with the electrode, at which time the plasma arc will start.
19. Immediately release the downward force on the plasma torch to let the cutting nozzle return to its normal position. It is not necessary to keep the cutting nozzle in contact with the work once the plasma arc has started; however, it makes cutting easier. While keeping the cutting nozzle in contact with the work, put very little, if any, downward pressure on the plasma torch. It is only necessary to drag it lightly on the work surface.
20. Begin to move the torch where the metal is to be cut. If the torch is moved too fast, it will not cut all the way through the workpiece. If moved too slowly, it will put too much heat into the workpiece and might also extinguish the plasma arc.
21. After completing the cut, lift the torch away from the workpiece. To disassemble the equipment, begin at step 14 and work backward, reversing the instructions.

TABLE 5-5: CONDITIONS FOR PLASMA ARC CUTTING OF MILD STEEL

Thickness (Inches)	Speed (Ipm)	Orifice Diameter (Inches)	Current (DCSP) (Amps)	Power (kW)
1/4	200	1/8	275	55
1/2	100	1/8	275	55
1	50	5/32	425	85
2	25	3/16	550	110

FIGURE 5-40 Position the plasma cutting torch perpendicular to the work surface. *(Courtesy of Century Mfg. Co.)*

Other pointers that should be remembered when using a plasma arc cutter are:

- When piercing materials that are 1/8-inch thick or more, angle the plasma torch at a 45-degree angle until the arc pierces the material. This will allow the stream of sparks to shoot off at a 45-degree angle away from the gas diffuser. If the plasma torch is held perpendicular to the work when piercing heavy-gauge material, the sparks will shoot back at the gas diffuser. The molten metal will then collect on the diffuser, plugging the air holes and greatly shortening the life of the diffuser.

CAUTION: When angling the torch, be aware of the fact that the sparks will shoot off as far as 20 feet away. Be sure that there are no combustibles or bystanders in the area that might be ignited or hurt by the sparks. Cover any nearby upholstery and wiring to protect them from flying sparks.

- When making long straight cuts, it might be easier to use a metal straightedge as a guide. Simply clamp it to the work to be cut.
- When cutting 1/4-inch materials, it is beneficial to start the cut at the edge of the material.
- When making rust repairs, it is possible to piece the new metal over the rusted area and then cut the patch panel at the same time the rust is cut out. This process also works when splicing in a quarter panel.
- Be aware of the fact that the sparks from the cutting arc can damage painted surfaces and will also pit glass. Use a welding blanket to protect these surfaces.

Accessories

Two common accessories are available for mechanized plasma cutting machines to aid in fume and noise control. One is the water table, which is simply a cutting table filled with water up to the bottom surface of the plates. The high-speed gases emerging from the plasma jet produce turbulence in the water. Most of the fume particles are trapped, similar to the operation of a scrubber in an air pollution control system.

The second device is a water muffler designed to reduce noise. The muffler is a nozzle attached to the torch body that produces a curtain of water around the front of the torch; it is always used in conjunction with a water table. Water from the table is pumped through the nozzle, and the combination of the water curtain at the top of the plate and the water contacting the plate bottom encloses the plasma arc in a sound-deadening shield. The noise output is reduced by roughly 20 dBA. This equipment should not be confused with water injection or water shielding PAC variations.

REVIEW QUESTIONS

1. TIG weld speeds typically reach ____________.
   a. 25 to 30 ipm
   b. 15 to 20 ipm
   c. 10 to 15 ipm
   d. 5 to 10 ipm

2. Alternating current is used to TIG weld ____________.
   a. cast iron
   b. aluminum
   c. steel
   d. all of the above

3. In a particular TIG welding application, maximum heat is desired. Because of this, Welder A uses direct current straight polarity. Welder B uses direct current reverse polarity. Who is correct?
   a. Welder A
   b. Welder B
   c. Both A and B
   d. Neither A nor B

4. Which of the following statements is incorrect?
   a. When installing a tungsten electrode in a TIG gun, the color tip should be at the front end of the gun.
   b. When electrode size is matched with the correct current, the electrode tip will become a molten hemisphere.
   c. Too much current will cause excessive tungsten erosion.
   d. All of the above

5. What does a dirty, rough electrode usually mean?
   a. There was air leakage in the gas supply system of the torch.
   b. Inert gas was shut off before the electrode cooled.
   c. The electrode tip was contaminated by touching metal.
   d. All of the above

6. What is the best method for locating cracks in aluminum cylinder heads?
   a. pressure testing
   b. penetrating dye
   c. magnetic detection equipment
   d. both a and b

7. The typical acetylene pressure for oxyacetylene welding is ____________.

a. 15 to 100 psi
b. 3 to 12 psi
c. 3 to 25 psi
d. 30 to 120 psi

8. Mixing slightly more acetylene than oxygen will obtain what type of flame?
   a. neutral
   b. standard
   c. carburizing
   d. oxidizing

9. While making an oxyacetylene weld, a popping noise is heard. Welder A says this is normal and continues welding. Welder B says it means that insufficient gas is reaching the tip and opens both needle valves a bit more. Who is right?
   a. Welder A
   b. Welder B
   c. Both A and B
   d. Neither A nor B

10. Which of the following is a possible cause of an oxyacetylene flame cutting off in the middle of the procedure?
    a. clogged flame outlet
    b. oxygen pressure too low
    c. insufficient acetylene supply
    d. all of the above

11. The best method for cutting modern car metals is ______________.
    a. TIG
    b. plasma arc
    c. oxyacetylene
    d. both a and b

12. Dross is a direct result of plasma arc torch speed that is ______________.
    a. too fast
    b. too slow
    c. both a and b
    d. neither a nor b

13. Which of the following statements concerning plasma arc cutting is incorrect?
    a. It is not necessary to keep the cutting nozzle in contact with the work once the arc has started.
    b. The gas used is almost exclusively nitrogen or argon with hydrogen added.
    c. The nozzle must come in contact with an electrically conductive part of the work before the arc can start.
    d. When piercing material that is more than 1/8-inch thick, hold the torch perpendicular to the work.

14. Sparks from a plasma cutting arc can ______________.
    a. pit glass
    b. damage painted surfaces
    c. both a and b
    d. neither a nor b

15. What is the one disadvantage of using direct current straight polarity for TIG welding?
    a. minimal heat produced
    b. low degree of weld penetration
    c. cleaning action not provided
    d. weld speed slows considerably

CHAPTER SIX

RESISTANCE SPOT WELDING

Objectives

After reading this chapter, you should be able to

- Describe how a resistance spot welder works.
- List and explain the function of the various resistance spot welding components.
- List the different spot welder adjustments.
- Describe the safety and setup procedures for resistance spot welding.
- List and describe the factors affecting a spot welding operation.
- State and explain the three general methods for determining a spot weld's integrity.
- Explain the causes of weak welds.
- Describe some other spot welding functions.

Resistance welding is a high-speed method that can join metals in a fraction of a second. Pressure and electric heating under accurate control are used to join the metals together without the addition of any consumable welding material. The welding force is applied by the welder, eliminating extra tack welds or holding clamps.

Although squeeze-type resistance spot welding (S-TRSW) is just one of several resistance welding processes, it is the most common. Discovered more than 100 years ago, it is now one of the most widely used of all welding processes in the metal fabricating industry.

The process is used to weld thousands of different industrial and consumer products, from nuclear piping heat shields and railway cars to major home appliances, automobiles, and trucks. In this country, it is also widely used in the automotive aftermarket for sunroof installations and vehicle conversions, including recreational vehicles and stretch limousines. S-TRSW is fast, easy to use, and produces strong welds.

Although resistance spot welding has been used in the European and Japanese unibody collision repair industry for more than 25 years, it is a relatively new process in this country's repair industry. This is regrettable since the squeeze-type resistance spot welder (Figure 6-1) is ideal for welding many of the unibody's thin-gauge sections that call for good strength and no distortion. Since the early 1980s, a growing number of car manufacturers have published factory collision repair manuals that specify the use of squeeze-type resistance spot welding to repair weld many areas of their unibody cars (Figure 6-2). These areas include rocker panels, radiator core supports, window and door opening flanges, roofs, many exterior panels, and pinch weld areas (Figure 6-3).

These same manuals also specify those areas for which the MIG welding process should be used. Typical MIG weld areas usually include the heavier

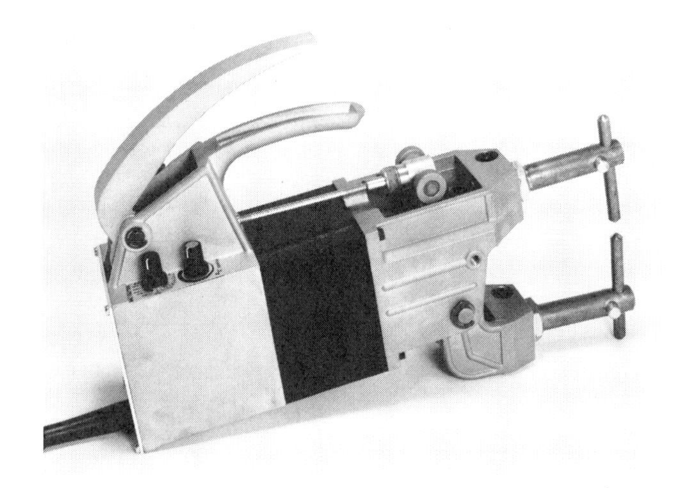

FIGURE 6-1 Typical squeeze-type resistance spot welder *(Courtesy of Lors Machinery, Inc.)*

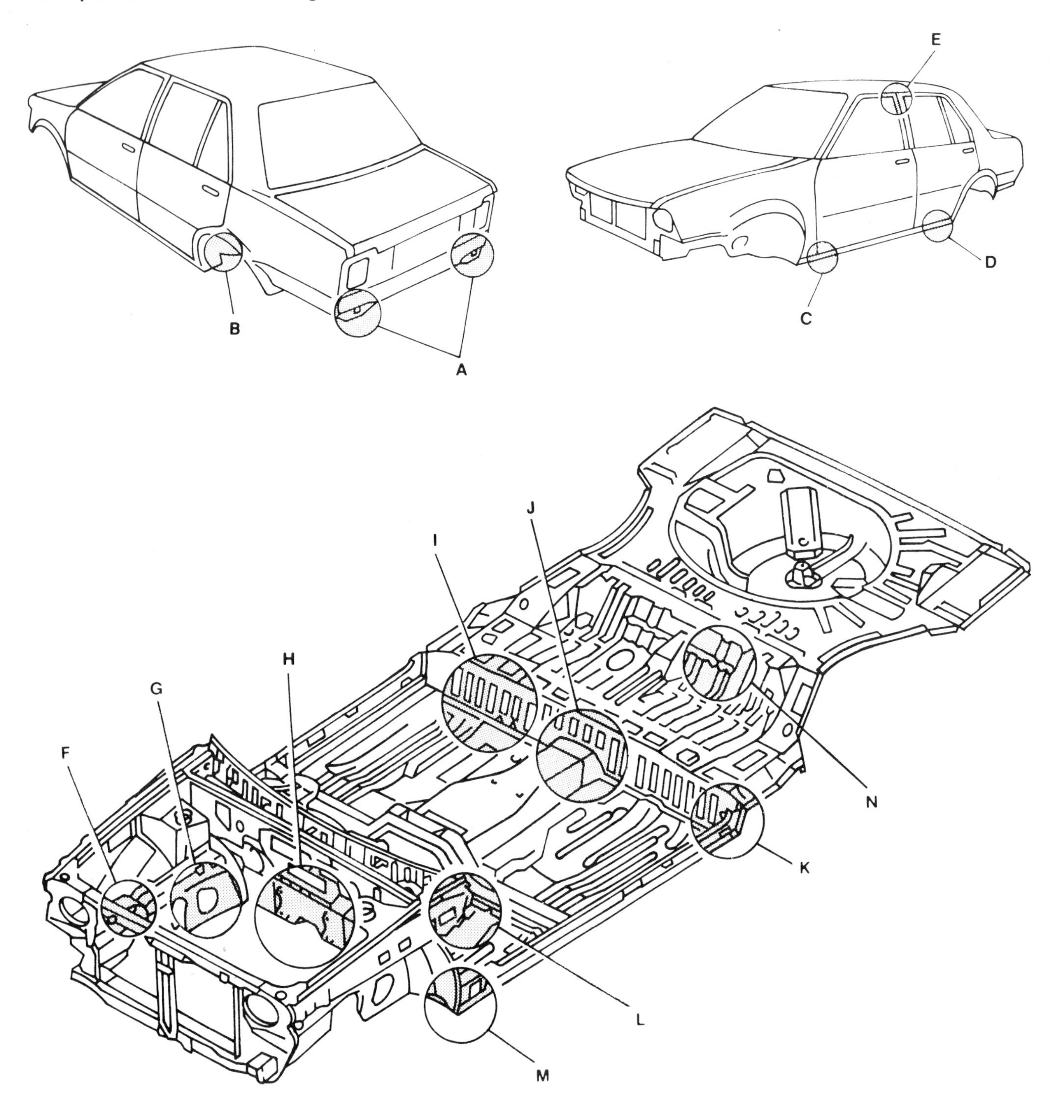

Construction	Location
Suspension mounting	G
Steering gear mounting	H
Fuel tank mounting	N
Engine, transmission mounting	F

Construction	Location
Belt anchor	E
Jack-up point	C, D
Major construction portions	A, I, J, K, L, M

FIGURE 6-2 Typical auto body locations in which spot welding is used during vehicle production *(Courtesy of Nissan Motor Corp.)*

A

B

C

FIGURE 6-3 Typical applications of squeeze-type resistance spot welders: (A) welding a side quarter window pinch weld flange; (B) welding the right quarter panel to the rear side member, trunk floor pan, and lower back panel; and (C) welding the left rear quarter panel to the rear valence *(Courtesy of Lors Machinery, Inc.)*

rail sections and a number of other areas where the metal thickness or restricted accessibility to both sides of the repair joint would not be suitable for a hand-held squeeze-type resistance spot welder normally used for unibody work.

SHOP TALK

Never assume that one welding process is correct for all repairs or car models. Due to vehicle design differences, one automaker might recommend a different repair procedure for a given damaged area than would another automaker. The welder should refer to the automaker's factory collision repair manual to determine which process and procedures are to be used.

This should make it clear that while 90 to 95 percent of all factory welds in a unibody structure are resistance spot welds, the MIG welder is still needed. The MIG welder and the resistance spot welder are two different machines designed for different applications. They are complementary in today's body shop.

MIG welders work best on inner structures, framework, gussets, and brackets. Conversely, resistance spot welders are designed for thin outer skins and cosmetic panels. On thin material like the 22- and 24-gauge sheet metal used on today's cars, resistance spot welding has the following advantages:

- Uses half as much heat as a MIG welder, which means less finishing time because of less heat distortion.
- Finishing time is further reduced because there is no weld metal to grind down.
- Duplicates OEM factory weld appearance.
- Clean; no smoke or fumes.
- Allows use of weld-through conductive zinc primers to restore corrosion protection to repair joints.
- Fast weld times of 1 second or less, which produce strong welds on HSS steel, HSLA steel, and mild steels.
- Low initial cost and no consumables such as wire, rods, or gas are required, so it is economical.

Keep in mind that these two processes, resistance spot welding and MIG, should not be viewed as competitive, but as complementing each other. The properly equipped welding shop should use both spot welders and MIG welders.

The need for squeeze-type resistance spot welders in automotive repair is growing, as much unibody automobiles are being manufactured and older unibody cars are being kept in service longer by their owners due to higher new car costs. The squeeze-type resistance spot welder will become an increasingly important time and cost saver, as well as necesary tool for repair welding today's unibody autos in compliance with the vehicle manufacturer's recommended repair procedures. For this reason, the competitive welder must know how to use a resistance spot welder.

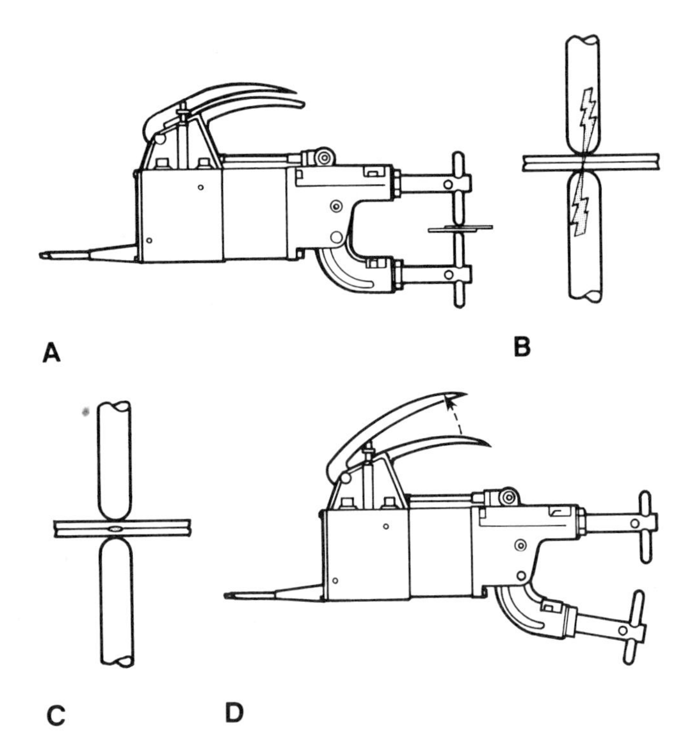

FIGURE 6-5 A resistance spot welding gun in the (A) squeeze position; (B) welding position; (C) hold position; and (D) off position

HOW RESISTANCE SPOT WELDING WORKS

Resistance spot welding relies on the resistance heat generated by low voltage electric current flowing through two pieces of metal held together, under pressure, by the squeeze force of the welding electrodes. One of the things that makes resistance spot welding unique is that the actual weld nugget is formed internally with relation to the surface of the base metal. Figure 6-4 shows a resistance spot weld nugget compared to a MIG spot weld. The MIG spot weld is made from one side only. The resistance spot weld is normally made with electrodes on each side of the workpiece. Resistance spot welds can be made with the workpiece in any position.

The actual process of resistance spot welding is fairly simple. It has been compared to the forge or hammer welding done by blacksmiths. The blacksmith used a forge of hot coals to heat the metal parts to be welded. When the metal was white hot, the blacksmith would overlap the parts on the anvil and quickly forge them together under the pressure of the hammer blows. The smith depended on heat, pressure, and time to weld parts together.

A squeeze-type resistance spot welder also depends on heat, pressure, and time to make strong welds. The heat of the weld comes from the resistance of the metal to the passage of heavy current. In all cases, of course, the current must flow or the weld cannot be made. The pressure of the electrode tips on the workpiece holds the parts together in close and intimate contact. It is at this location where a pulse of current forms the weld. The force applied before, during, and after the current flow forges the heated parts together so that coalescence will occur. The welding time is very short, usually 1/2 second (30 cycles) or less. Of course, the correct combination of welding current intensity, weld force, and weld time must be selected to make consistently strong welds. This is easy to do once it is understood how a resistance spot welding gun works (Figure 6-5).

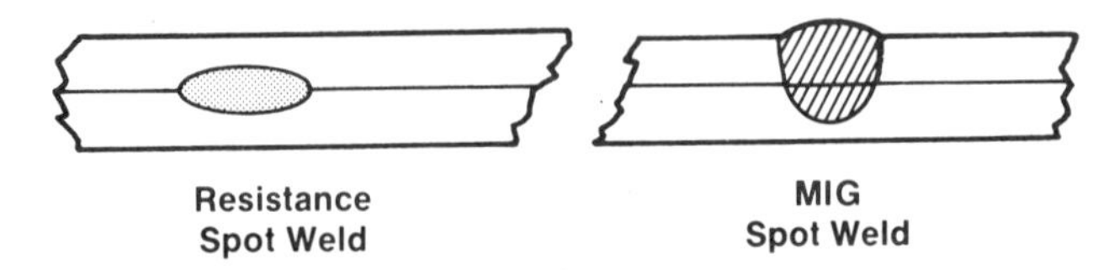

FIGURE 6-4 Resistance and MIG spot weld comparison

- *Pressurization.* The mechanical welding bond between two pieces of sheet metal is directly related to the amount of force exerted on the sheet metal by the welding tips. As the tips squeeze the sheet metal together, an electrical current flows from the tips through the base metal, causing the two pieces to melt and fuse together. Weld spatter internal or external) is the result of light pressure on the tip or excessive current flow. Heavy tip pressure causes a small spot weld (Figure 6-6) and a reduced mechanical bond. In other words, as the tip pressure increases, the electrical current and subsequent heat is distributed over a wider area, thus reducing the diameter and the penetration of the weld.

- *Current Flow.* When pressure is applied to the metal, a large surge of electric current flows through the electrodes and through the two pieces of metal. The temperature rises rapidly at the joined portion of the metal where the resistance is greatest (Figure 6-7A). If the current continues to flow, the metal melts and fuses together (Figure 6-7B). If the current becomes too great or the pressure too low, internal spatter will result.
- *Holding.* If the current flow is stopped, the melted portion begins to cool and forms a round, flat bead of solidified metal known as a nugget. This structure becomes very dense due to the pressurization force, and its subsequent mechanical bonding is excellent (Figure 6-8). Pressurization time is very important; do not use less time than specified in the operator's manual.

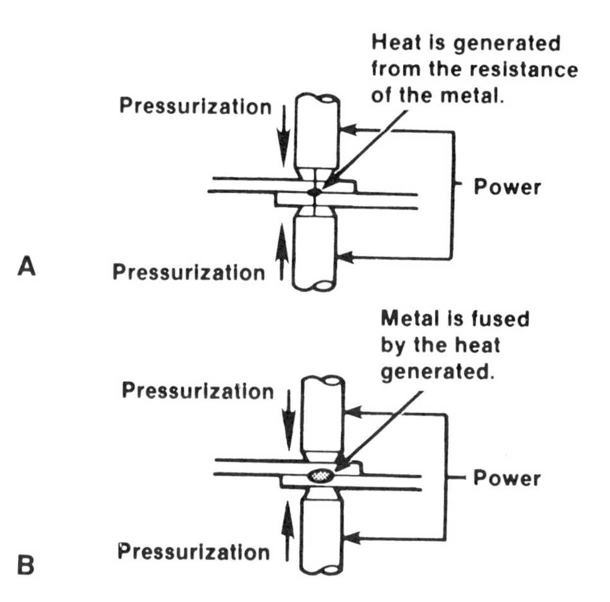

FIGURE 6-7 Electrical current (amperage)

HEAT

When current is passed through a conductor, the electrical resistance of the conductor to current flow will cause heat (or power) to be generated. The basic formula for heat generation is:

$H = I^2R$ where H = Heat
I^2 = Welding current squared
R = Resistance

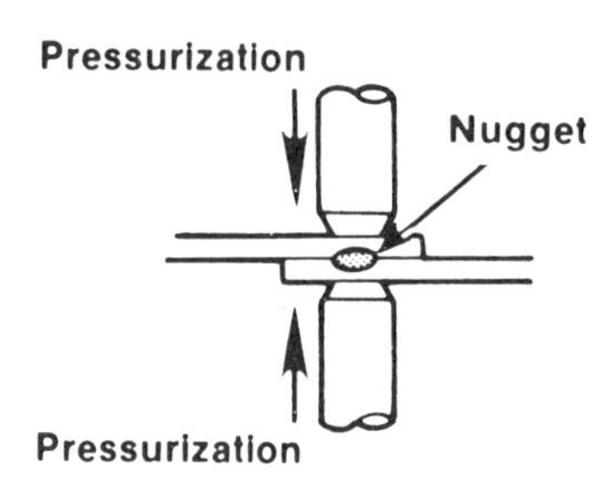

FIGURE 6-8 Electrical current (holding) flow time

The secondary portion of a resistance spot welding circuit, including the parts to be welded, is actually a series of resistances. The total additive value of this electrical resistance affects the current output of the resistance spot welding gun and the heat generation of the circuit.

Although current value is the same in all parts of the electrical circuit, the key fact is the resistance values can vary considerably at different points in the circuit. The heat generated is directly proportional to the resistance at any point in the circuit. It is at the interface of the top and bottom workpieces where the weld is to be made that the greatest relative resistance is required. (The term *relative* means with relation to the rest of the actual welding circuit.)

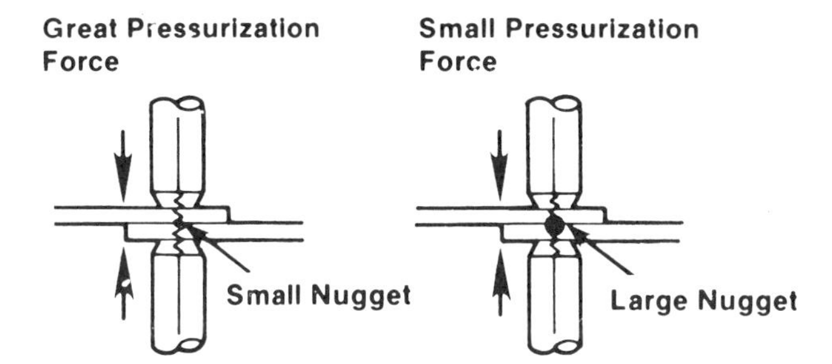

FIGURE 6-6 Electrode (tip) pressure

There are six major points of resistance in the work area.

1. Contact point between the electrode and top workpiece
2. Top workpiece
3. Interface of the top and bottom workpieces
4. Bottom workpiece
5. Contact point between the bottom workpiece and electrode
6. Resistance of electrode tips

The resistances are in series (which means the resistances are cumulative), and each point of resistance will retard current flow. The amount of resistance at the interface of the workpieces will depend on the heat transfer capabilities of the material, its electrical resistance, and the combined thickness of the materials at the weld joint. It is at this part of the circuit that the nugget of the weld is formed.

PRESSURE

The primary purpose of pressure is to hold the parts to be welded in intimate contact at the joint interface. This action assures consistent electrical resistance and conductivity at the point of weld. The tong and electrode tips should not be used to pull the workpiece together for welding. The resistance spot welder is not designed to be an electrical C-clamp. The parts to be welded should be in intimate contact before pressure is applied.

Investigations have shown that high pressures exerted on the weld joint decrease the resistance at the point of contact between the electrode tip and the workpiece surface. The greater the pressure, the lower the resistance factor.

Where intimate contact of the electrode tip and the base metal exists, proper pressures will tend to conduct heat away from the weld. Higher currents are necessary with greater pressures and, conversely, lower pressures require less amperage from the resistance spot welding gun. This fact should be carefully noted, particularly when using a heat control with a resistance spot welder.

TIME

Resistance spot welding depends on the resistance of the base metal and the amount of current flowing to produce the heat necessary to make the spot weld. Another important factor is time. In most cases, several thousands of amperes are used in making the spot weld. Such amperage values, flowing through a relatively high resistance, will create a lot of heat in a short time. To make good resistance spot welds, it is necesary to have close control of the time the current is flowing.

Previously, the formula for heat generation was used. With the addition of the time element, the formula is completed as follows:

$H = I^2RTK$ where H = Heat
I^2 = Current squared
R = Resistance
T = Time
K = Heat losses

Welding heat is proportional to the square of the welding current. If the current is doubled, the heat generated is quadrupled. Welding heat is proportional to the total time of current flow; thus, if current is doubled the time can be reduced considerably. The welding heat generated is directly proportional to the resistance and is related to the material being welded and the pressure applied. The heat losses should be held to a minimum. It is therefore an advantage to shorten welding time.

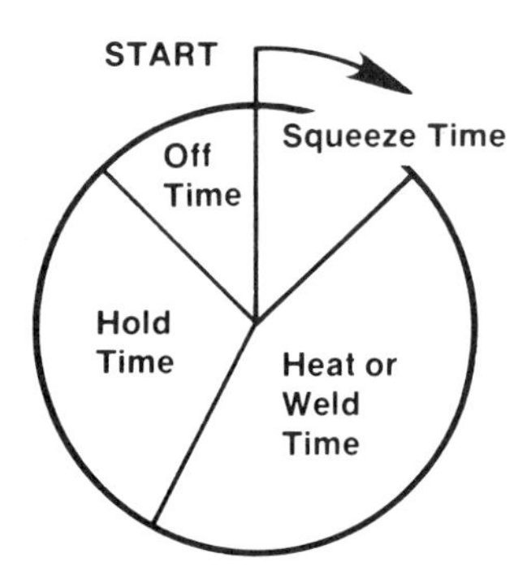

FIGURE 6-9 Spot welding time cycle

Most resistance spot welds are made in very short time periods. Since alternating current is normally used for the welding process, procedures may be based on a 60-cycle time (60 cycles = 1 second). Figure 6-9 shows the resistance spot welding time cycle. The squeeze time is the time between pressure application and welding; the weld or heat time is measured in cycles; the hold time is the time that pressure is maintained after the weld is made; and the off time is the time wherein the electrodes separate to permit movement to the next spot.

Control of time is important. If the weld time element is too long, the base metal in the joint can exceed the melting point (and possibly the boiling) of the material. This could cause faulty welds due to gas porosity. There is also the possibility of expulsion of molten metal from the weld joint (also called weld spatter), which could decrease the cross section of the joint and weaken the weld. Shorter weld times also decrease the possibility of excessive heat transfer in the base metal. Distortion of the welded parts is minimized and the heat affected zone around the weld nugget is substantially smaller. Not only is control of the welding time important, the holding time must be controlled also. As mentioned earlier, the specified holding times given in the operator's manual should not be shortened.

RESISTANCE SPOT WELDING COMPONENTS

The components of a resistance spot welder (Figure 6-10) are the transformer, the welder control, and the gun (with interchangeable arm sets and electrode tips).

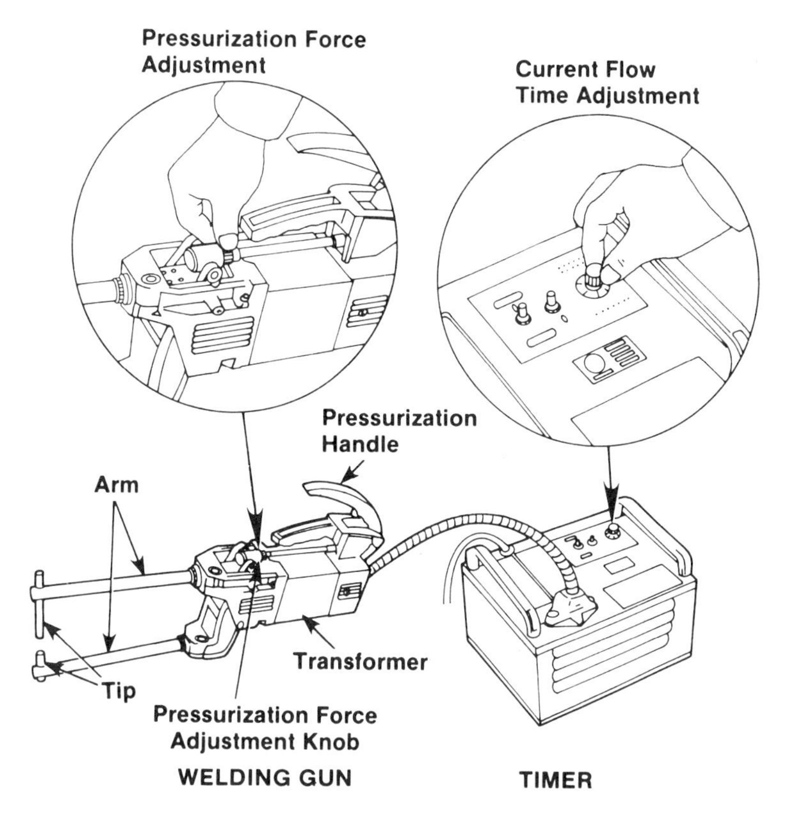

FIGURE 6-10 Components of a resistance spot welding system *(Courtesy of Toyota Motor Corp.)*

TRANSFORMER

The heart of a resistance spot welder is the transformer. It converts the shop line's high voltage and low amperage current (usually 220 volts and 30 amps) into a low secondary voltage and a very high secondary amperage current (2 to 5 volts and more than 8000 amps), safe from electrical shock. It can be either built into the welding gun or mounted remotely and connected to the gun with cables. A built-in transformer is electrically more efficient, since there is little or no loss of welding current between the transformer and the gun. A remote transformer must be larger and draw more shop line current to compensate for power losses through the long cables connecting it to the gun. Remember that this high weld current will decrease when long reach or wide gap arm sets are used. A high weld current output can be adjusted to a lower intensity by use of the welder control.

WELDER CONTROL

The brain of a resistance spot welder is the welder control (also called the *control unit*). By means of a solid-state weld timer and a phase shift weld current regulator, the control unit adjusts the transformer's weld current output and permits precise adjustment of the weld time.

The welder control must be capable of providing a full range of welding current adjustments. Weld current settings vary, depending upon the thickness of the steel to be welded and the length and gap of the arm sets needed to reach into the weld area. It might be necessary to decrease the weld current when welding with short-reach arm sets or increase the weld current when using long-reach or wide-gap arm sets.

Weld timers are usually adjustable from 1/30 of a second to 1 second (two 60 cycles). Typical weld current adjustment range is from 30 to 100 percent of the maximum available weld current. A repeatable accuracy of at least 1/10 of a second is recommended for consistent weld quality.

SHOP TALK

Some manufacturers of resistance spot welders designed for unibody repair work offer additional control features that compensate for small amounts of surface scale or rust on the metal. Such features make it possible to determine when a poor weld condition exists.

Welders that do not have adjustable weld time and weld current capabilities or that use mechanical or manual timing are not satisfactory for consistent quality welding and thus are not recommended for unibody repair welding.

GUN

The muscle of a resistance spot welding operation is the gun that applies the squeeze force and delivers the welding current through the arms and electrodes to the weld area. It usually consists of a spring-assisted operating lever to generate upward of 250 pounds of active clamping force when its short arms are fitted to the welder. Most resistance spot welders are designed with a force-multiplying mechanism to produce the high electrode force required for consistent weld quality. These force-multiplying mechanisms can be spring or pneumatically assisted. Squeeze-type resistance welders that do not use a force-multiplying mechanism, and instead rely solely on the operator's manual grip for pressure, are not recommended for welding unibody structures.

This active clamping or squeeze force holds together the parts to be welded to maintain a good metal contact. This permits the welding current to flow easily through the repair joint to prevent expulsion or weld spatter of the molten weld nugget, which could weaken the weld. Active clamping force applies a follow-up pressure to forge the weld as the welding current is turned off and the weld nugget contracts.

A toggle clamp or static force, like vise-grip pliers, cannot follow up; it loses clamping pressure as the weld nugget contracts, resulting in expulsion and weak welds. The squeeze pressure mechanism should be adjusted to develop sufficient clamping and follow-up force at the electrodes for the metal thickness being welded (Figure 6–11).

The majority of welding guns in auto body shops have a maximum capacity of up to two times 5/64-inch-thick steel when equipped with short-reach arm sets of 5 inches or less. Capacity with long reach or wide gap arm sets is at least two times 1/32-inch-thick steel. These capacities comply with the specifications listed in most factory body repair manuals.

Arms

Resistance spot welders used for unibody repair welding are available with a full range of interchangeable arm sets. Standard arm sets (Figure 6–12) are designed to reach difficult areas on most makes of cars, such as wheel well flanges, drip rails, taillight openings, and other tight pinch weld areas, as well as floor pan sections, rocker panels, and window and door openings. When working with new cars, be sure to check the factory repair manual, and look for special arm sets for the hard-to-reach areas.

Electrodes

Copper is the base metal normally used for resistance spot welding tongs and tips. The purpose of the electrode tips is to conduct the welding current to the workpiece, to be the focal point of the pressure applied to the weld joint, to conduct heat from the work surface, and to maintain their integrity of shape and characteristics of thermal and electrical conductivity under working conditions.

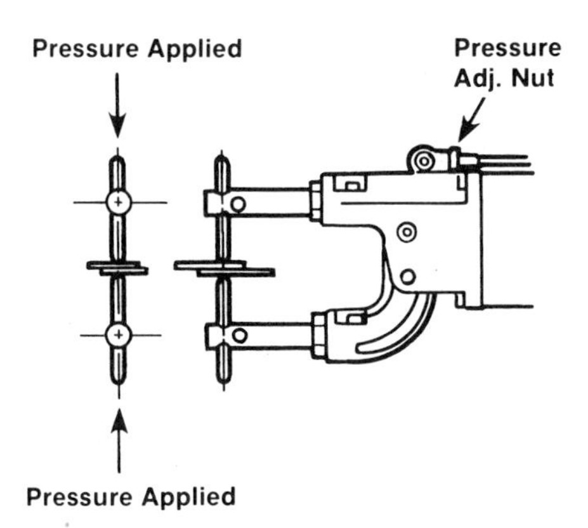

FIGURE 6–11 Adjustment of the squeeze pressure mechanism

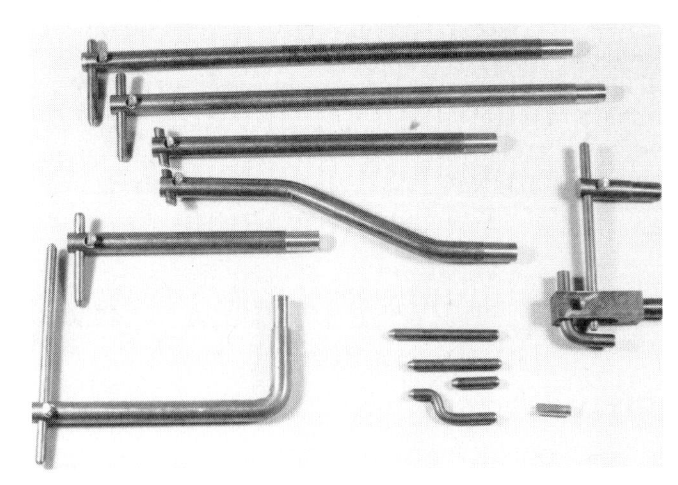

FIGURE 6–12 Standard arm sets for resistance spot welders *(Courtesy of Henning Hansen, Inc.)*

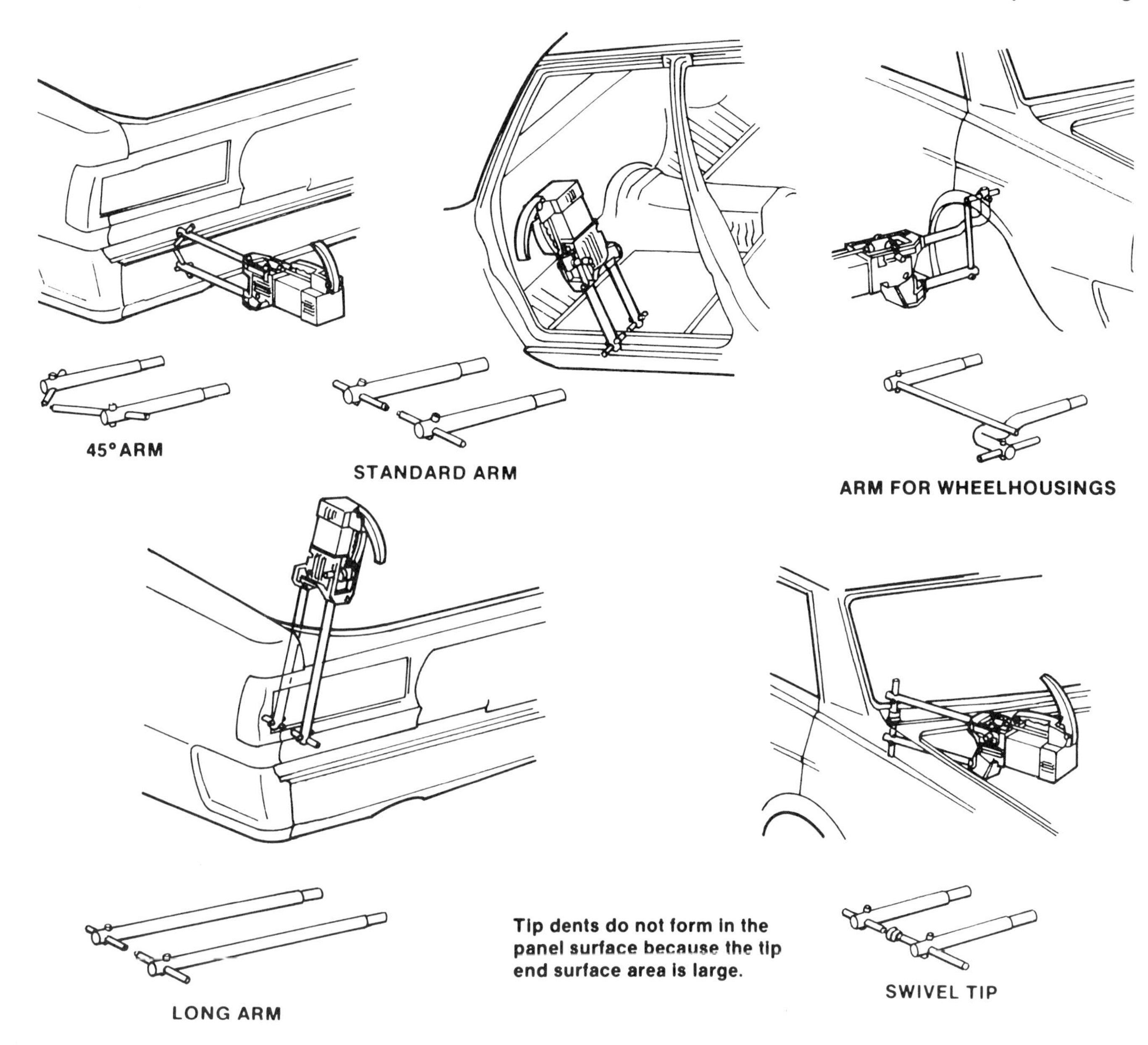

FIGURE 6-13 Select the proper arms for the job. *(Courtesy of Toyota Motor Corp.)*

SPOT WELDER ADJUSTMENTS

To obtain sufficiently strong spot-welds, perform the following checks and adjustments on the squeeze-type gun before starting the operation:

- *Arm Set Selection.* It is important to select the appropriate arm set for the area to be welded (Figure 6-13).
- *Arm Adjustment.* Keep the gun arm as short as possible to obtain the maximum pressure for welding (Figure 6-14). Tighten the gun

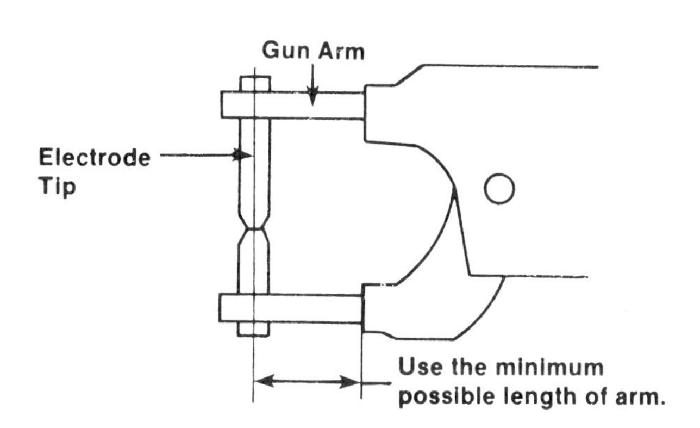

FIGURE 6-14 Adjusting the gun arm

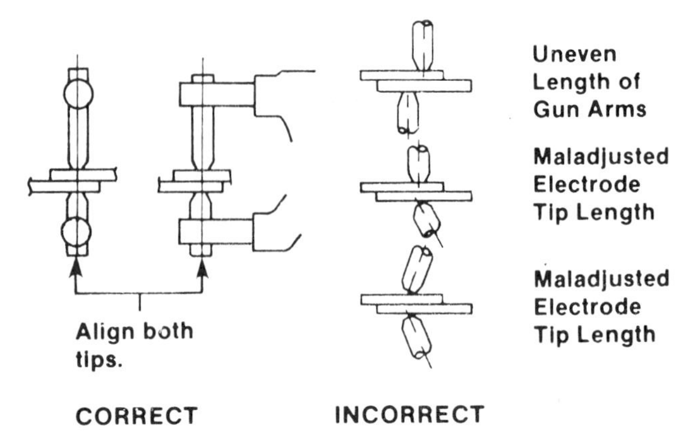

FIGURE 6-15 Correct and incorrect alignment of the electrode tip *(Courtesy of Nissan Motor Corp.)*

arm and tip securely so that they will not become loose during the operation.

- *Electrode Tip Alignment.* Align the upper and lower electrode tips on the same axis (Figure 6-15). Poor alignment of the tips causes insufficient pressure, which results in insufficient current density and reduced strength at the welded portions.

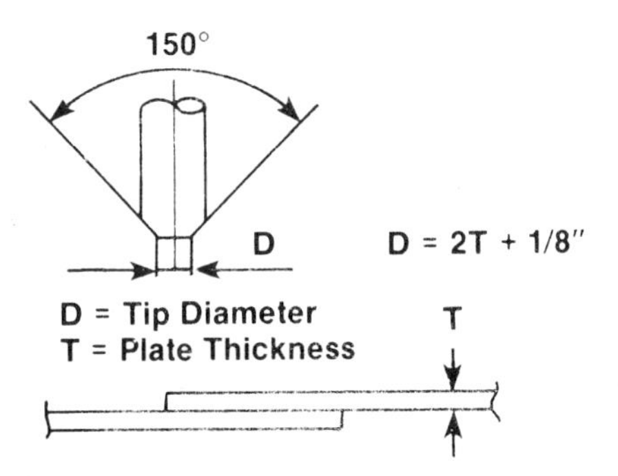

FIGURE 6-16 Determining tip diameter *(Courtesy of Nissan Motor Corp.)*

- *Electrode Tip Diameter.* The diameter of the spot weld decreases as the diameter of the electrode tip increases. However, if the electrode tip is too small, the spot weld will not increase in size. The tip diameter (Figure 6-16) must be properly controlled to obtain the desired weld strength. Before starting the operation, make sure that the tip diameter is kept the proper size and file it cleanly to remove burnt or foreign matter from its surface. As the amount of dirt on the tip increases, the resistance at the tip also increases; this reduces both the current flow through the base metal and the weld penetration, resulting in an inferior weld. If the tips are used continuously over a long period of time, they will not dissipate heat properly and will become red hot. This will cause premature tip wear, increased resistance, and a drastic reduction in the welding current. If necessary, let the tips cool down after every five or six welds. If they are worn, use a tip dressing tool to reshape them (Figure 6-17).
- *Electrical Current Flow Time.* When the electrical current flow time increases, the resulting heat increases the weld diameter and penetration. The amount of heat that is dissipated at the weld increases as the current flow time increases. Since the weld temperature will not rise after a certain amount of time, even if current flows longer than that, the spot weld size will not increase. However, tip pressure marks and heat warping might result.

The pressurization force and welding current cannot be adjusted on many spot welders, and the

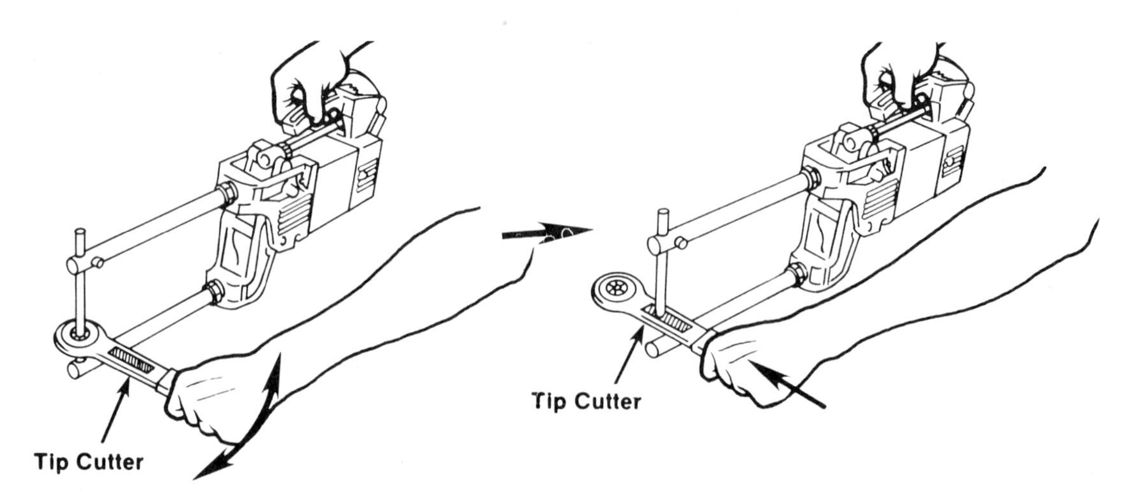

FIGURE 6-17 Reshaping the ends of the tips *(Courtesy of Toyota Motor Corp.)*

current value might be low. However, weld strength can be assured by lengthening the current flow time (in other words, letting the low current flow for a long time).

The best results can be obtained by adjusting the arm length or welding time according to the thickness of the panels. Although the welder instruction manual has these values listed inside, it is best to test the quality of the weld using the methods described later in this book as the adjustments are being done.

SHOP TALK

When spot welding antirust steel panels, offset the drop in current density by raising the current value 10 to 20 percent above that for ordinary steel panels. Since the current value cannot be adjusted in spot welders ordinarily used for body repairs, lengthen the current flow time a little. It is important to differentiate between antirust sheet metal and ordinary sheet metal, since the protective zinc coating on the antirust panels must be removed along with the paint (when sanding prior to welding).

SAFETY AND SETUP

Read the welder manufacturer's installation and operating instructions. Check the electrical service voltage in the shop to see if it matches the voltage requirements on the welder nameplate. Make sure the welding equipment is connected to a positive ground and have a qualified technician check for ground. Never connect a ground wire of single-phase electrical equipment to a leg of a three-phase electrical supply because it could permanently damage the equipment and injure the operator. Check for the correct fuse and wire size recommended for the welder. As with all electrical equipment, be sure local electrical codes and legal requirements are met.

Unlike other welding processes, a helmet and dark eye lens are not required with resistance spot welders. However, clear safety glasses should be worn at all times and light shop gloves are usually recommended. Standard shop clothing in accordance with accepted shop safety practices is recommended.

If the welder has a separate control unit, connect the welding gun to the control following the manufacturer's instructions. If the welding gun is self-contained with a built-in control unit, this step will not be necessary.

Take two pieces of metal the same thickness as the sheets to be repaired. Adjust the pressure mechanism to assure sufficient weld force on the two thicknesses of sample metal. Make sure the welder arms are parallel and the tips make full contact when force is applied. If they are not, adjust the tips up or down until the arms are parallel. Secure the tips in place and check to see that enough squeeze force can be applied to the metal piece being welded.

Now you are ready to connect the control unit or the self-contained gun, following the manufacturer's instructions, to a correct voltage electrical supply that is fused with a long-delay type fuse. Remember, all electrical connections should comply with all applicable codes.

Adjust the weld current regulator to its lowest value. This is usually indicated as a percent of current (30 to 100 percent). Adjust the weld timer to about 5 or 10 cycles (1/12 to 1/6 second).

Next, make sure that the metal used for sample welds is identical to the metal to be used on the vehicle. Scrap pieces cut from the repair area of the car are ideal. Be sure that the sample metal reaches as far into the welder throat (arm length) as the actual repair will.

Remember, the amount of steel placed in the throat of a welder will affect the welder's electrical efficiency and capacity. Squeeze the operating lever to apply pressure to the parts being welded. As pressure is being applied, a built-in switch turns on the welding current for the amount of time preset by the weld timer adjustment. The timer will automatically turn off the current at the end of the weld time. Since the weld time is less than 1 second, the entire process is very fast. Keep pressure on the weld for a moment to forge the weld before releasing the operating lever.

Test the sample weld for strength. The easiest and best method is the peel test. Simply peel the ends of the top sheet of metal upward, away from the bottom sheet. Then, peel the ends of the bottom sheet downward, away from the top sheet until the weld sample looks like two letter U's joined at the base (Figure 6-18).

Continue to rock the pieces apart until a hole is pulled in one of them. An alternate version of the peel test can be made using a vise, hammer, and chisel. Clamp the welded sample upright in a vise, gripping just below the weld. Place the chisel alongside the weld (never over the weld) and force the two welded pieces apart. A hole should be pulled in one of the two pieces of metal. The weld should always be stronger than the parent metal.

If the weld is not stronger, increase the weld current setting and make sample welds until expul-

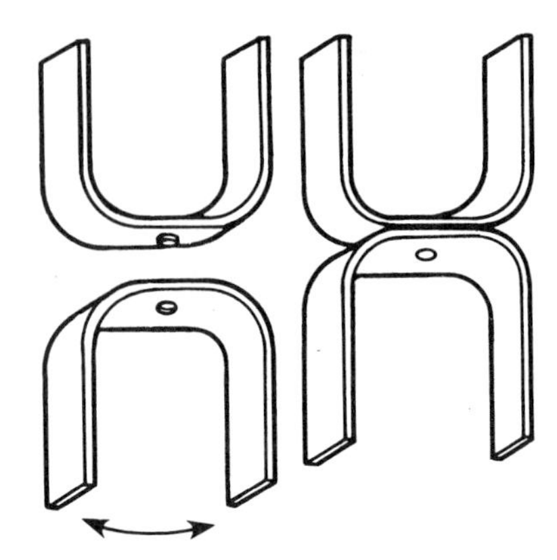

FIGURE 6-18 Weld sample

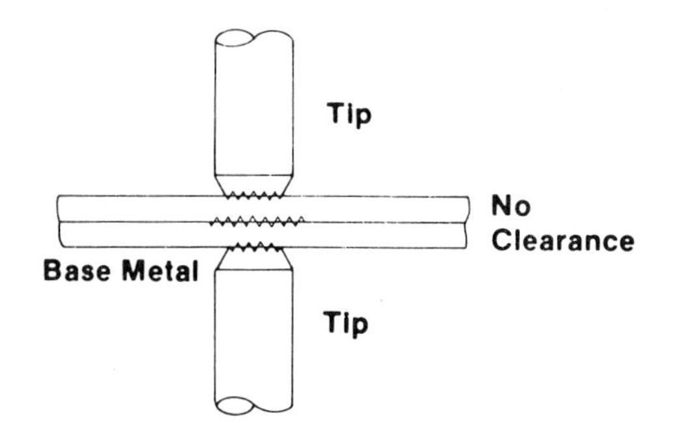

FIGURE 6-20 Condition of base metal surfaces

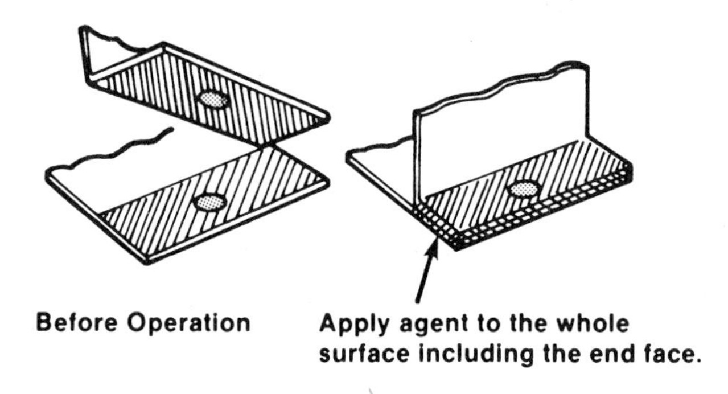

FIGURE 6-21 Areas to be protected with anticorrosion agent

sion or weld spatter is encountered. Expulsion may be visible as sparks shooting away from the weld. When expulsion is encountered, slightly reduce the weld current intensity setting and slightly increase the weld time. Make another sample weld and test it for strength. When a solid weld nugget can be peeled away leaving a hole in one of the two sheets of metal, the weld is good. The actual repair work can now be started.

FACTORS AFFECTING OPERATION

Important operational procedures that should be considered when using a squeeze-type resistance spot welder are:

- *Clearance between Welding Surfaces.* Any clearance that exists between the surfaces to be welded causes poor current flow (Figure 6-19). Even if the weld can be made without removing the gap, the weld area will become smaller, resulting in insufficient strength. Flatten the two surfaces to remove the gap and clamp them tightly before welding.
- *Surfaces to be Welded.* Paint film, rust, dust, or any other foreign matter on the metal surfaces to be welded can cause insufficient current flow and poor results, and must be removed (Figure 6-20). Coat the surfaces to be welded with an anticorrosion agent that has higher conductivity. Apply the agent uniformly, even to the end face of the panel (Figure 6-21).

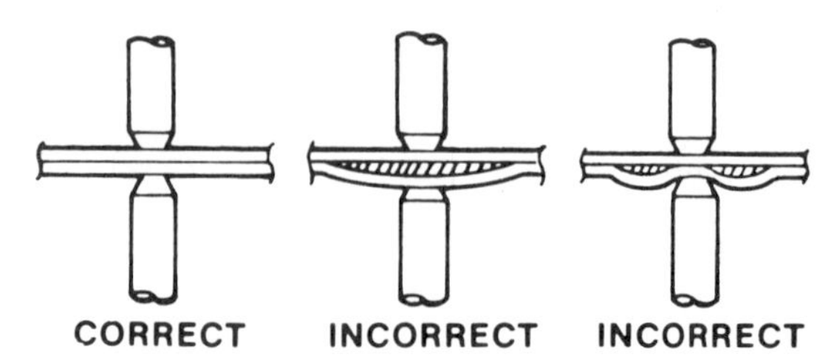

FIGURE 6-19 Correct and incorrect clearance between welding surfaces

- *Performance of Spot Welding Operations.* When performing spot welding operations, be sure to:
  —Use the direct welding method. For the portions to which direct welding cannot be applied, use MIG plug welding.
  —Apply electrodes at a right angle to the panel (Figure 6-22A). If the electrodes are not applied at right angles, the current density will be low, resulting in insufficient weld strength.
  —Spot weld twice (Figure 6-22B).
- *Number of Welding Points.* Generally, the capacity of spot welding machines available in an auto repair shop is less than that of welding machines at the factory. Accordingly, the number of points of spot welding should be increased by 30 percent in a shop (Figure 6-23).

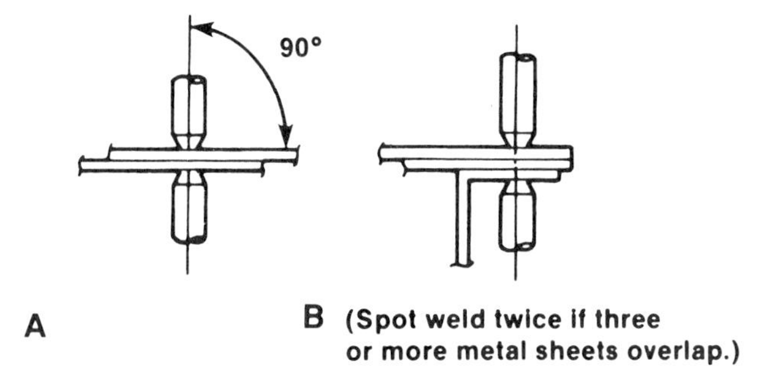

FIGURE 6-22 Precautions in performing spot welds

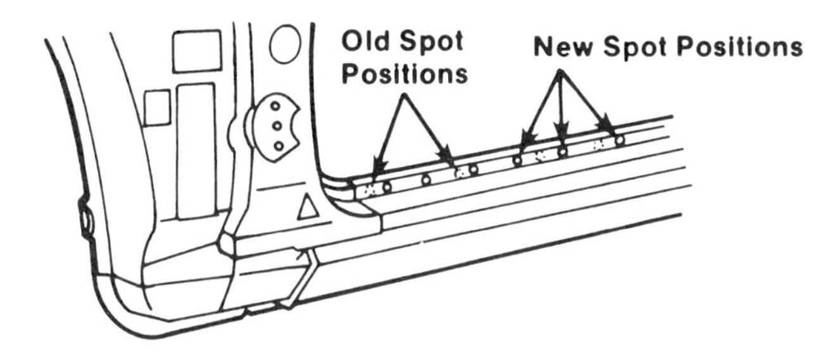

FIGURE 6-23 Number of points to spot weld

- *Minimum Welding Pitch.* The strength of individual spot welds is determined by the spot weld pitch (the distance between the welds) and edge distance (the distance from the spots to the panel edge). The bond between the panels becomes stronger as the weld pitch is shortened. However, past a certain point, the metal becomes saturated and further shortening of the pitch will not increase the strength of the bond; this is because the

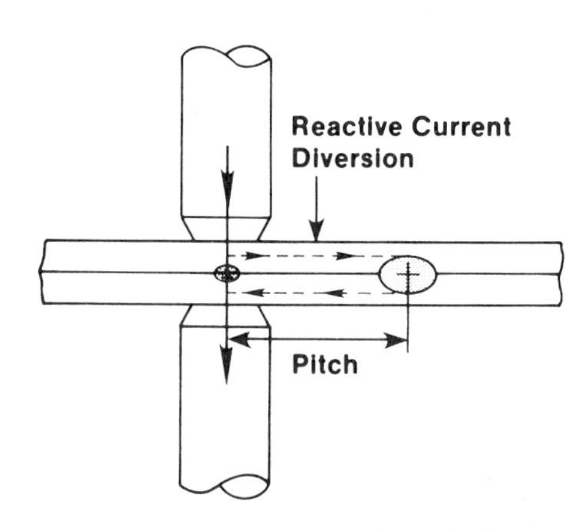

FIGURE 6-24 Minimum welding pitch

current will flow to the spots that have previously been welded. This reactive current diversion increases as the number of spot welds increases, and the diverted current does not raise the temperature at the welds (Figure 6-24). The distance of the weld pitch must be beyond the area influenced by the reactive current diversion. In general, the values presented in Table 6-1 should be observed.
- *Position of Spot Weld from the End of the Panel.* The edge distance is also determined by the position of the welding tip. Even if the spot welds are normal, the welds will not have sufficient strength if the edge distance is insufficient. When welding near the end of a panel, observe the values presented in Table 6-2.
- *Spotting Sequence.* Do not spot continuously in one direction only. This method pro-

TABLE 6-1: SPOT WELDING POSITION

Panel Thickness	Pitch (S)	Edge Distance (P)
1/64″	7/16″ or more	13/64″ or more
1/32″	9/16″ or more	13/64″ or more
Less Than 3/64″	11/16″ or more	1/4″ or more
3/64″	7/8″ or more	9/32″ or more
1/16″	1-9/64″ or more	5/16″ or more

TABLE 6-2: POSITION OF WELDING SPOT FROM THE END OF PANEL

Thickness (t)	Minimum pitch (★)
1/64"	7/16" or over
1/32"	7/16" or over
Less than 3/64"	15/32" or over
3/64"	9/16" or over
1/16"	5/8" or over
5/64"	11/16" or over

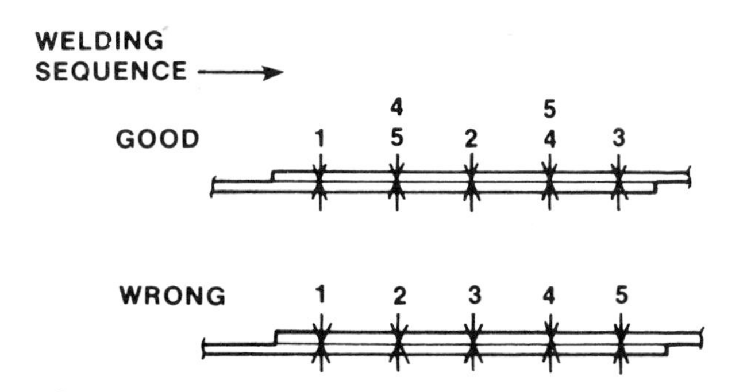

FIGURE 6-25 Proper welding sequence *(Courtesy of Nissan Motor Corp.)*

motes weak welds due to the shunt effect of the current (Figure 6-25). If the welding tips become hot and change their color, stop and allow the tips to cool.

- *Welding Corners.* Do not weld the corner radius portion (Figure 6-26); this results in concentration of stress that leads to cracks. The following locations require special consideration: the upper corner of the front and center pillars, the front upper portion of the rear fender, and the corner portion of the front and rear windows.

SPOT WELD INTEGRITY

The quality of spot welds can be determined largely through inspection and nondestructive and destructive testing. Inspection is used to judge the quality of the weld by its outward appearance and texture. The nondestructive and destructive tests are used to measure the strength of the weld.

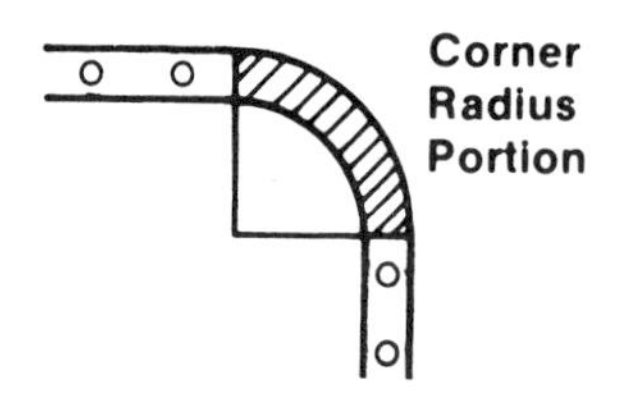

FIGURE 6-26 Proper method of welding corners *(Courtesy of Nissan Motor Corp.)*

INSPECTION

Check the finish of the weld visually and by touching. The items to check include:

- *Spot Position.* The spot weld position should be in the center of the flange, with no tip holes and no welds overriding the edge. As a rule, the old spot position should be avoided.
- *Number of Spots.* There should be at least 1.3 times the number made by the manufacturer. (For example, 1.3 times four original factory welds equals roughly five new repair welds.)
- *Pitch.* It should be a little shorter than that of the manufacturer, and the spots should be uniformly spaced. The minimum pitch should be at a distance where reactive current diversion will not occur.
- *Dents.* There should be no dents, or tip bruises, on the surfaces that exceed half the thickness of the panel.
- *Pinholes.* There should be no pinholes large enough to see.
- *Spatter.* A spatter is too large if a glove can catch on the surface when rubbed across it.

NONDESTRUCTIVE TESTING

To conduct a nondestructive test, use a chisel and hammer and proceed as follows:

1. Insert the tip of the chisel between the welded plates (Figure 6-27) and tap the end of the chisel until a clearance of 1/8 to 5/32 inch (when the plate thickness is approximately 1/32 inch) is formed between the plates. If the welded portions remain normal, it indicates that the welding has been done properly. This clearance varies depending on the location of the welded spots, the length of the flange, plate thick-

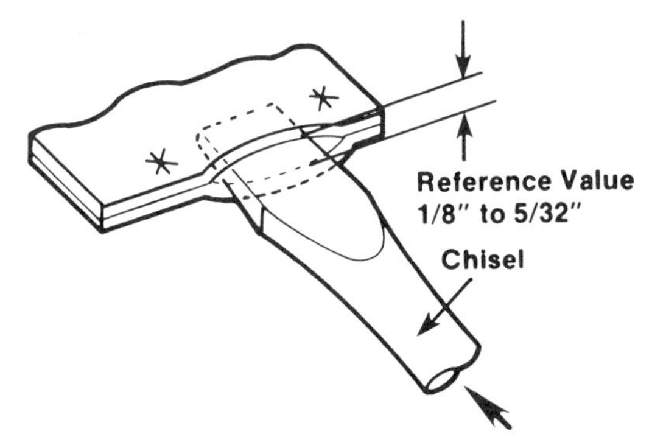

FIGURE 6-27 Performing the nondestructive test *(Courtesy of Nissan Motor Corp.)*

ness, welding pitch, and other factors. The values given here are only reference values.

2. If the thickness of the plates is not equal, the clearance between the plates must be limited to 1/16 to 5/64 inch. Any further opening of the plates can become a destructive test.
3. Be sure to repair the deformed portion of the panel after inspection.

DESTRUCTIVE TESTING

Most destructive tests require the use of sophisticated equipment, a requirement that most auto body shops are unable to meet. For this reason, simpler methods have been developed for use in body shops.

One method of destructive testing was given earlier in this chapter. Here is another method. A test piece using the same metal as the welded piece is made; it should also be the same panel thickness and be welded in the positions shown in Figure 6-28. Next, force is applied in the direction of the arrow to separate the spots. If the weld pulls out cleanly (like a cork from a bottle), it is good. It should be noted that since the weld performance cannot be exactly duplicated by this test, these results should only serve as a reference.

CAUSES OF WEAK WELDS

The three most common causes of weak welds in auto body repair are failure to properly clean the areas to be welded, use of a nonconductive (insulating) primer at the repair joint, and mushrooming of the welder tips.

The first can be corrected by cleaning all sides of the repair joint with a wire brush (a wire cat-type

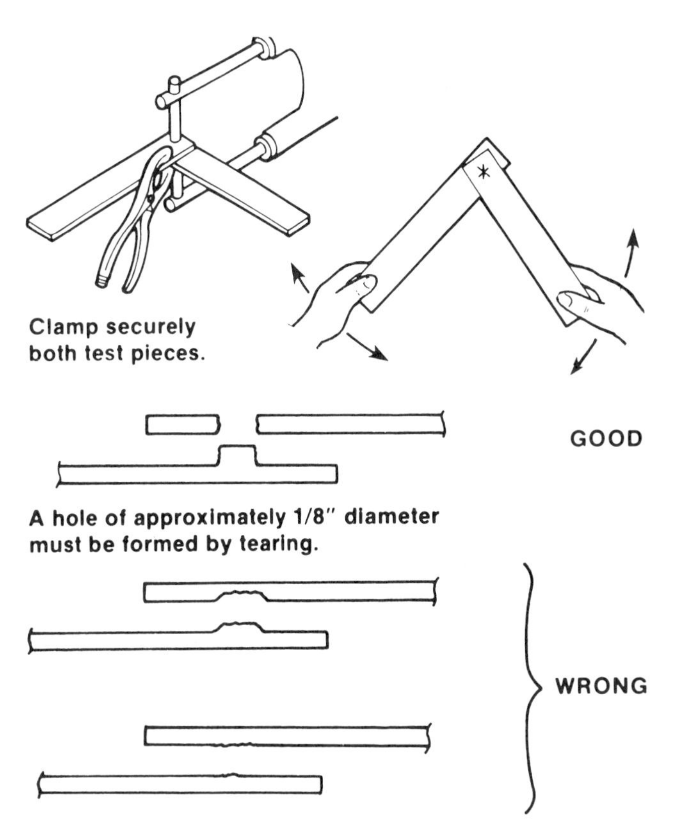

FIGURE 6-28 Performing the destructive test *(Courtesy of Nissan Motor Corp.)*

brush in a power drill will clean two sides of a flange at the same time) or a recovery-type sandblaster. Use of only electrically conductive zinc-based weld-through primers on repair joint areas will eliminate the second cause.

Mushrooming of the welder tips is easily avoided by using the correct weld time and weld current for the job. Excessive weld times do not make better welds and will essentially "cook" the tips, causing them to mushroom.

If a tip mushrooms to double its original size, its area has quadrupled. At that point, four times the weld current and four times the squeeze force previously needed with the correct size tip face are now required because the weld current and force are spread out over a greater area. Mushroomed tips can be reshaped with a file or cutter tool, which can be fitted to an ordinary power drill.

The troubleshooting guide shown in Table 6-3 lists the common resistance welding problems encountered in auto body repair.

OTHER SPOT WELDING FUNCTIONS

Although the squeeze-type welding gun is used most often in the repair shop, there are other types of

TABLE 6-3: RESISTANCE SPOT WELDING DEFECTS AND THEIR CAUSES

Possible Cause of Weld Defect		Type of Weld Defect				
		Weak Weld	Excessive Expulsion	Electrode Mushrooming	Excessive Weld Marking	No Weld
Weld Current	Low	Primary			Secondary	Primary
	High		Primary		Secondary	
Weld Time	Short	Primary				Primary
	Long		Primary	Primary	Primary	
Low weld force		Secondary	Primary	Primary		
Electrode face dia.	Small	Primary				
	Large	Primary				Secondary
Poor metal Fit-up		Primary	Primary		Primary	Primary
Too close to edge		Secondary				
Dirty metal		Secondary	Primary			Primary
Weld spacing too close		Primary				Secondary

guns used with spot welding equipment. With the proper gun attachment, the spot welder can be used as a panel spotter, stud welder, and mold rivet welder.

PANEL SPOTTING

When operating a panel spotter (Figure 6-29), the two electrode guns are placed on the nonstructural replacement panel. Figure 6-30 shows how

FIGURE 6-29 Typical panel spotter *(Courtesy of Lenco Inc.)*

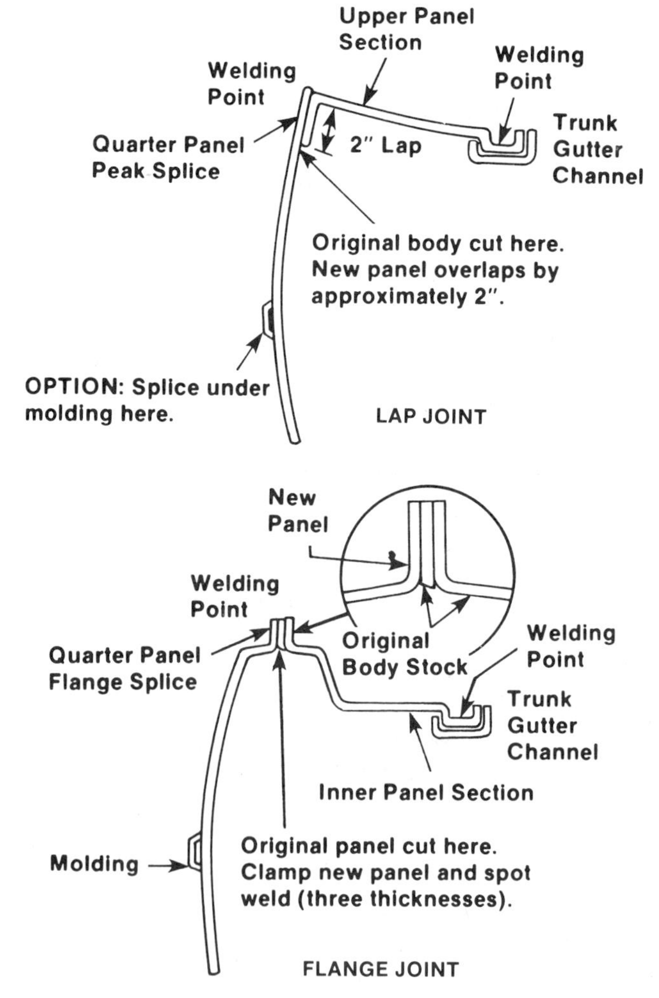

FIGURE 6-30 Panel spotting lap and flange joints

both lap and flange joints can be made with a spliced or full panel installation. After the adjustments are made following the manufacturer's directions, push both electrodes against the panel and apply moderate pressure to close any gaps. Press the weld button on the switch handle and hold it down until the welding cycle stops automatically. When finger pressure is released from the weld button, the electrodes are moved to the next welding location.

Here are some other panel spotter operational tips:

- As in any spot welding operation, thoroughly clean the surfaces along the weld seam. If a replacement panel has been primed, strip off this coat on both sides and along the weld seam with a coarse abrasive paper. If the panel has a rust preventive film coating instead of a primer, it is sufficient to wipe off both sides of the seam with a clean rag and solvent.
- Use vise grips on all flange joint and drip rail applications (Figure 6-31). Weld near the vise grip jaws where the fit-up is tight.
- A few sheet metal screws can be used on lap joints to position the panel for spot welding. Make sure that the paint has been removed from the joints.
- On long splice jobs, start in the middle of the panel and spot weld in one direction only; for example, from the middle of the panel to the door post. Start in the middle again and complete the panel welds to the taillight area. This helps in eliminating distortion.
- Remove burrs on the newly cut panel to insure good metal-to-metal contact when body pressure is applied to the electrodes. Burrs and dents cause air space between mating parts and prevent positive metal contact.

The twin electrodes of the panel spotter permit spot welding in many places where the squeeze-type has difficulty operating. In addition, the panel spotter can be converted in a squeeze type spot

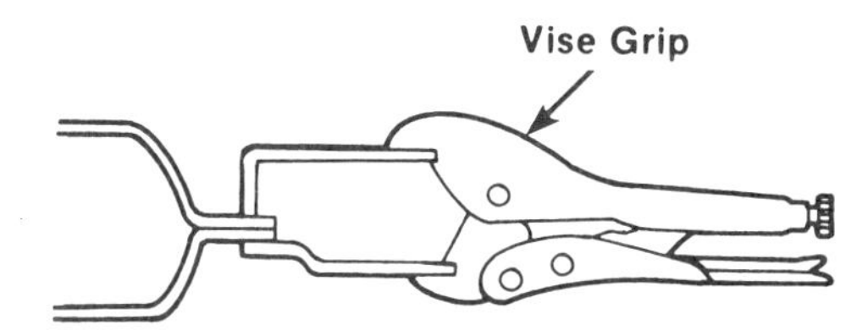

FIGURE 6-31 Using vise grips to hold a flange joint together

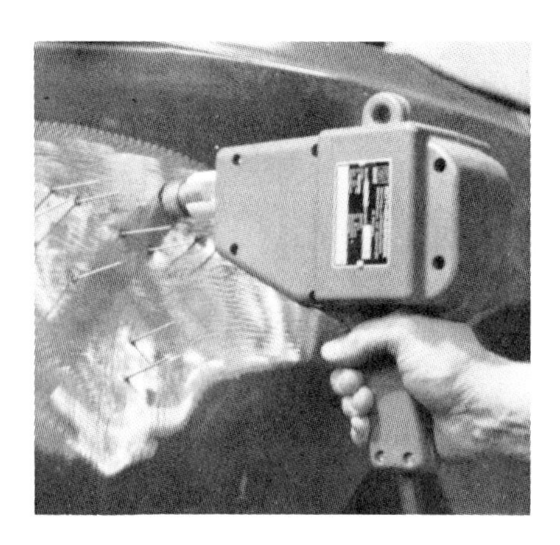

FIGURE 6-32 Typical stud welder for removing dents *(Courtesy of Henning Hansen, Inc.)*

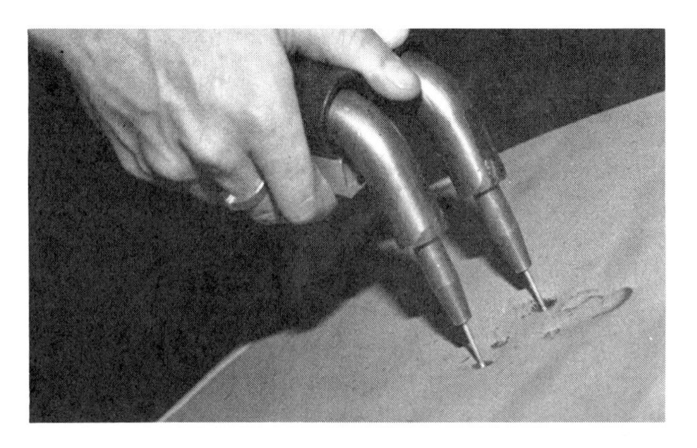

FIGURE 6-33 Using a panel spotter to install spot studs *(Courtesy of Lenco Inc.)*

welder with a gun attachment. However, this arrangement should be used only on nonstructural parts—never on structural parts.

STUD SPOT WELDING FOR DENT REMOVAL

Studs used in dent removal can be welded with a stud spot welder (Figure 6-32) or panel spotter (Figure 6-33). When using the latter, a stud pulling kit (Figure 6-34) that contains a slidehammer and other necessary items must be used.

To remove a dent with either a stud or stud spot welder, a good quality stud is necessary. The stud should offer the necessary combination of pull strength and tensile strength, while remaining extremely flexible. Flexibility allows the stud to be bent out of the way when working on adjacent studs, then bent back when needed. The importance of this stud is to minimize the heat required and, therefore, to maintain the flexibility of the steel when being applied and removed. This technique avoids drilling or

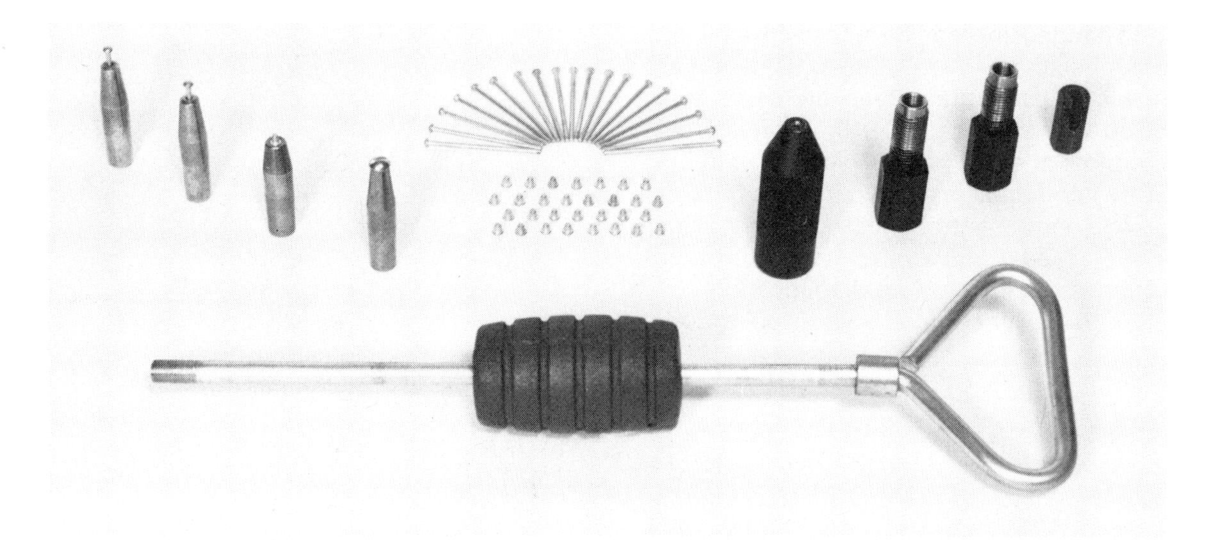

FIGURE 6-34 Typical panel spotter stud kit *(Courtesy of Lenco Inc.)*

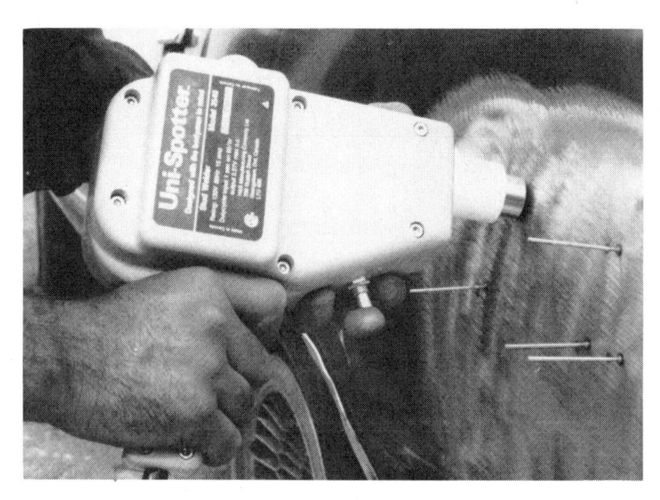

FIGURE 6-35 Welding the stud to the dented panel

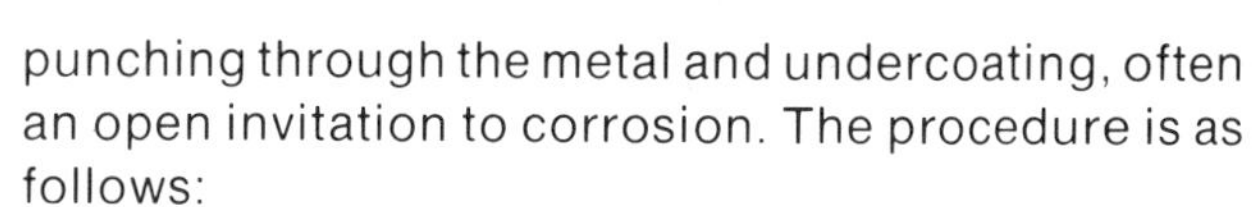

punching through the metal and undercoating, often an open invitation to corrosion. The procedure is as follows:

1. Fuse the stud to the dented area (Figure 6-35).
2. Use a dent puller or power jack to pull the stud (Figure 6-36).
3. When the dent is removed, snip off the stud with cutters and grind it flush with the panel (Figure 6-37). The entire procedure takes very little time, with no damage at all to the panel.

MOLD RIVET WELDING

Although some decorative strips are applied with adhesive, many moldings are still applied with

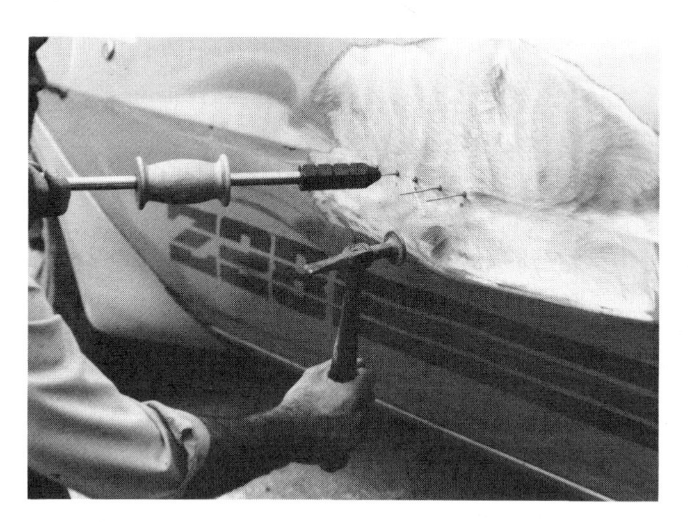

A

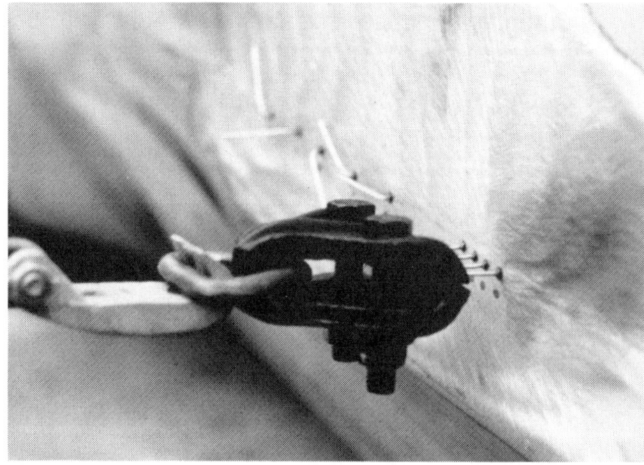

B

FIGURE 6-36 Pulling a dent (A) with a puller and hammer and (B) with power equipment

FIGURE 6-37 Grinding off the welded studs

mold rivets and clips; for example, chrome strips on rocker panels and window and vinyl roof moldings.

When patching or refinishing areas that are susceptible to moisture, salt, or high humidity, a technician is usually apprehensive about drilling holes exposing inner panels. Mold rivet welding with a stud or spot welder is a logical alternative. (As shown in Figure 6-38, one electrode has the mold rivet welding tip, thc other has the ground tip.) No holes are made; rivets can be relocated or replaced while not exposing vulnerable areas to outside elements. This one-step operation achieves a factory replica and is ideal for placing rivets on new skins. If rivets need to be removed or relocated, they require very little grinding.

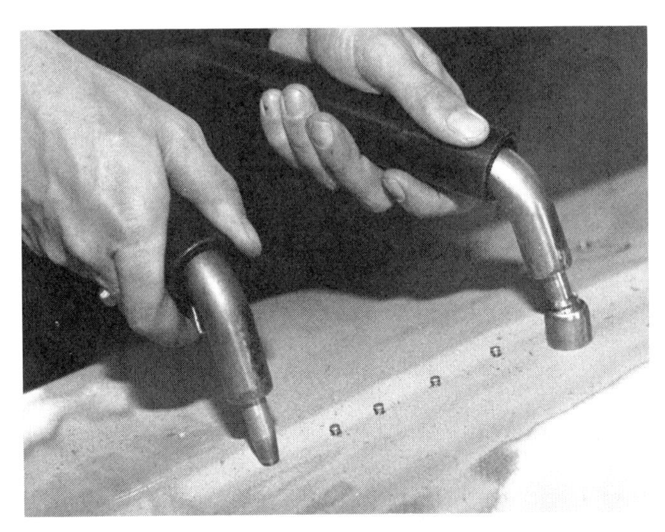

FIGURE 6-38 The panel spotter also welds molding rivets. *(Courtesy of Lenco Inc.)*

REVIEW QUESTIONS

1. Resistance spot welding depends on which of the following?
   a. heat
   b. pressure
   c. time
   d. both a and b
   e. all of the above

2. A round, flat bead of solidified metal is known as a(n) ____________.
   a. expulsion
   b. weld spatter
   c. nugget
   d. both a and b
   e. all of the above

3. When current passes through a conductor, the ____________.
   a. voltage causes heat to be generated
   b. resistance of the conductor causes heat to be generated
   c. conductivity of the conductor causes heat to be generated
   d. none of the above

4. For the purposes of resistance spot welding, where is the greatest relative resistance required?
   a. top workpiece
   b. bottom workpiece
   c. interface of the top and bottom workpieces
   d. both a and b
   e. all of the above

5. Increasing the pressure on the weld joint ____________.
   a. increases the resistance
   b. decreases the resistance
   c. has no effect on the resistance
   d. either a or b, depending on other factors

6. Which component is the heart of the resistance spot welder?
   a. transformer
   b. welder control
   c. gun
   d. none of the above

7. Which of the following squeeze mechanisms is recommended for welding unibody structures?
   a. spring
   b. pneumatic
   c. manual
   d. both a and b
   e. all of the above

8. Which of the following base metal is normally used for resistance spot welding tongs and tips?
   a. cast iron
   b. copper
   c. aluminum
   d. tungsten

9. Welder A installs a larger diameter electrode tip to increase the diameter of the spot weld. Welder B installs a smaller diameter electrode tip to increase the diameter of the spot weld. Who is right?
   a. Welder A
   b. Welder B
   c. Both A and B
   d. Neither A nor B

10. When spot welding antirust steel panels, Welder A raises the current value 10 percent; Welder B raises the current value 20 percent. Who is right?
    a. Welder A
    b. Welder B
    c. Both A and B
    d. Neither A nor B

11. When expulsion is encountered, what should be done?
    a. slightly reduce the weld current intensity setting
    b. slightly increase the weld time
    c. both a and b
    d. neither a nor b

12. How many spot welds should be made in a panel?
    a. a number equal to the original factory welds
    b. 20 percent more than the number of original factory welds
    c. 30 percent more than the number of original factory welds
    d. none of the above

13. Which of the following judges the quality of a weld by its outward appearance and texture?
    a. inspection
    b. nondestructive testing
    c. destructive testing
    d. both a and b
    e. all of the above

14. Which of the following causes weak welds?
    a. dirty welding surface
    b. nonconductive primer at the repair joint
    c. mushrooming
    d. all of the above

15. Which of the following is a spot welding function?
    a. panel spotting
    b. stud spot welding for dent removal
    c. mold rivet welding
    d. all of the above

CHAPTER SEVEN

BRAZING

Objectives

After reading this chapter, you should be able to

- Explain the difference between soft and hard brazing.
- List the advantages of brazing.
- Describe the three functions of flux in the brazing process.
- Perform the general brazing procedure.
- Explain the differences involved in brazing aluminum.
- Identify the unique characteristics of torch, induction, resistance, and diffusion brazing.
- Perform the general soldering procedure.

Brazing is still a popular method of joining metal in the automotive trade. It is a form of welding in which a nonferrous metal, whose melting point is above 840 degrees Fahrenheit but lower than that of the base metal, is melted without melting the base metal (Figure 7-1). Brazing can be compared to joining two objects with adhesives; the melted material spreads between the base metals to form a strong bond. The bending intensity against impact is less than that of the base metal, but the same as the melted material. Under no circumstances should brazing be applied to any part of an automobile other than those to which it was applied at the factory.

Brazing is classified as either soft (also known as soldering) or hard. Hard brazing can take the form of nickel or brass; the latter is used frequently on automotive bodies. Keep in mind that when the general term brazing is used, it most often refers to hard brazing.

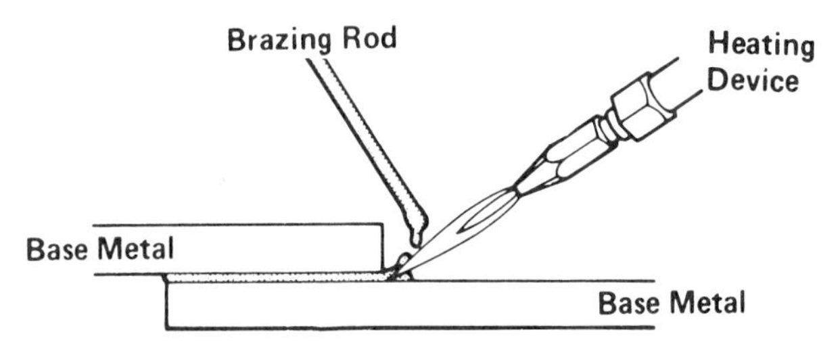

FIGURE 7-1 The basic brazing process *(Courtesy of Toyota Motor Corp.)*

ADVANTAGES OF BRAZING

There are several reasons for the widespread use of brazing in the auto body repair shop:

- Because the pieces of base metal are joined at a relatively low temperature at which the base metal does not melt, there is less risk of distortion and stress in the base metal.
- Because of its excellent flow characteristics, brazing penetrates well into narrow gaps and is therefore very convenient for filling gaps in body seams.
- Because the base metal does not melt, it is possible to join otherwise incompatible metals, such as copper to steel or stainless steel to mild steel (Figure 7-2).
- The brazing technique takes relatively little time and skill to learn.
- Parts can be easily repositioned or even disassembled later by reheating the joint.
- Parts of varying thickness can be joined without burning or overheating.

BRAZING PRINCIPLES

More often than not, brazing utilizes brass or bronze filler metal. However, silver solder, aluminum, copper, and many other materials fit just as well into the category of brazing. Under the proper conditions, the filler metal is distributed between the properly fitted surfaces of the joint to be brazed. This process is very similar to that of soldering copper pipes with wire solder.

Although many brazed joints are as strong as welded steel, there is one notable exception: a butt

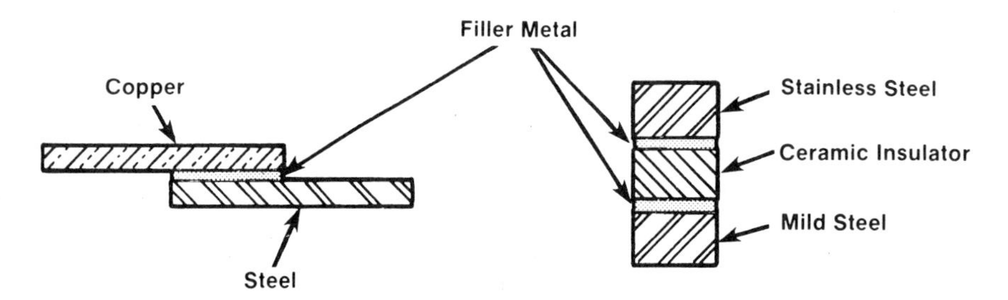

FIGURE 7-2 Joining dissimilar metals by brazing

joint on light-gauge metal. Because of its high incidence of failure, this joint should be avoided if at all possible. To repair a brazed butt joint, the brass deposited on the base metal must be completely removed and rewelded with steel filler rod. Many inexperienced welders fail to do this when trying to repair old fenders with cracks caused by stress or vibrations. Because of the constant expansion and contraction due to extreme heat, brazed butt joints should also be avoided when repairing the exhaust system; use the lap joint technique instead (Figure 7-3).

When working with sheet metal or light-gauge metal, heat can cause warping because of the large surface area of the panels. Because the metal is very thin, it reacts to any significant amount of heat; the temperature for mild steel oxyacetylene welding is relatively high, in the 2200- to 2400-degree-Fahrenheit range. This is why the brazing method is preferred over standard welding methods. The controlled temperature of brazing is sufficiently low as to practically eliminate warping.

Automobile assembly plants use arc brazing to join the roof and quarter panels (Figure 7-4). Arc brazing utilizes essentially the same principles as MIG welding, the major difference is that argon is used as the shielding gas (Figure 7-5). Since the amount of heat applied to the base metal is low, the chances of overheating are greatly minimized and there is little distortion or warpage of the base metal. When compared to the previously used methods of depositing brass on the base metal, arc brazing shortens both the time for making the weld and the time for finishing. Also, there is no danger of lead poisoning.

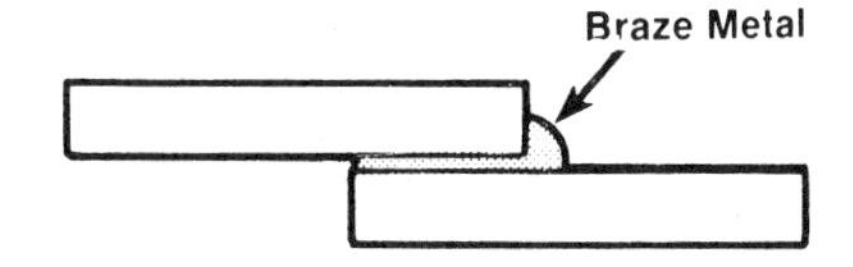

FIGURE 7-3 Brazed lap joint

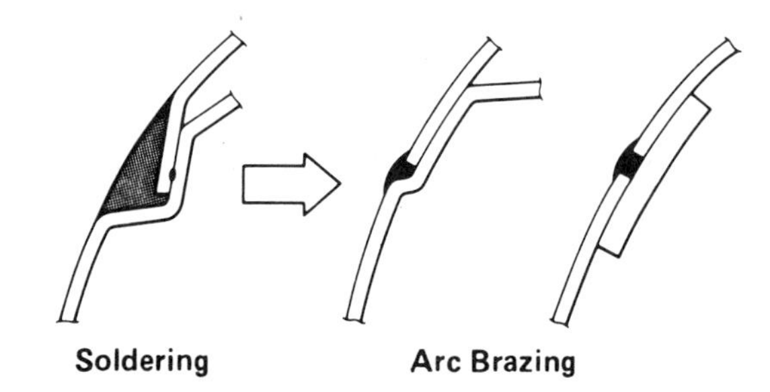

FIGURE 7-4 Typical body construction using arc brazing *(Courtesy of Toyota Motor Corp.)*

In today's body shop, the brazing equipment is usually about the same as for oxyacetylene welding: an oxyacetylene torch, filler rods, flux welding goggles, gloves, and a torch lighter. Although an oxyacetylene torch can be used in soft brazing, it is best to use one specifically designed for soldering.

BRAZING RODS

To achieve good flow characteristics, melting temperature, compatibility with the base metal, and strength, brazing rods are made of two or more metals that form an alloy (Table 7-1). There are many different types of rods available; the most common is the brass rod containing 70 percent copper and 30 percent zinc. This ratio varies considerably according to the strength of the joint desired and the necessary temperature. Brazing rods can also contain small amounts of tin, manganese, iron, lead, or silver. A lower melting temperature can be obtained by adding tin and silver. Nickel bronze rods are made of copper and tin alloy, with other compounds added for strength.

Some manufacturers of brass or bronze rods make brazing rods to specifications. Brass and bronze are available either plain or flux covered; common sizes range from 1/16 inch to 1/8 inch in diameter and larger, and lengths of 18 or 36 inches. Some technicians prefer to use a bare brazing rod; a flux-coated rod can produce too much molten flux in the brazing area that can interfere with the proper

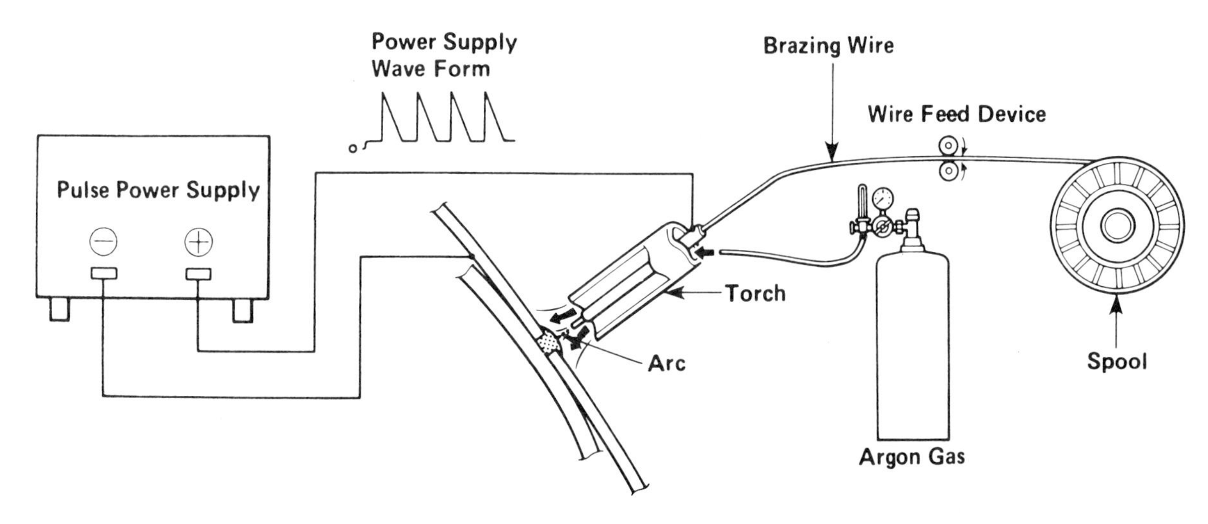

FIGURE 7-5 The arc brazing process *(Courtesy of Toyota Motor Corp.)*

TABLE 7-1: BRAZING ROD INGREDIENTS

Brazing Material	Main Ingredients
Brass brazing metal	Copper, Zinc
Silver brazing metal	Silver, Copper
Phosphor copper brazing metal	Copper, Phosphorus
Aluminum brazing metal	Aluminum, Silicon
Nickel brazing metal	Nickel, Chrome

flow of the brass. Using a powdered flux is an even better way to control the amount of flux.

FLUXES

Fluxes used in brazing have three major functions:

1. Remove any oxides that form as a result of heating the parts
2. Promote wetting
3. Aid in capillary action

When heated to its working temperature, the flux must be able to flow through the tap provided at the joint. As it does, the flux reduces oxides, allowing the molten filler metal to be pulled in behind it (Figure 7-6). After the joint is complete, the flux residue should be easily removable.

Fluxes are available in many forms including solids, powders, pastes, liquids, sheets, rings, and washers. They can be mixed with the filler metal, inside the filler metal, or on the outside of the filler metal. Sheets, rings, and washers can be placed within the joints of an assembly before heating so that total coverage inside the joint will be assured.

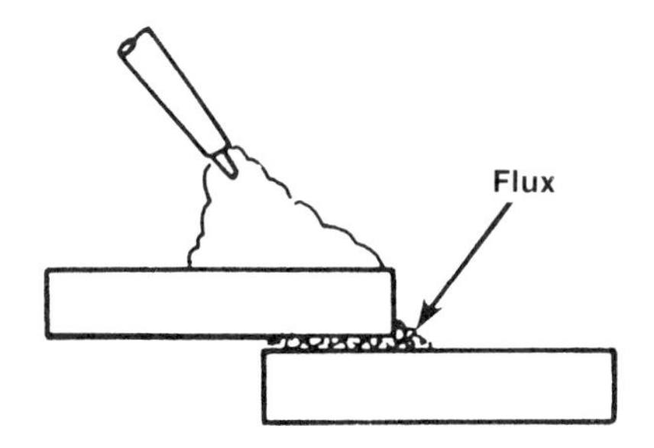

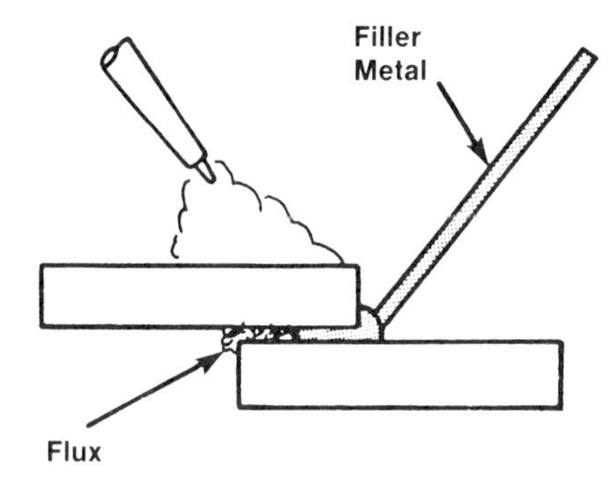

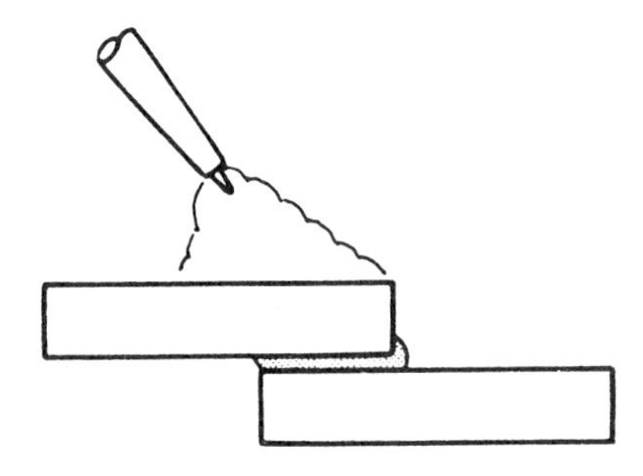

FIGURE 7-6 Flux flowing into a joint reduces oxides and causes the filler metal to flow behind it.

Paste, powders, and liquids can be brushed on the joint either before or after the material is heated. Paste and powders can also be applied to the end of the brazing rod by heating it and dipping it in the flux. Most powders can be made into paste, which can be thinned by adding distilled water or alcohol. Never use tap water because it contains minerals that weaken the flux.

Some liquid fluxes can also be added to the gas when using an oxyacetylene torch for soldering or brazing. The flux is picked up by the fuel gas as it is bubbled through the flux container and carried to the torch, where it becomes part of the flame.

Flux and filler metal combinations are convenient and easy to use. It is often necessary to stock more than one type of flux/filler metal combination for each type of filler metal needed, and these combinations are more expensive than other types of flux. In cases where the flux covers the outside of the filler metal, it might be chipped off during storage. Keep in mind that using more flux than is needed to perform the soldering or brazing job can result in excessive amounts of flux being trapped in the joint, weakening of the joint, and future joint failure.

Fluxing Action

An important point to keep in mind is that the use of fluxes does not eliminate the need for good joint preparation. Flux will not remove heavy oxide layers, oils, dirt, or any other surface contaminants. Chemical compounds such as muriatic acid (hydrochloric acid), sal ammoniac (ammonium chloride), or rosin are used for soldering fluxes and borates; fluorides, chlorides, boric acids, and alkalies are used in brazing fluxes. These chemical compounds react to dissolve, absorb, or break up surface oxides that are formed as a result of heating. They must be active through the entire temperature range of the solder or braze filler metal. The chemicals in the flux react with the oxides as either acids or bases.

The activity or speed of a flux can be greatly affected by temperature. As the parts are heated to the soldering or brazing temperature, the flux becomes more active. Some fluxes are completely inactive at room temperature, and the majority have a temperature range at which they are most effective. Care should be taken to avoid overheating fluxes. If this is allowed to happen, the flux will stop working and thus prevent the solder or braze from wetting and flowing. In the event of overheating, stop and clean off the flux before continuing.

Fluxes that are active at room temperature must be neutralized or washed off after the job is completed. If these fluxes are left on the joint, premature failure can result due to flux-induced corrosion. Fluxes that are inactive at room temperature do not have to be cleaned off. However, if the part is to be finished in any way, the flux must be completely removed.

CAUTION: Some fluxes give off a very toxic fume when heated. Use them in a well-ventilated place. When removing flux residue, wear safety glasses to prevent eye infection or other damage.

Flux not only removes harmful oxide film, it also prevents the metal surface from reoxidizing. The importance of flux is illustrated in Figure 7-7. If brazing material is melted over a surface with oxidized film and other foreign matter adhering to it, there will not be adequate bonding to the base metal. This is because the surface tension causes the brazing material to ball up and not stick.

BRAZING JOINT STRENGTH

Since the strength of the brazing material is less than that of the base metal, the shape and the clearance of the joint are extremely important. Figure 7-8 shows basic brazing joints. Joint strength is dependent on the surface area of the pieces to be joined, so make the joint overlap as wide as possible.

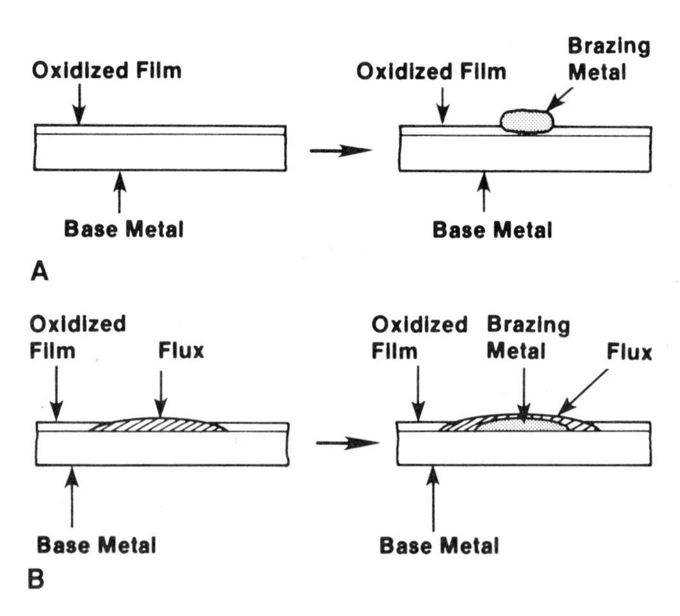

FIGURE 7-7 Comparing the effects on the brazing metal (A) when flux is not used and (B) when flux is used *(Courtesy of Toyota Motor Corp.)*

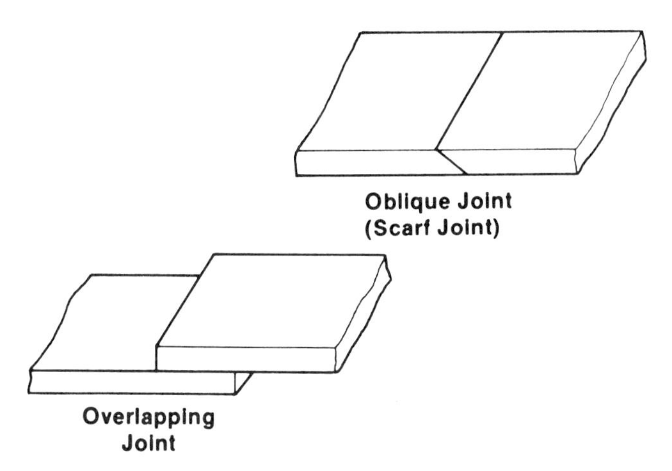

FIGURE 7-8 Two basic brazing joints *(Courtesy of Toyota Motor Corp.)*

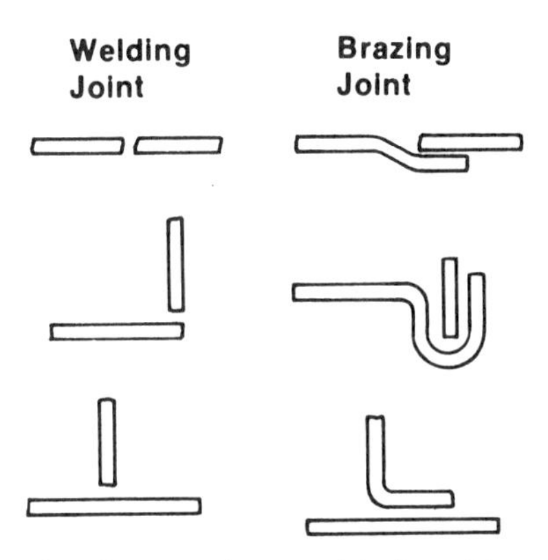

FIGURE 7-9 Comparing welded and brazed joints *(Courtesy of Toyota Motor Corp.)*

Even when the items being joined are of the same material, the brazed surface area must be larger than that of a welded joint (Figure 7-9). As a general rule, the overlapping portion must be at least three times wider than the panel thickness.

The spacing between the parts being joined greatly affects the tensile strength of the finished part. Table 7-2 lists the spacing requirements at the joining temperature for the most common alloys. As the parts are heated, the initial space might increase or decrease, depending upon the joint design and fixturing. The changes due to expansion can be calculated, but trial and error also works. Parts that are 1/4-inch thick or less should not be considered for brazing or soldering.

TABLE 7-2: BRAZING JOINT CLEARANCES

Filler Metal	Joint Spacing	
	Inches	**Millimeters**
BAlSi	0.006-0.025	0.15-0.61
BAg	0.002-0.005	0.05-0.12
BAu	0.002-0.005	0.05-0.12
BCuP	0.001-0.005	0.03-0.12
BCuZn	0.002-0.005	0.05-0.12
BNi	0.002-0.005	0.05-0.12

Some joints are designed so that the flux and filler metal can be preplaced. When this is possible, visually checking for filler metal around the outside of the joint is easy. Evidence of filler metal around the outside is a good indication of an acceptable joint.

As mentioned earlier, joint preparation is very important to a successful soldered or brazed part. The surface must be free of all oil, dirt, paint, oxides, or any other contaminants. The surface can be either mechanically cleaned (wire brushed, sanded, ground, scraped, or filed) or chemically cleaned (with an acid, alkaline, or salt bath). Soldering or brazing should start as soon as possible after the parts are cleaned to prevent any additional contamination of the joint.

BRAZING OPERATIONS

In discussing the general brazing technique, the procedure will be geared toward auto body work on 20- to 24-gauge sheet metal. It is wise to remember that the great majority of brazing jobs will not take place under ideal conditions; the metal will likely be covered with paint, rust, undercoating, galvanizing, and other assorted materials. And while it might be very tempting not to bother stripping the surface, the time and effort should be taken to ensure a high-quality brazing job.

The general brazing procedure is as follows:

1. Clean the surface thoroughly with a wire brush, grinder, or scraper.
2. Apply flux uniformly to the brazing surface. If a brazing rod with flux is being used (Figure 7-10), this step can be omitted.
3. Adjust the flame of the welding torch to a slight carburizing flame. Use the melting of the flux to estimate the proper temperature of the brazing material.
4. Heat the base metal to a uniform temperature capable of accepting the brazing material (Figure 7-11).
5. When the base metal has reached the proper temperature, melt the brazing material onto the base metal (Figure 7-12), letting it flow naturally.

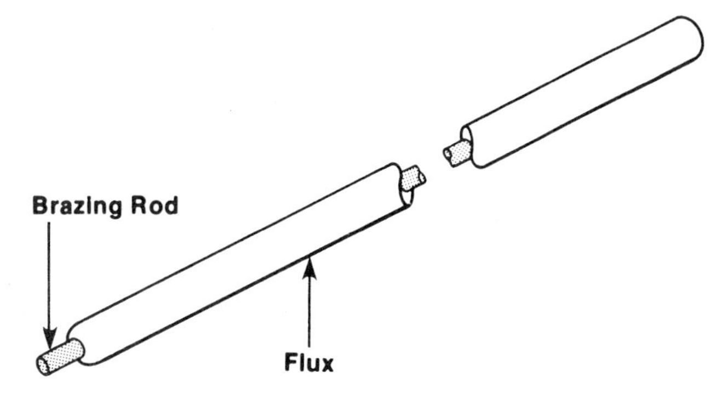

FIGURE 7-10 Brazing rod with a flux coating

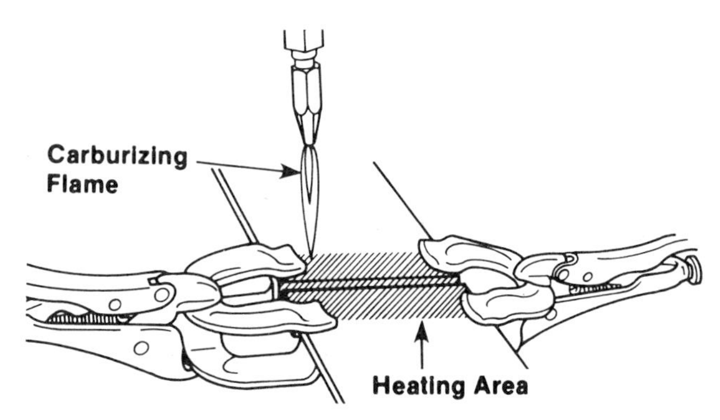

FIGURE 7-11 Base metal heating *(Courtesy of Toyota Motor Corp.)*

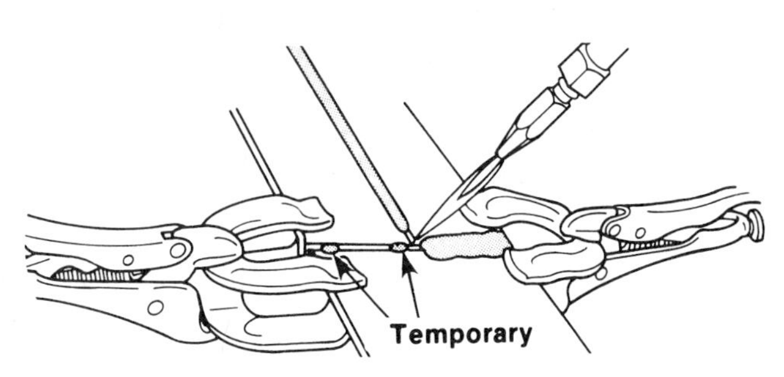

FIGURE 7-12 Base metal brazing operation *(Courtesy of Toyota Motor Corp.)*

6. Stop heating when the brazing material has flowed into the gaps of the base metal. If the surface temperature of the base metal is allowed to get too high, the flux will not clean it properly; the result will be a poor brazing bond and inferior joint strength.

The following additional precautions should be taken when brazing:

- The brazing temperature must be higher than the melting point of brass (or whatever brazing material is being used) by 50 to 190 degrees Fahrenheit.
- The size of the torch tip must be slightly larger than the thickness of the panel.
- Preheat the panel until it becomes bright red; this will promote more effective depositing of the brazing filler.
- Secure the panel to prevent the base metal from moving and the brazing zone from breaking.
- Heat the portion to be welded evenly, without melting the base metal.
- Make sure the torch feed angle is vertical to the base metal and tilted 10 to 15 degrees in the welding direction.
- Brazing time must be as short as possible to prevent the joint strength from being reduced.
- Avoid brazing the same spot twice.

Once the brazed portion has cooled sufficiently, rinse off the remaining flux sediment with water and scrub the surface with a stiff wire brush. Baked-on flux can be removed with a sander. If all of the flux is not removed, the paint will not adhere properly, and corrosion and cracks might form in the joint.

ALUMINUM BRAZING

Aluminum can be brazed with an oxyacetylene torch in a way that is very similar to iron. It shares with iron a great affinity for oxygen. Once aluminum is cleaned, it immediately reacts with oxygen to produce a thin, glass-like film of aluminum oxide on the surface. Immediately before brazing aluminum, the oxides must be removed by scraping, wire brushing, sandpapering (Figure 7-13), or other means. The oxides can also be removed chemically by the use of a flux.

An important factor that makes aluminum different from steel is its high heat conductivity; aluminum conducts heat almost five times that of steel or iron depending on the type of alloy. The result is that aluminum requires more heat to achieve its proper brazing temperature.

For a good look at how aluminum reacts to the heating process, take a piece of steel sheet metal and a piece of aluminum of the same size and thickness and heat one corner of both pieces for a few seconds. On the pieces of sheet metal, the heat will be concentrated at the corner, while the heat will travel across the aluminum very rapidly. Another key difference is the color change. The steel will gradually turn redder, then a bright red until it melts; the aluminum will not change color until the melting point is reached, and then it will simply fall away. A good method to determine the proper temperature

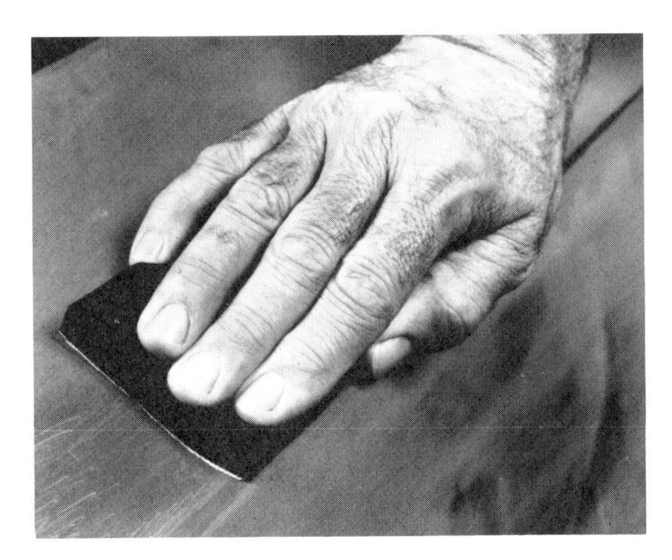

FIGURE 7-13 Sandpaper is effective in removing oxides from aluminum.

to braze aluminum is by using a special temperature indicating crayon that will melt or become liquid at a certain temperature.

When brazing aluminum, use a flux to obtain the proper adhesion. Apply the flux to the surface or, if necessary, to the brazing rod. Applying the flux directly to the rod is especially useful if the rod has been in the shop for a long time. There are many types of aluminum brazing rods available, the most common ones being the 1100 and the 4043. Some manufacturers produce a rod with a low melting point that can be used for general aluminum repairs.

When applied to the surface of the repair area, the flux will dry out as the temperature rises, then turn to a glossy liquid. Rub or scrape the aluminum brazing rod in the heated area at 3- to 5-second intervals, removing the rod from the flame each time. When the right temperature is reached, the rod will flow. Keep the torch moving on and off the surface to avoid overheating. Do not melt the rod with the torch before the surface is hot enough. The torch should be held at a flatter angle than when working with steel.

The hardest part of brazing aluminum is its tendency to melt away the edges. By moving the torch continually, there is less chance of this happening. Aluminum is more difficult to braze than brass, but in time it can be mastered.

Cleaning the area after brazing is very important because the flux is corrosive. If left on the aluminum, the flux will cause oxidation or corrosion. Also, when the component is painted, any remaining flux residue will cause the paint to peel (Figure 7-14). There are several different chemicals used by manufacturers to remove flux, but boiling hot water and a scrubbing brush will remove most of it. Finally, rinse the aluminum with cold water.

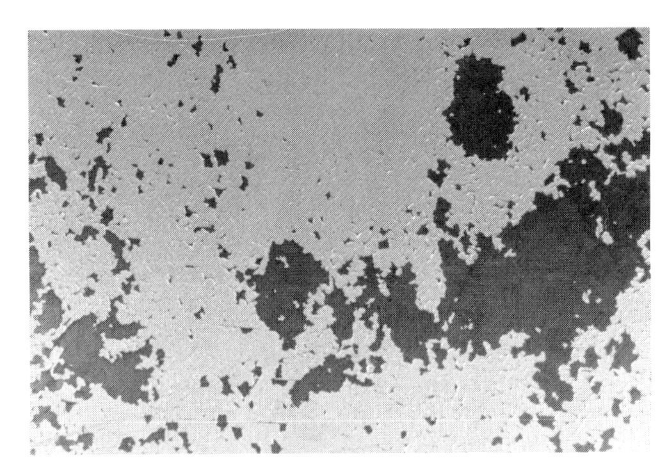

FIGURE 7-14 Flux allowed to remain on the aluminum will lead to paint failure in the form of peeling.

CAUTION: Aluminum flux might give off a toxic fume when heated. Always work in a well-ventilated area. If using powdered flux, be very careful to avoid getting it in the eyes.

SPECIAL BRAZING PRACTICES

A quality brazing job depends on several factors, as discussed earlier. Joint design, filler metal selection, uniform heating, and adequate protective covering are all extremely important. Each of the brazing methods that follow satisfactorily meets these requirements.

Torch Brazing

In this process, the parts to be brazed are heated with an oxyfuel gas torch and one of several gas fuels (Figure 7-15). While torch brazing is almost always manual in the auto body shop, automated machines are available as well. The most common application of torch brazing is in lap joining thin sections of steel ranging from 0.01 to 0.25 inch. In this case, the process is very economical, but its efficiency drops dramatically with thicker sections.

Induction Brazing

Induction brazing obtains the necessary heat from the resistance of a high-frequency current induced in the parts to be brazed; the workpiece does not become part of the electrical circuit. The current is passed through a copper tubing working coil spe-

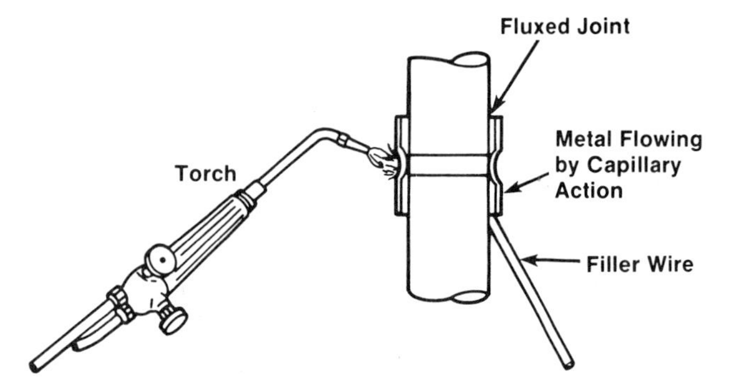

FIGURE 7-15 The torch brazing process

cially designed to fit around the workpiece (Figure 7-16). This coil does not touch the workpiece, but is joined to it by an electrical field. Heating of the workpiece usually occurs within 10 to 60 seconds.

The brazing filler metal is usually preplaced, and careful joint design and coil setup is necessary to assure that the surfaces of all joint members reach the brazing temperature at the same time. Flux and shielding atmospheres are the two methods of providing protection, with flux being the most common. The thickness of the sheets that can be brazed using this method does not normally exceed 1/8 inch, and production speeds are high.

Resistance Brazing

Resistance brazing obtains its heat from the resistance to the flow of an electrical current passed through the pieces being brazed. Unlike induction brazing, the workpiece does become part of the electric circuit. Resistance welding equipment can be used for resistance brazing, but it must be set at a lower power output. Specially designed resistance brazing machines are also available.

The pieces to be brazed are held between two electrodes while the correct current and pressure are applied. Pressure is maintained until the weld has solidified. Filler metal is normally preplaced in the joints but can also be face fed. Fluxes are usually employed to provide shielding, but shielding gases can also do the job. If flux is used, it must be electrically conductive.

The thickness of the base metals can range between 0.005 and 0.50 inch. The lap joint is the most common joint design used in resistance brazing. With the proper equipment the speed can be very fast. Resistance brazing is ideally suited for special application joints where the goal is to confine the heat to a very localized area and avoid overheating the surrounding parts.

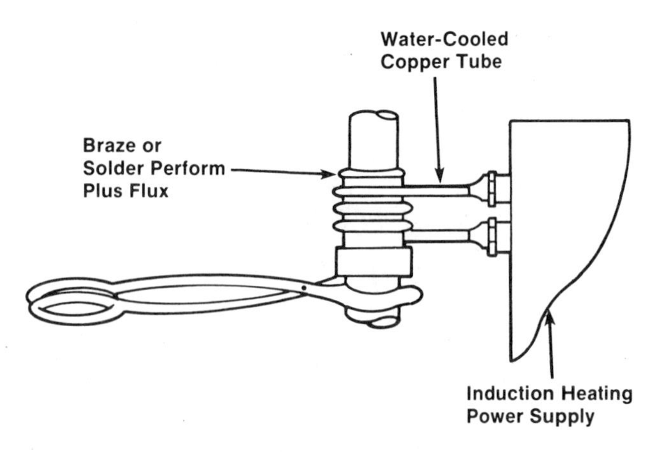

FIGURE 7-16 The induction brazing process

Diffusion Brazing

The principles of diffusion brazing closely resemble those of diffusion welding. The brazed joint is formed by holding the pieces together under pressure at a suitable temperature until diffusion occurs between the base metal and the filler metal. The resulting joint has a composition uniquely different from either the base or filler metal. Diffusion brazing produces joints consideraly stronger than any other brazing process. The joint will remelt at temperatures approaching the melting point of the base metal.

The normal thickness range of the base metals ranges up to about 2 inches, but much heavier pieces can be diffusion brazed because thickness has very little effect on this process. Many parts that are difficult to work with using other processes can be diffusion brazed. Both butt and lap joints with superior mechanical properties are produced, with the parts usually fixtured mechanically or tack welded together.

SOLDERING

Soldering defines a group of joining processes that are closely related to the brazing operations discussed previously. Like brazing, soldering unites the metals by heating them to a suitable temperature below their melting points and using a nonferrous filler metal (Figure 7-17). In the case of soldering, the filler has a melting temperature below 840 degrees Fahrenheit—this is the major difference between soldering and brazing.

Once the filler or solder is melted, it is distributed between the fitted surfaces of the joint by capillary action. Like brazing, successful soldering involves closely fitting the parts, cleaning and fluxing, and proper application of the heat and filler

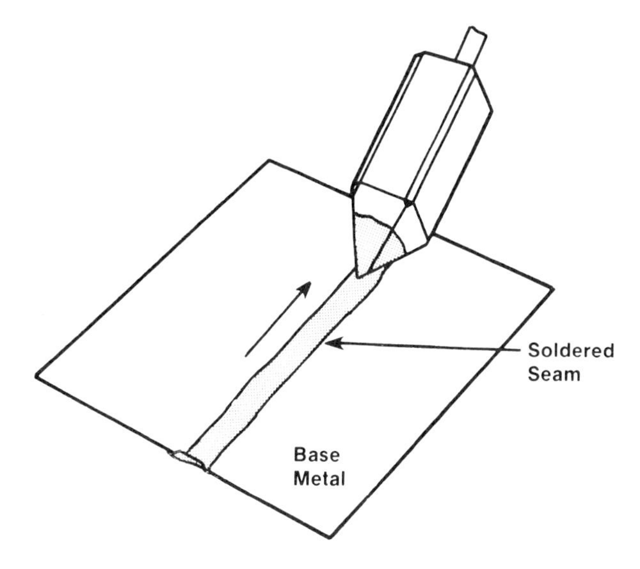

FIGURE 7-17 The soldering process

metal. Solder fillers are alloys of zinc and tin, with ratios ranging from 50/50 to 95 tin/5 zinc. They can be used to join similar and dissimilar metals in a wide range of thicknesses. Soldered joints have excellent heat and electrical conductivity, as well as minimum warpage.

Soldering is not used to reinforce panel joints; it is not as strong as brazing or welding. Therefore, it should never be used as a substitute for welding joints and seams. It is only used as a filler for final finishing, such as in leveling the panel surface and correcting the surface of the welded joints. Probably the biggest advantage of soldering is its outstanding sealing ability. Before attempting to solder a joint, be sure to remove any paint, rust, oil, and other foreign substances from the built-up area and the vicinity.

SOLDER FLUXES

The most common solder flux is a mixture of zinc chloride (71 percent) and ammonium chloride (29 percent). This is known as an acid flux and has corroding tendencies. It must, therefore, be thoroughly cleaned from the joint after the soldering operation. Some fluxes contain predominantly zinc chloride, with just a trace of ammonium chloride or stannous chloride. These fluxes are usually dissolved in water, although some are available in paste form.

As solder fluxes are heated, they melt and partially decompose to form a hydrochloric acid. All fluxes have some moisture content; the initial boiling of a flux as it is heated is the moisture vaporizing. The acid dissolves the oxides and permits the solder metal to adhere to the base metal. The flux fuses as it cools and covers the solder to prevent or minimize further oxidation.

Be sure the flux container is sealed when not in use in order to keep the flux clean. Remove only that quantity of flux needed for the particular job and apply it with a clean brush or paddle. A soldered joint should be completed as soon as possible once the flux is heated, because delaying will cause harmful flux salts to form.

GENERAL SOLDERING PROCEDURE

The metals to be soldered must be thoroughly cleaned. A chemical cleaner is good to use (Figure 7-18), provided it is rinsed and dried off completely. The metals can also be cleaned by filing or wire brushing; steel wool also works well. After the surface has been cleaned, proceed as follows:

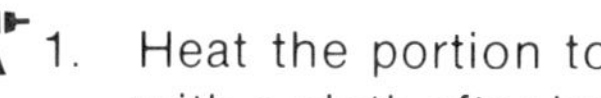

1. Heat the portion to be soldered. Wipe it with a cloth after heating.
2. Stir the solder paste well. Apply it with a brush to an area 1 to 1-1/2 inches larger than the built-up area.
3. Heat it from a distance.
4. Wipe the solder paste from the center to the outside.
5. Make sure the soldered portion is silver gray. If it is bluish, it has overheated.
6. If any spot is not adequately soldered, reapply the paste for additional soldering.

FIGURE 7-18 Using a chemical cleaner prior to soldering

When soldering, keep the following additional points in mind:

- It is wise to use a special torch for soldering. If a gas welding torch is used, the oxygen and acetylene gas pressures must be 4.3 to 5.0 psi.
- To maintain the appropriate temperature, move the torch so that the flame evenly heats the entire portion to be soldered. When the solder begins to melt, remove the flame and finish with a spatula.
- When additional solder is required, the previously built-up solder must be reheated.

If either piece of metal moves while it is cooling from its flow temperature to its solidification temperature, the solder will form cracks and likely fail. It is therefore necessary to firmly support the base metals with a fixture, clamps, or the like to make sure they do not move while the soldering operation is in progress.

SHOP TALK

The use of plastic is not recommended for automotive soldering because it is prone to cracking and rusting, requires a long time for drying, does not finish smoothly, and cannot be recycled as ordinary solder can.

A common soldering error is to use too much solder; this is wasteful because it does not add any strength to the joint. If too much solder is used, a neat looking joint can still be achieved by wiping off the excess molten solder using a clean, thick cloth. The solder should be wiped away while it is at its flowing temperature rather than at its melting temperature. It is very important to avoid overheating the metal.

A popular use for soldering is to build up irregular surfaces to secure a smooth finish. This method is used extensively in auto body repair work. The irregular surface is mechanically cleaned, then chemically cleaned with a weak acid. A wood paddle is sometimes used to apply the solder to the torch-heated surfaces. Figure 7-19 shows a joint being filled in by soldering. The solder is then dressed by filing and sanding to match the sheet metal surface.

COPPER SOLDERING

Soldering copper produces a very concentrated heat, while at the same time acting as a means of

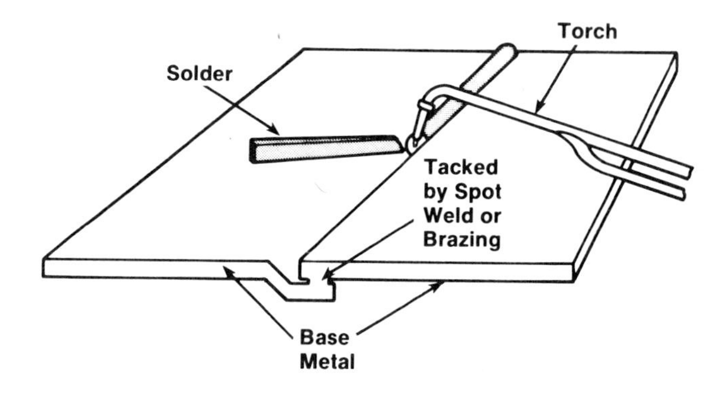

FIGURE 7-19 Filling a joint using the soldering method

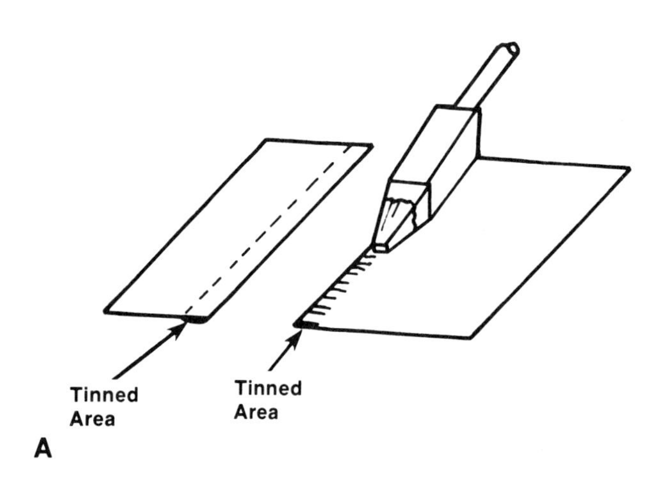

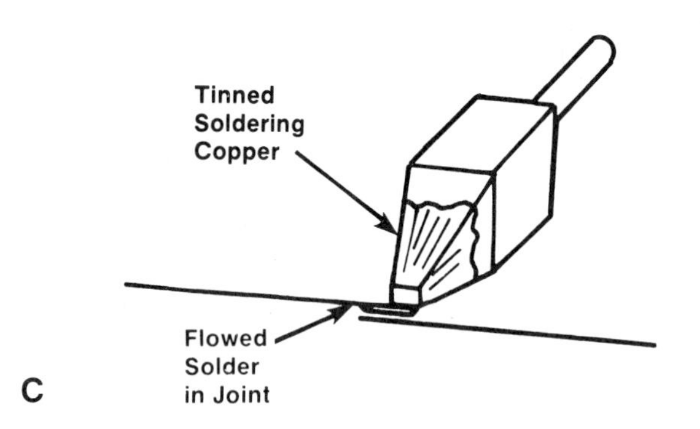

FIGURE 7-20 Steps required to make a sweated soldered joint: (A) Solder is applied in a thin film; (B) metal surfaces are lapped to form the joint; (C) soldering copper is moved along the joint to flow the solder on the previously tinned surfaces.

spreading the solder as it adheres to the base metals. In some parts of the United States, it is known as soldering iron. It has the disadvantage of requiring reheating quite frequently. However, electrically heated soldering coppers and internal flame heated coppers do not have this problem. The tip of the soldering copper must be kept clean at all times.

One of the best applications of soldering copper is to use it for sweating a soldered joint. This means that the two metals to be soldered together are lapped at their joint with a previously applied film of solder on the two contacting surfaces. (This application of film is known as *tinning.*) These edges are then lapped, and the copper is slowly moved along the seam, permitting the heat from the soldering copper to penetrate through the metal and fuse the solder films together, as shown in Figure 7-20. The resulting joint is strong and neat; this method is recommended when a high-quality, leakproof joint is desired and for difficult-to-solder joints.

REVIEW QUESTIONS

1. Which of the following is not characteristic of brazing?
   a. relatively high strength
   b. can join parts of varying thickness
   c. greater risk of distortion in the base metal
   d. can join otherwise incompatible metals

2. What type of brazed joint should be avoided because of its high incidence of failure?
   a. butt joint on light-gauge metal
   b. butt joint on heavy-gauge metal
   c. lap joint on light-gauge metal
   d. lap joint on heavy-gauge metal

3. The most common brazing rod contains ______________.
   a. 70 percent copper and 30 percent zinc
   b. 30 percent copper and 70 percent zinc
   c. 100 percent copper
   d. 100 percent zinc

4. Brazing flux is available in what form?
   a. liquid
   b. powder
   c. solid
   d. all of the above

5. Powdered fluxes can be thinned by adding ______________.
   a. alcohol
   b. distilled water
   c. both a and b
   d. neither a nor b

6. After working with a flux that is active at room temperature, Welder A lets the flux remain on the finished joint. After working with the same flux, Welder B washes the flux completely off. Who is right?
   a. Welder A
   b. Welder B
   c. Both A and B
   d. Neither A nor B

7. As a general rule, what is the minimum width the overlapping portion of a brazed joint must be?
   a. twice as wide as the panel thickness
   b. three times as wide as the panel thickness
   c. four times as wide as the panel thickness
   d. half as wide as the panel thickness

8. Which of the following statements concerning brazing is incorrect?
   a. The size of the torch tip should be slightly smaller than the thickness of the panel.
   b. Make brazing time as short as possible to prevent the joint strength from being reduced.
   c. Avoid brazing the same spot twice.
   d. Baked-on flux can be removed with a sander.

9. The thickness of the parts being brazed has very little effect on which brazing process?
   a. torch
   b. induction
   c. resistance
   d. diffusion

10. Soldering is used to ______________.
    a. reinforce panel joints
    b. correct the surface of welded joints
    c. level panel surfaces
    d. both b and c

11. After soldering, the work area is bluish in color. Welder A says that this is normal. Welder B says it is a sign that the work area was overheated. Who is right?
    a. Welder A
    b. Welder B

c. Both A and B
d. Neither A nor B

12. Excess solder should be wiped away while it is at its ______________.
 a. flowing temperature
 b. melting temperature
 c. both a and b
 d. neither a nor b

13. The application of a thin film of solder is known as ______________.
 a. dressing
 b. sweating
 c. tinning
 d. none of the above

14. Which type of brazing is frequently used on automotive bodies?
 a. nickel
 b. brass
 c. lead
 d. all of the above

15. Which of the following can be used to clean metal prior to soldering?
 a. chemical cleaner
 b. steel wool
 c. wire brush
 d. all of the above

CHAPTER EIGHT

PLASTIC REPAIR MATERIALS AND EQUIPMENT

Objectives

After reading this chapter, you should be able to

- List some of the growing number of plastic automotive applications.
- Explain the difference between the two types of automotive plastics.
- Identify automotive plastics through the use of national identification symbols, the burn test, and by making a trial-and-error weld.
- Describe the basic differences between welding metal and welding plastic.
- Describe the setup and shutdown procedures for a typical hot-air welder.
- Explain the ways airless plastic welding compares favorably with the hot-air method.
- Explain the principles of ultrasonic welding.

In 1975, plastic represented only about 3 percent of the total weight of the average automobile. In the past few years, more and more plastic has been used in various parts of car and light truck bodies, particularly in the front end: in bumper and fender extensions, soft front fascia, fender aprons, grille opening panels, stone shields, instrument panels, trim panels, and elsewhere. The reason is simply that many of the new reinforced plastics are nearly as strong and rigid as steel, and some are even more stable dimensionally. Many such plastics are also extremely corrosion resistant. Figure 8-1 illustrates the difference between today's automobiles and those made just ten years ago. A recent study by Market Search, Inc., of Toledo, Ohio, suggests that this growth is only the beginning. By 1998, the combined use of plastics in vehicle manufacturing is predicted to rise to nearly 3.3 billion pounds, up from 2.2 billion pounds in 1988 (Figure 8-2).

The General Motors APV van, with its fiberglass-reinforced plastic body and urethane fenders, leads the way in the current "plastics boom." The popular Pontiac Fiero features all of its exterior body panels made from various plastics. Other expected automotive growth opportunities in this technology include gas tanks, intake manifolds, composite springs, bumpers, energy absorber systems, and even entire frames.

Because plastic parts are much lighter in weight than sheet metal, they have become an important part of every American car manufacturer's fuel saving, weight reduction program. And because of the high strength-to-weight ratio of plastic, the weight decrease does not mean a decrease in strength. Every indication is that plastic body parts are here to stay, with new applications constantly being found. Therefore, plastic repairs can be expected to be a permanent and increasingly prominent part of life in every body shop. Figure 8-3 shows where plastic parts are currently being used in a typical automobile.

This increasing use of plastic in the automobile industry has resulted in new approaches to collision damage repair. Many damaged plastic parts can be repaired more economically than they can be replaced, especially if the part does not have to be removed from the vehicle. Cuts and cracks, gouges, tears, and punctures are all repairable; and, when

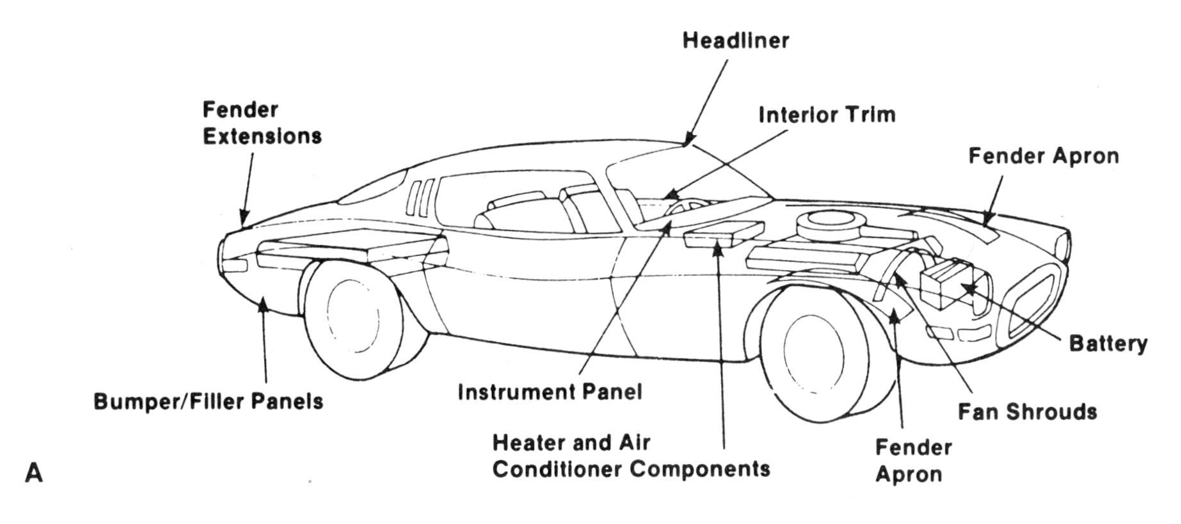

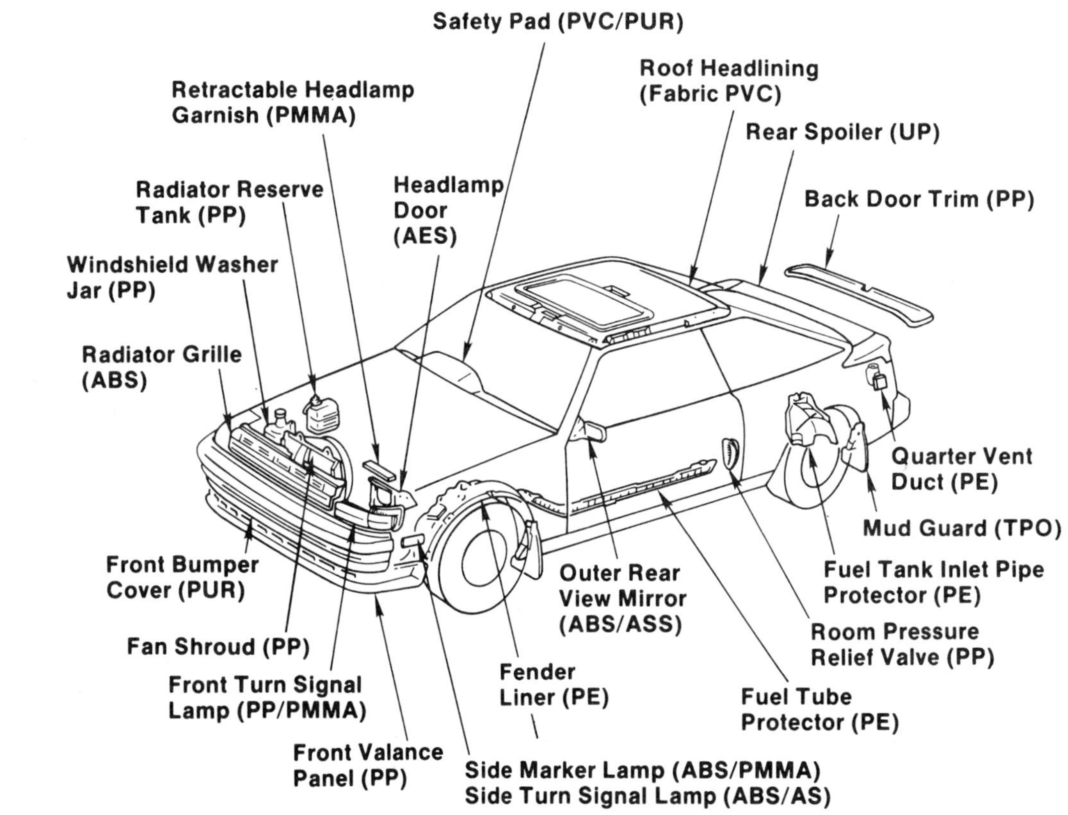

FIGURE 8-1 The use of plastics in automobiles has grown greatly from (A) 10 years ago to (B) today.

necessary, some plastics can also be re-formed after distortion from their original shape. Repair is quicker as well, since replacement parts are not always available; this means less downtime for the vehicle.

Another factor underlining the importance of learning plastic repair technology is the changing attitude of insurance companies. Simply put, they are much less inclined nowadays to pay for operations such as dash pad removal and replacement because of the considerable time and expense involved. This means that the body shop technician faced with a badly torn or dented plastic dash has one of two choices: either repair it or make the replacement at his or her own expense.

TYPES OF AUTOMOTIVE PLASTICS

Two types of plastics are used today in automotive production:

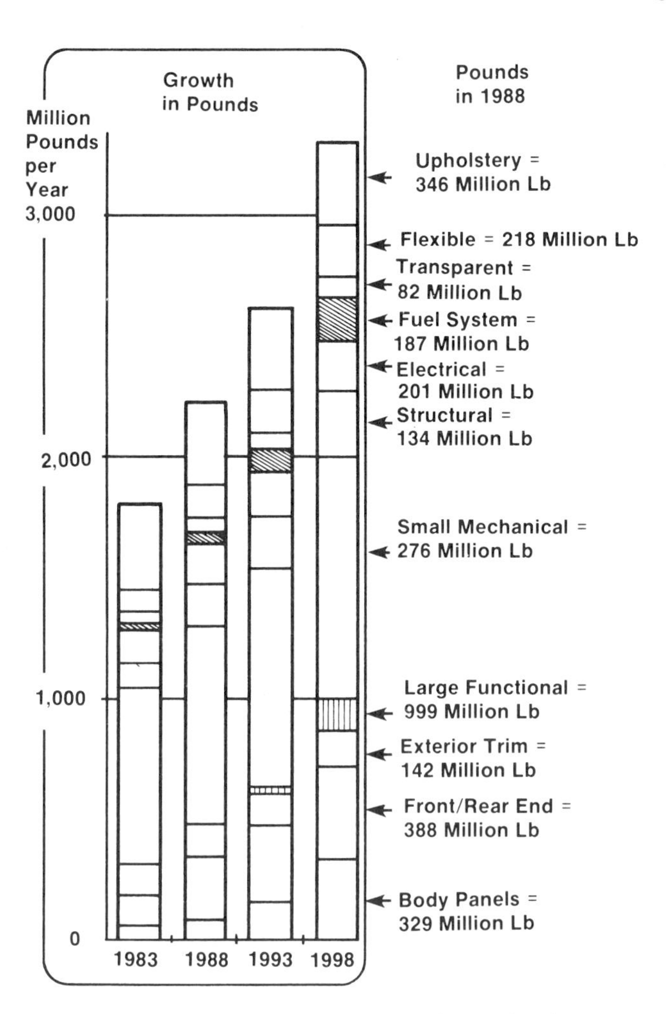

FIGURE 8-2 Projected growth of plastics in automotive applications

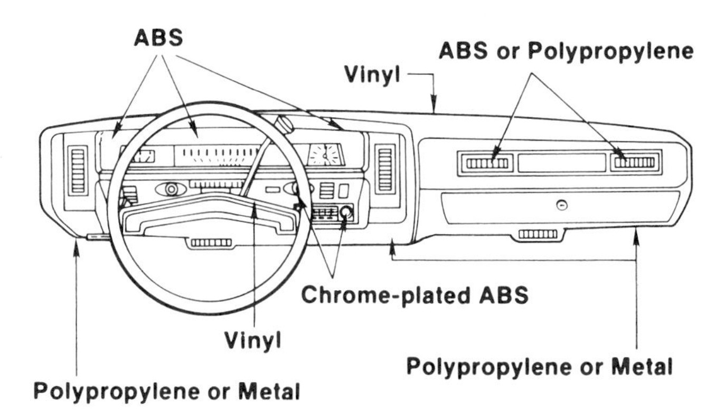

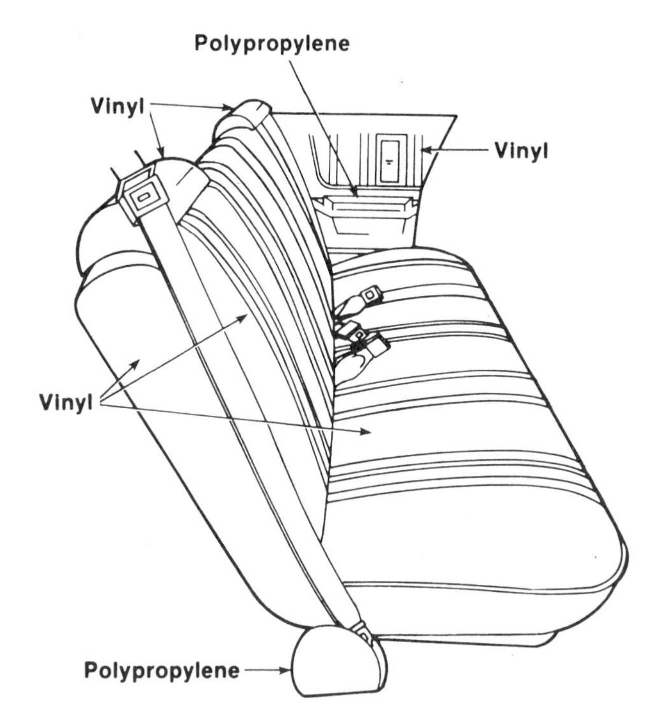

FIGURE 8-3 Common uses of plastic in today's automobile

1. *Thermoplastics.* These plastics are capable of being repeatedly softened and hardened by heating or cooling, with no change in their appearance or chemical makeup. They soften or melt when heat is applied to them, and therefore are weldable with a plastic welder.
2. *Thermosetting Plastics.* These plastics undergo a chemical change by the action of heating, a catalyst, or ultraviolet light leading to an infusible state. They are hardened into a permanent shape that cannot be altered either by reapplying heat or catalysts. Thermosets are not weldable, although they can be "glued" using an airless welder.

Figure 8-4 explains the relationship of heat to plastics more fully. In general, the recommended repair techniques are chemical adhesive bonding for thermosetting plastics and welding for thermoplastics. Without a doubt, the ever-growing use of both of these materials has helped to transform plastic welding from a curious technique practiced by a select few into an important skill that has been accepted as an integral part of today's auto body shop. Following are brief descriptions of the most common automotive plastics.

ABS

Acrylonitrile/butadiene/styrene, or ABS, is a thermoplastic that has excellent forming properties for most applications. ABS is available in normal and high-temperature types, either of which can be welded. Conventional ABS should be welded with air or inert gas at approximately 350 to 400 degrees

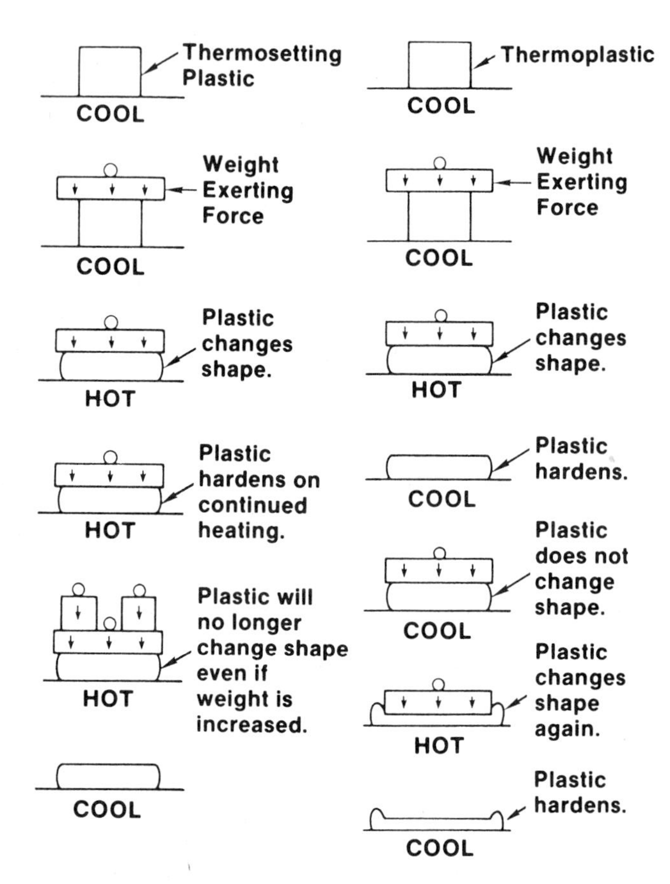

FIGURE 8-4 The effect of heat on plastics

Fahrenheit, and high-temperature ABS at 500 to 550 degrees Fahrenheit.

ABS does not have a sharp melting point, but softens gradually after reaching its heat distortion point. Normal welding rod pressure is sufficient for holding an approximate 60-degree welding angle. Visual observation of the wavy flow pattern along the edges of the deposited rod is the best way to monitor ABS weld quality.

POLYETHYLENE

Oxidation is an important consideration when working with this thermoplastic. For this reason, nitrogen is the recommended gas for welding polyethylene; it also helps achieve the maximum weld strength. Because a very thin coating of oxidized film can adversely affect the final welded product, best results are obtained by welding immediately after cutting the polyethylene, then removing the oxidized film from the welding rod with fine sandpaper or by scraping with a knife.

The rod and base material should be of the same composition, since this affects the weld bond. Use large-sized rods for polyethylene wherever possible, and inspect for poor tolerance and stress. The term *stress cracking*, often included in discussions about polyethylene, refers to cracking or splitting of the material under certain conditions, including chemical reaction, heat, or stress. It results from welding materials of slightly different composition, welding at improper temperatures, and subjecting materials to undue stress or chemical attack (Figure 8-5). Correct weld speeds should always be used to avoid stress; maximum polyethylene weld strength is achieved 10 hours after the weld is completed.

POLYVINYL CHLORIDE

Polyvinyl chloride, or PVC, is another thermoplastic whose use in automobile applications is on the rise. Three important factors influence the welding of this plastic: the type of PVC used (normal or high impact), the amount of plasticizer used, and the quality of the welding rod. Plasticizers are liquids or compounds added during extrusion to make PVC plastic more flexible. They are also used in some welding rods to improve weld quality; under normal conditions, a 10 percent plasticized rod yields improved performance and strength. However, at high temperatures, a plasticized rod decreases in strength, but an unplasticized rod will not.

POLYPROPYLENE

A higher temperature thermoplastic that is quite similar to polyethylene, polypropylene is susceptible to stress cracking and oxidation. It requires the use of nitrogen to obtain maximum weld strength. Some splash of molten plastic will occur when welding polypropylene, but this does not affect the weld and can be eliminated by throttling down the airflow.

POLYURETHANE

Polyurethane is unique in that it is available as both a thermoplastic and thermosetting plastic. Because it is extremely lightweight and formable, it has

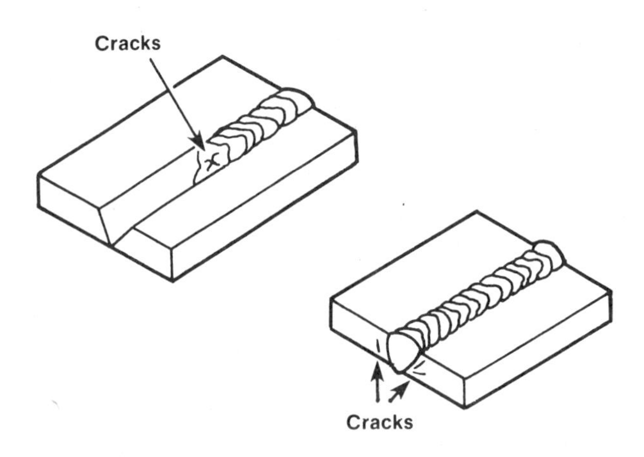

FIGURE 8-5 Examples of stress cracks

excellent aerodynamic properties. Polyurethane is also very durable and noncorrosive, thus making it popular with both domestic and foreign carmakers (Figure 8-6).

PLASTIC IDENTIFICATION

In 1980, General Motors published the names of seventy-nine different plastics used, to one extent or another, in the auto industry. With such a variety of plastics in existence—and new ones constantly being tested and refined—it is easy for the uninformed body shop technician to be reduced to playing a guessing game when trying to do a repair. This, in turn, jeopardizes the entire job, because a repair that is done based on incorrect identification will likely yield unsatisfactory results. Not only do the chances of using the wrong repair products and techniques increase greatly, but selection of the proper refinishing procedure also hinges on accurate identification of the plastic substrate material (Figure 8-7). A repair that initially appears to be sound can quickly delaminate, crack, or discolor if the job was performed based on incorrect identification of the plastic.

There are several ways to identify an unknown plastic. One possibility is the use of national identification symbols, which can often be found on the parts to be repaired. When parts are not identified by these symbols, refer to the manufacturer's technical literature for plastic identification (Figure 8-8). Domestic manufacturers are using identification symbols more and more; unfortunately, there are still some who do not. Keep in mind that one problem with this system is that it is usually necessary to remove the part to read the letters.

One of the best ways to identify an unknown plastic is to become familiar with the different types

FIGURE 8-6 Foreign car manufacturers have begun to use polyurethane extensively.

FIGURE 8-7 Incorrect plastic identification can even affect the final finishing.

FIGURE 8-8 In some cases, it is possible to refer to the manufacturer's literature for plastic identification.

of plastics and where they are commonly used. Table 8-1 gives the identification symbol, chemical and common names, and applications of common automotive plastics. Plastic applications information can often be found in shop manuals or in special manufacturer's guides.

Another technique for identifying various types of plastic is the so-called burn test or melt test. Different plastics have different burn characteristics and some produce unique odors.

To do a burn test, scrape off a small sliver of the material from the backside or underside of the part, hold it with pliers or on the end of a piece of wire, and try to ignite it with a match or propane torch (Figure

TABLE 8-1: STANDARD SYMBOL, CHEMICAL NAME, TRADE NAME, AND DESIGN APPLICATIONS OF MOST COMMONLY USED PLASTICS

Symbol	Chemical Name	Common Name	Design Applications	Thermosetting or Thermoplastic
ABS	Acrylonitrile-butadiene-styrene	ABS, Cycolac, Abson, Kralastic, Lustran, Absafil, Dylel	Body panels, dash panels, grilles, headlamp doors	Thermoplastic
ABS/MAT	Hard ABS reinforced with fiberglass	—	Body panels	Thermosetting
ABS/PVC	ABS/Polyvinyl chloride	ABS Vinyl	—	Thermoplastic
EP	Epoxy	Epon, EPO, Epotuf, Araldite	Fiberglass body panels	Thermosetting
EPDM	Ethylene-propylene-diene-monomer	EPDM, Nordel	Bumper impact strips, body panels	Thermosetting
PA	Polyamide	Nylon, Capron, Zytel, Rilsan, Minlon, Vydyne	Exterior finish trim panels	Thermosetting
PC	Polycarbonate	Lexan, Merlon	Grilles, instrument panels, lenses	Thermoplastic
PRO	Polyphenylene oxide	Noryl, Olefo	Chromed plastic parts, grilles, headlamp doors, bezels, ornaments	Thermosetting
PE	Polyethylene	Dylan, Fortiflex, Marlex, Alathon, Hi-fax, Hosalen, Paxon	Inner fender panels, interior trim panels, valances, spoilers	Thermoplastic
PP	Polypropylene	Profax, Olefo, Marlex, Olemer, Aydel, Dypro	Interior moldings, interior trim panels, inner fenders, radiator shrouds, dash panels, bumper covers	Thermoplastic
PS	Polystyrene	Lustrex, Dylene, Styron, Fostacryl, Duraton	—	Thermoplastic
PUR	Polyurethane	Castethane, Bayflex	Bumper covers, front and rear body panels, filler panels	Thermosetting
TPUR	Polyurethane	Pellethane, Estane, Roylar, Texin	Bumper covers, gravel deflectors, filler panels, soft bezels	Thermoplastic
PVC	Polyvinyl chloride	Geon, Vinylete, Pliovic	Interior trim, soft filler panels	Thermoplastic
RIM	"Reaction injection molded" polyurethane	—	Bumper covers	Thermosetting
R RIM	Reinforced RIM-polyurethane	—	Exterior body panels	Thermosetting

TABLE 8-1: STANDARD SYMBOL, CHEMICAL NAME, TRADE NAME, AND DESIGN APPLICATIONS OF MOST COMMONLY USED PLASTICS (CONTINUED)

Symbol	Chemical Name	Common Name	Design Applications	Thermosetting or Thermoplastic
SAN	Styrene-acrylonitrite	Lustran, Tyril, Fostacryl	Interior trim panels	Thermosetting
TPR	Thermoplastic rubber	—	Valance panels	Thermosetting
UP	Polyester	SMC, Premi-glas, Selection Vibrin-mat	Fiberglass body panels	Thermosetting

TABLE 8-2: BURN CHARACTERISTICS OF COMMON PLASTICS

Plastic	Burn Characteristic
Polypropylene (PP)	Burns with no visible smoke and continues to burn once the flame source is removed. Produces a smell like burned wax. Bottom of flame is blue and top is yellow.
Polyethylene (PE)	Also smells like burned wax, makes no smoke, and continues to burn once the flame source is removed. Bottom of flame is blue and top is yellow.
ABS	Burns with a thick, black, sooty smoke and continues to burn when the flame source is removed. Produces a sweet odor when burned. Flame is yellowish orange.
PVC	Only chars and does not support a flame when you try to burn it. Gives off gray smoke and an acid-like smell. End of flame is yellowish green.
Thermoplastic Polyurethane (TPUR) and Thermosetting Polyurethane (PUR)	Burns with a yellow-orange sputtering flame and gives off black smoke. The thermoset version of polyurethane, however, will not support a flame.

8-9). The burn characteristics of common automotive plastics are given in Table 8-2.

The burn test has fallen out of favor as a technique for identifying plastics for several reasons. For one, having an open flame in a body shop environment creates a potential fire hazard. And although the burn test does not generally pose a health threat, it is a good idea to avoid inhaling the fumes. Another reason the burn test is not as popular as it was once is because it is not always reliable. For example, it is difficult to determine the difference between polypropylene and polyethylene because both burn with the same characteristics. Furthermore, some parts are now being manufactured from "hybrids," which are blended plastics that use more than one ingredient. A burn test is of no help here.

The best means of identifying an unknown plastic (assuming it is probably a thermoplastic that is potentially weldable) is to make a trial-and-error weld (Figure 8-10). Try several different filler rods until one sticks. Most suppliers of plastic welding equipment offer only half a dozen or so different types of plastic filler rods, so the range of possibilities is not that great. The rods are color coded, so once the rod that works is found, the base material is identified. For more specifics on working with plastics, see Table 8-3.

PRINCIPLES OF PLASTIC WELDING

The welding of plastics is not unlike the welding of metals. Both methods use a heat source, welding rod, and similar types of finished welds (butt joints, fillet welds, lap joints, and the like). Joints are prepared in much the same manner, and similarly evaluated for strength. Due to differences in the physical characteristics of each material, however, there are notable differences between welding metal and welding plastics.

When welding metal, the rod and base metal are made molten and puddled into a joint. And while metals have a sharply defined melting point, plastics have a wide melting range between the temperature

TABLE 8-3: HANDLING PRECAUTIONS FOR PLASTICS

Code	Material Name	Heat Resisting Temperature* °F	Resistance To Alcohol or Gasoline	Notes
AAS	Acrylonitrile Acrylic Rubber Styrene Resin	176	Alcohol is harmless if applied only for short time in small amounts (example, quick wiping to remove grease).	Avoid gasoline and organic or aromatic solvents.
ABS	Acrylonitrile Butadiene Styrene Resin	176	Alcohol is harmless if applied only for short time in small amounts (example, quick wiping to remove grease).	Avoid gasoline and organic or aromatic solvents.
AES	Acrylonitrile Ethylene Rubber Styrene Resin	176	Alcohol is harmless if applied only for short time in small amounts (example, quick wiping to remove grease).	Avoid gasoline and organic or aromatic solvents.
EPDM	Ethylene Propylene Rubber	212	Alcohol is harmless. Gasoline is harmless if applied only for short time in small amounts.	Most solvents are harmless, but avoid dipping in gasoline, solvents, etc.
PA	Polyamide (Nylon)	176	Alcohol and gasoline are harmless.	Avoid battery acid.
PC	Polycarbonate	248	Alcohol is harmless.	Avoid gasoline, brake fluid, wax, wax removers, and organic solvents.
PE	Polyethylene	176	Alcohol and gasoline are harmless.	Most solvents are harmless.
POM	Polyoxymethylene (Polyacetal)	212	Alcohol and gasoline are harmless.	Most solvents are harmless.
PP	Polypropylene	176	Alcohol and gasoline are harmless.	Most solvents are harmless.
PPO	Modified Polyphenylene Oxide	212	Alcohol is harmless.	Gasoline is harmless if applied only for quick wiping to remove grease.
PS	Polystyrene	140	Alcohol and gasoline are harmless if applied only for short time in small amounts.	Avoid dipping or immersing in alcohol, gasoline, solvents, etc.

at which they soften and the temperature at which they char or burn. Also unlike metals, plastics are poor conductors of heat, and thus difficult to heat uniformly. Because of this, the plastic filler rod and the surface of the plastic will char or burn before the material below the surface becomes fully softened. The decomposition time at welding temperature is shorter than the time required to completely soften many plastics for fusion welding. The result is that a plastic welder must work within a much smaller temperature range than the metal welder.

Because a plastic welding rod does not become completely molten and appears much the same before and after welding, a plastic weld might appear incomplete to the shop technician who is used to welding only metal. The explanation is simple: since only the outer surface of the rod has become molten and the inner core has remained hard, the welder is able to exert pressure on the rod to force it into the joint and create a permanent bond. When heat is taken away, the rod reverts to its original form. Thus, even though a strong and permanent bond has been obtained between the rod and base material, the appearance of the rod is much the same as before the weld was made, except for molten flow patterns on either side of the bead.

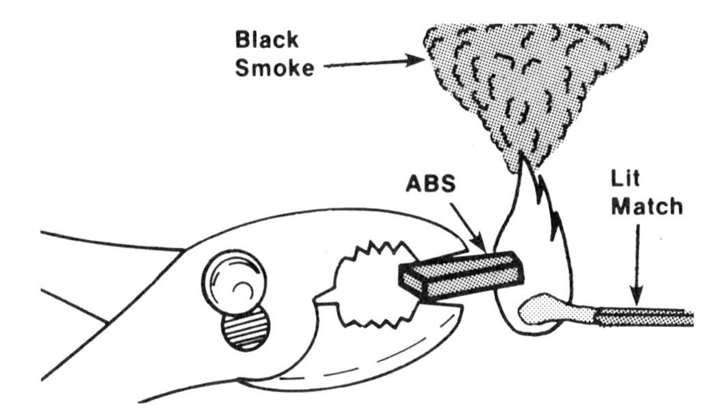

FIGURE 8-9 Conducting a burn test on ABS plastic

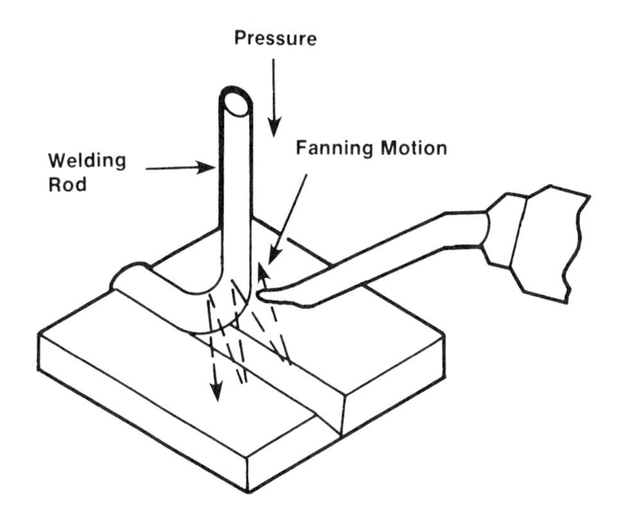

FIGURE 8-11 Successful plastic welding requires the proper combination of heat and pressure.

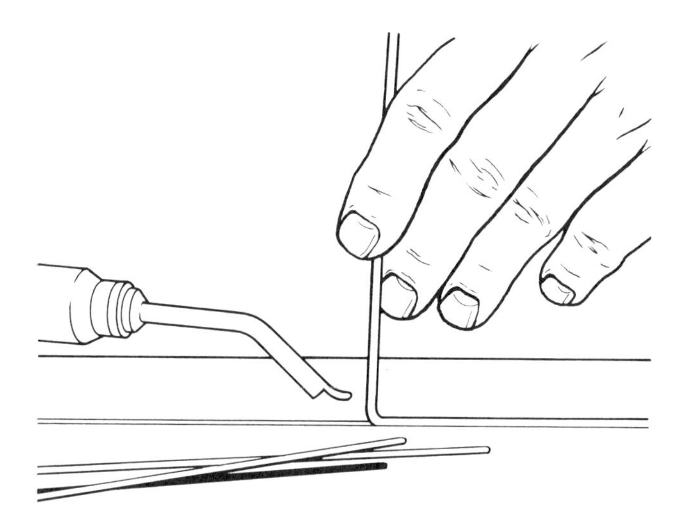

FIGURE 8-10 Making a trial-and-error weld

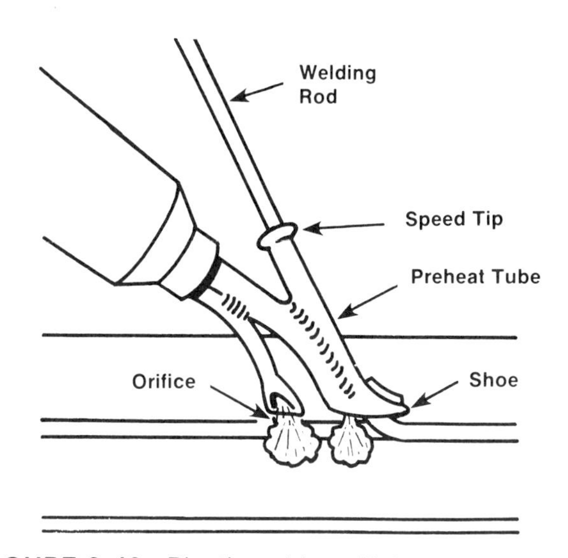

FIGURE 8-12 Plastic welder utilizing a high-speed tip

When welding plastics, the materials are fused together by the proper combination of heat and pressure. With the conventional hand welding method, this combination is achieved by applying pressure on the welding rod with one hand, while at the same time applying heat and a constant fanning motion to the rod and base material with hot gas from the welding torch (Figure 8-11). Successful welds require that both pressure and heat be kept constant and in proper balance. Too much pressure on the rod tends to stretch the bead and produce unsatisfactory results; too much heat will char, melt, or distort the plastic. With practice, plastic welding can be mastered as completely as metal welding.

HIGH-SPEED WELDING

High-speed welding incorporates the basic methods utilized in hand welding. Its primary difference lies in the use of a specially designed and patented high-speed tip (Figure 8-12), which enables the welder to produce more uniform welds and work at a much higher rate of speed. As with hand welding, constant heat and pressure must be maintained.

The increased efficiency of high-speed welding is made possible through preheating of both the rod and base material before the point of fusion. The rod is preheated as it passes through a tube in the speed tip; the base material is preheated by a stream of hot air passing through a vent in the tip ahead of the fusion point. A pointed shoe on the end of the tip applies pressure on the rod, thus eliminating the need for the operator to apply pressure. At the same time it smooths out the rod, creating a more uniform appearance in the finished weld.

In high-speed welding, the conventional two-hand method becomes a faster and more uniform one-hand operation. Once started, the rod is fed automatically into the preheating tube as the welding torch is pulled along the joint. High-speed tips are designed to provide the constant balance of heat and pressure necessary for a satisfactory weld. The average welding speed is about 40 inches per minute.

High-speed welding does have its disadvantages. Because increased speeds must be maintained in order to achieve the best possible weld, the high-speed welding torch is not suited for small, intricate work. Also, when the operator is new to this technique, the position in which the welder is held might seem clumsy and difficult. However, experience will enable the operator to successfully make all welds that can be made with a hand welder, including butt welds, V-welds, corner welds, and lap joint welds. Speed welds can be made on circular as well as flat work. In addition, inside welds on tanks can be speed welded, provided the working space is not too small to manipulate the torch.

HOT-AIR PLASTIC WELDING

There are a number of manufacturers who make plastic hot-air welding equipment available for the body repair industry, and all use the same basic technology. A ceramic or stainless steel electric heating element is used to make hot air (450 to 650 degrees Fahrenheit), which in turn blows through a nozzle and onto the plastic. The air supply comes from either the shop's air compressor or a self-contained portable compressor that is mounted in a carrying case that comes with the unit. Most hot-air welders use a working pressure at the tip of around 3 psi. A pair of pressure regulators is required to reduce the air pressure first to around 50 psi, and then finally to the working pressure of 2-1/2 to 3-1/2 psi. A typical hot-air welder is illustrated in Figure 8–13.

The barrel of the torch itself gets sufficiently hot so that skin contact could cause a burn if the hot air is directed against the skin long enough. The torch is used in conjunction with the welding rod, which is normally 3/16 inch in diameter and made from the same material as the plastic being repaired. This will ensure that the strength, hardness, and flexibility of the repair is the same as the damaged part. Use of the proper welding rod is very important; an adequate weld is impossible if the wrong rod is used.

One of the problems with hot-air welding is that the 3/16-inch-diameter rod is often thicker than the panel to be welded. This can cause the panel to overheat before the rod has melted. Using a 1/8" rod with the hot-air welder can often correct such warpage problems. (This works on the same principle as using a 1/64-inch wire instead of a 1/32 inch for MIG welding.)

Three shapes of welding tips are available for use with most plastic welding torches. They are:

- *Tacking Welding Tips.* These are used to tack broken sections together before welding (Figure 8–14). If necessary, tack welds can be easily broken apart for realigning.
- *Round Welding Tips.* These tips are used to fill small holes and make short welds, welds in hard-to-reach places, and on tight or particularly sharp corners (Figure 8–15).
- *Speed Welding Tips.* These are used for long, fairly straight welds. They hold the filler rod, automatically preheat it, and feed the rod into the weld, thus allowing for faster welding rates (Figure 8–16).

SETUP, SHUTDOWN, AND SERVICING

Naturally, no two hot-air welders are exactly alike; their design can vary from one manufacturer to another. The setup, shutdown, and service procedures that follow, while typical of all hot-air welders, should nonethless be regarded as general guidelines only. For specific instruction, always refer to the owner's manual and other material provided by the welder manufacturer. Keep in mind that some manufacturers advise against using their welder on plastic that is any thinner than 1/8 inch because of the likelihood of distortion. In other cases, it is acceptable to weld plastics as thin as 1/16 inch, provided they are supported from underneath during the operation. Again, it is very important to read and follow the specific directions for the welder being used.

Setup and Shutdown

To set up a typical hot-air welder, proceed as follows:

1. Close the air pressure regulator valve by turning the control handle counterclockwise until it is loose. This will prevent possible damage to the gauge from a sudden surge of excess air pressure.
2. Connect the regulator to a supply of either compressed air or inert gas. The standard rating for an air pressure regulator is 200

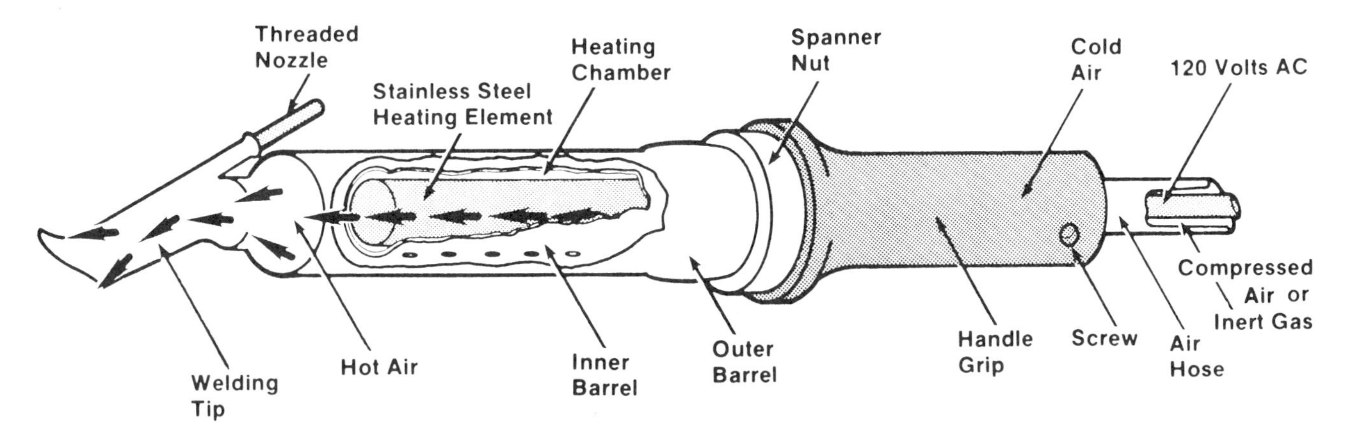

FIGURE 8-13 Typical hot-air welder

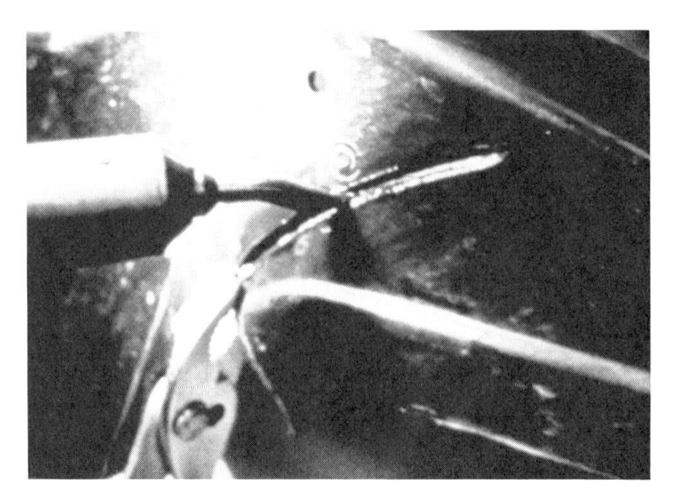

FIGURE 8-14 Tacking welding tip *(Courtesy of Seelye Inc.)*

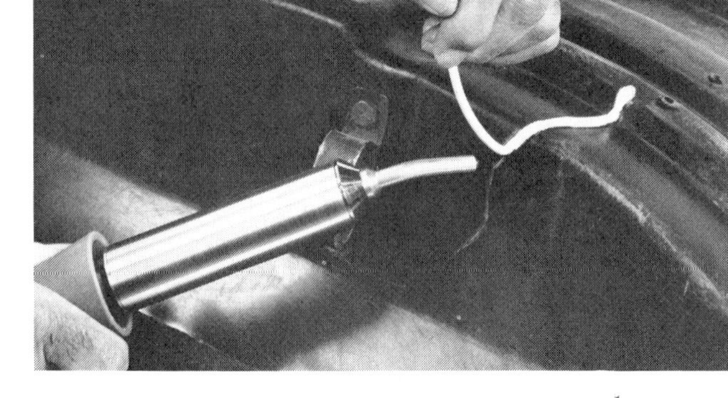

FIGURE 8-15 Round welding tip *(Courtesy of Seelye Inc.)*

pounds of line pressure. If inert gas is used, a pressure reducing valve is needed.

3. Turn on the air supply. The starting air pressure depends on the wattage of the heating element and the air pressure. The operating air pressure requires slightly less air.
4. Connect the welder to a common 120-volt AC outlet. A three-prong grounded plug or temporary adapter must be used with the welder at all times, as shown in Figure 8-17.
5. Allow the welder to warm up at the recommended pressure. It is essential that either air or inert gas flows through the welder at all times, from warm-up to cool down, to prevent burnout of the heating element and further damage to the gun.
6. Select the proper tip and insert it with pliers to avoid touching the barrel while hot.

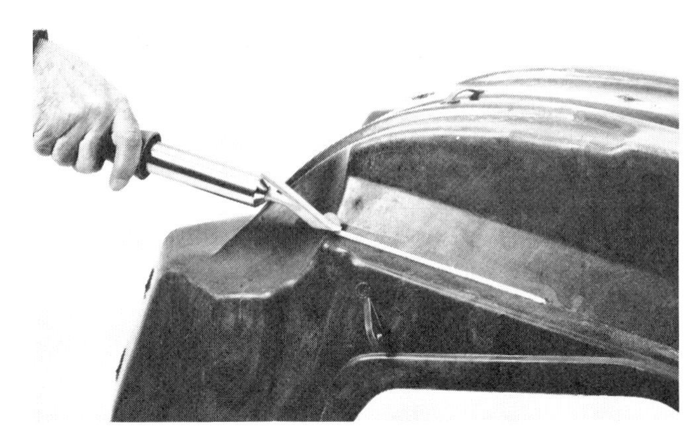

FIGURE 8-16 Speed welding tip *(Courtesy of Seelye Inc.)*

7. After the tip has been installed, the temperature will increase slightly due to back pressure. Allow 2 to 3 minutes for the tip to reach the required operating temperature.

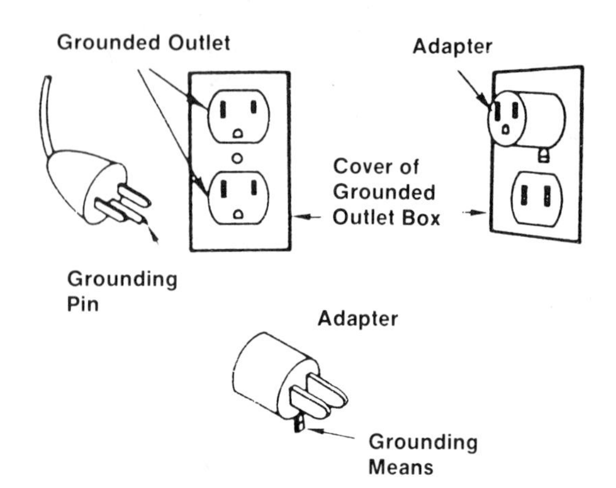

FIGURE 8-17 Methods of grounding a plastic welder

8. Check the air temperature by holding a thermometer 1/4 inch from the hot air end of the torch. For most thermoplastics, the temperature should be in the 450 to 650 degree Fahrenheit range. Information supplied with the welder usually includes a chart of welding temperatures.
9. If the temperature is too high to weld the material, increase the air pressure slightly until the temperature decreases. If the temperature is too low for the particular application, decrease the air pressure slightly until the temperature rises. When increasing and decreasing the air pressure, allow at least 1 to 3 minutes for the temperature to stabilize at the new setting.
10. Damage to the welder or heating element will not occur from too much air pressure; however, the element can become overheated by too little air pressure. When decreasing the air pressure, never allow the round nut that holds the barrel to the handle of the welder to become too hot to the touch. This is an indication of overheating.
11. A partial clogging of the dirt screen in the regulator or a fluctuation in the line voltage can also cause over or underheating. Watch for these symptoms.
12. If the threads at the end of the barrel become too tight, clean them with a good high-temperature grease to prevent seizing.
13. When the welding is finished, disconnect the electric supply and let the air flow through the welder for a few minutes or until the barrel is cool to the touch. Then disconnect the air supply.

Maintenance

If it becomes necessary to change the heating element, proceed as follows:

1. While pushing the end of the barrel against a solid object, hold the handle tightly and push in. The pressure on the barrel will compress the element spring.
2. Use a spanner wrench to loosen the spanner nut. Keep the pressure on the handle and back off the nut all the way by hand.
3. Hold the barrel and place the complete welder on a bench. Remove the barrel.
4. Gently pull the element out of the handle. At the same time, unwind the cable (which has been spiraled into the handle) until it is completely out of the handle.
5. Grasp the socket at the end of the wire tightly. Rock the element while pulling until the element is disconnected.
6. To install the new element, reverse the above procedure. Turn the element clockwise (about 1-1/2 turns) while pushing the wire gently back into the handle. This prevents kinking of the wire.

Other, more involved servicing procedures are best left to qualified repair technicians. Many manufacturers make it clear that disassembling the welder automatically invalidates the warranty.

AIRLESS PLASTIC WELDING

Although only in existence for a relatively short period of time, airless welding has become very popular with the auto body repair industry (Figure 8-18). Compared to the hot-air method, it is less expensive, easier to learn, simpler to use, and more versatile. Although the hot-air method utilizes 3/16-inch-diameter welding rods, 1/8-inch-diameter rods are used with the airless method. This not only provides a quicker rod melt, it also helps eliminate two troublesome problems: panel warpage and excess rod buildup.

When setting up the typical airless welder, the first and most important step is to put the temperature dial at the appropriate setting, depending on the specific thermosetting plastic being worked on. It is crucial that the temperature setting is correct; otherwise, the entire welding operation will be jeopardized. It will normally take about 3 minutes for the welder to fully warm up. As for the selection of the welding rod, there is another factor to consider besides size; namely, compatibility. Make sure the rod

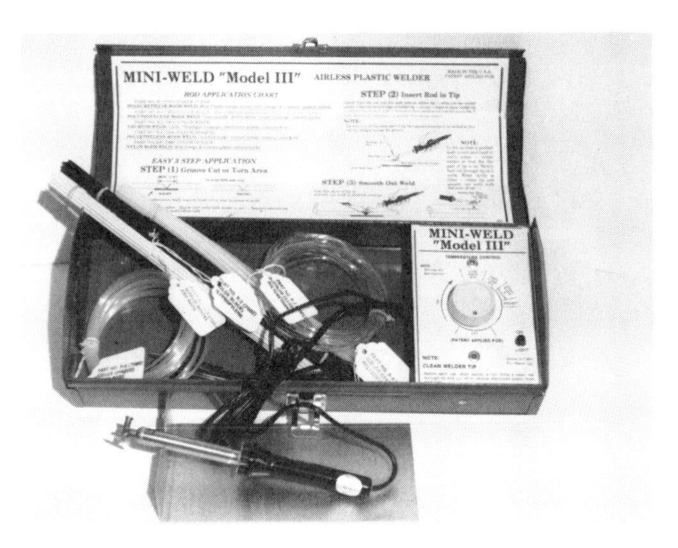

FIGURE 8-18 Typical airless welder

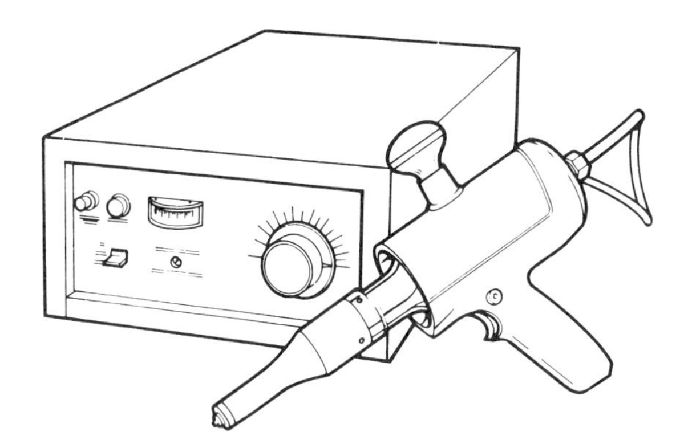

FIGURE 8-19 Typical hand-held ultrasonic welder and power supply

is the same as the damaged plastic or the weld will more than likely not hold. To this end, many airless welder manufacturers provide rod application charts. When the correct rod has been chosen, it is good practice to run a small piece of it through the welder to clean out the tip before beginning.

ULTRASONIC PLASTIC WELDING

The technology of ultrasonic plastic welding was originally developed solely as a means of assembly, but it is also suitable for making repairs. Hand-held systems (Figure 8-19) are available in 20 and 40 kHz; they are equally adept at welding large parts and tight, hard-to-reach areas. Welding time is controlled by the power supply. And although the use of ultrasonics in the auto body shop is still in its infancy, the strong, clean, precise welds the process produces will likely make it increasingly popular in the years to come.

Most commonly used injection-molded plastics can be ultrasonically welded without the use of solvents, heat, or adhesives. Ultrasonic weldability depends on the plastic's melting temperature, elasticity, impact resistance, coefficient of friction, and thermal conductivity. Generally, the more rigid the plastic, the easier it is to weld ultrasonically. Thermoplastics such as polyethylene and polypropylene are ideal for ultrasonic welding, provided the welder can be positioned close to the joint area.

ULTRASONIC STUD WELDING

Ultrasonic stud welding, a variation of the shear joint, is a reliable technique that can be used to join plastic parts at a single point or numerous locations. In many applications requiring permanent assembly, a continuous weld is not required. Frequently, the size and complexity of the parts seriously limit attachment points or weld location. With similar materials, this type of assembly can be effectively and economically accomplished using ultrasonic stud welding. The power requirement is low because of the small weld area, and the welding cycle is short, almost always less than half a second.

Figure 8-20 shows the basic stud weld joint before, during, and after welding. The weld is made along the circumference of the stud; its strength is a function of the stud diameter and the depth of the weld. Maximum tensile strength is achieved when the depth of the weld equals half the diameter of the stud. The radial interference (dimension A) must be uniform and should generally be 0.008 to 0.012 inch for studs having a diameter of 0.5 inch or less. The hole should be a sufficient distance from the edge to prevent breakout; a minimum of 0.125 inch is recommended.

In the joint, the recess can be on the end of the stud or in the mouth of the hole, as shown in the examples. When using the latter, a small chamfer can be used for rapid alignment. To reduce stress concentration, a good-sized fillet radius should be incorporated at the base of the stud. Recessing the fillet below the surface allows flush contact of the parts.

Other ways in which the ultrasonic stud weld can be used are illustrated in Figure 8-21. A third dissimilar material can be locked in place, as in view A. View B shows separate molded rivets in lieu of metal self-tapping screws. Unlike metal fasteners, they produce a relatively stress-free assembly.

Figure 8-22 shows a variation that can be used where appearance is important or an uninterrupted surface is required. The stud is welded into a boss,

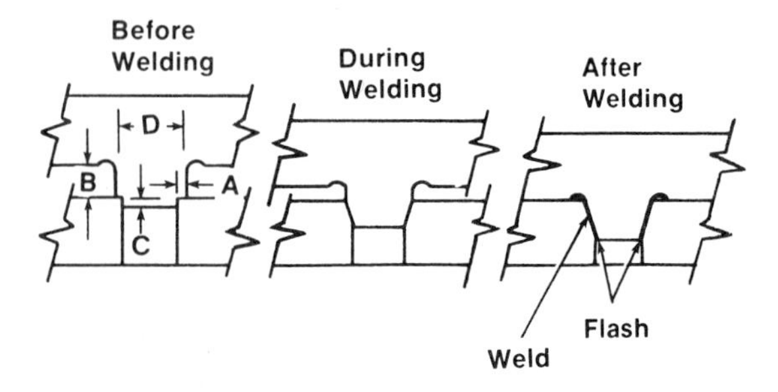

FIGURE 8-20 Basic ultrasonic stud welding joint

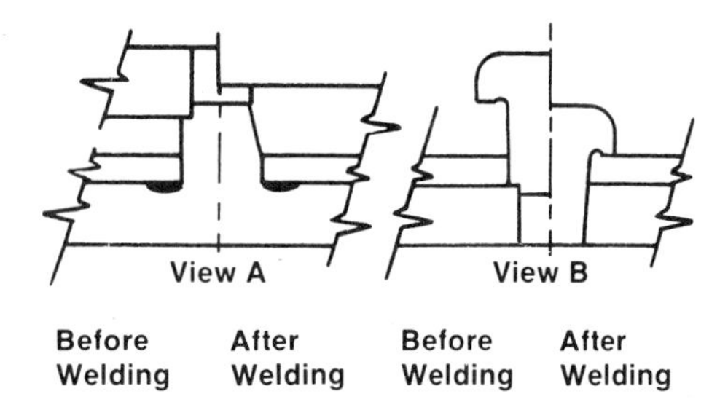

FIGURE 8-21 Variations of the basic stud joint

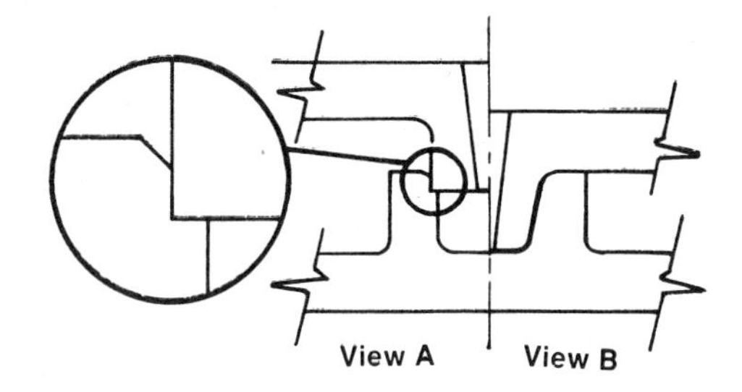

FIGURE 8-22 Welding a stud in a blind hole

whose outside diameter can be no less than twice the stud diameter. When welding into a blind hole, it might be necessary to provide an outlet for air. Two possibilities are shown: a center hole through the stud (View A), or a small, narrow slot in the interior wall of the boss (View B).

REVIEW QUESTIONS

1. In the past few years, more and more plastic has been used in ____________.
   a. bumper and fender extensions
   b. trim panels
   c. instrument panels
   d. all of the above

2. Which of the following statements applies to thermoplastics?
   a. They are weldable with a plastic welder.
   b. They undergo a chemical change by the action of heating, a catalyst, or ultraviolet light leading to an infusible state.
   c. Both a and b
   d. Neither a nor b

3. What is available as both a thermoplastic and a thermosetting plastic?
   a. Polyurethane
   b. ABS
   c. PVC
   d. Polypropylene

4. Welder A uses nitrogen for welding polyethylene; Welder B uses argon. Who is correct?
   a. Welder A
   b. Welder B
   c. Both A and B
   d. Neither A nor B

5. During a burn test, the plastic in question burns with a thick, black, sooty smoke and produces a sweet odor. Welder A says the plastic is polyethylene; Welder B says it is thermosetting polyurethane. Who is correct?
   a. Welder A
   b. Welder B
   c. Both A and B
   d. Neither A nor B

6. A burn test is of no use with ____________.
   a. automotive plastics
   b. thermoplastics
   c. thermosetting plastics
   d. hybrid plastics

7. Upon inspection of a plastic weld, the bead is found to be stretched. Welder A says a likely cause is excessive heat during the welding operation. Welder B says a likely cause is excessive pressure on the welding rod. Who is correct?
   a. Welder A
   b. Welder B
   c. Both A and B
   d. Neither A nor B

8. The heat resisting temperature of polystyrene is ____________.
   a. 212 degrees Fahrenheit
   b. 140 degrees Fahrenheit
   c. 248 degrees Fahrenheit
   d. 176 degrees Fahrenheit

9. In high-speed welding, what applies pressure on the welding rod?
   a. operator
   b. pointed shoe
   c. preheat tube
   d. none of the above

10. Most hot-air welders use a working pressure at the tip of around ______________.
    a. 50 psi
    b. 30 psi
    c. 10 psi
    d. 3 psi

11. Which type of welding tips are ideal for working in hard-to-reach places?
    a. round
    b. speed
    c. both a and b
    d. neither a nor b

12. A plastic welding operation is plagued by almost constant overheating. Welder A says a likely cause is fluctuation in the line voltage. Welder B says a likely cause is a partially clogged dirt screen in the regulator. Who is correct?
    a. Welder A
    b. Welder B
    c. Both A and B
    d. Neither A nor B

13. Compared to hot-air welding, the airless method is
    a. more expensive
    b. more versatile
    c. harder to learn
    d. all of the above

14. The recommended welding rod size for airless welding is ______________.
    a. 3/16-inch diameter
    b. 1/32-inch diameter
    c. 1/16-inch diameter
    d. 1/8-inch diameter

15. In ultrasonic stud welding, maximum tensile strength is achieved when the depth of the weld equals ______________.
    a. radial interference
    b. half the radial interference
    c. diameter of the stud
    d. half the diameter of the stud

16. Conventional ABS should be welded with air or inert gas at approximately ______________.
    a. 350 to 400 degrees Fahrenheit
    b. 400 to 450 degrees Fahrenheit
    c. 450 to 500 degrees Fahrenheit
    d. 500 to 550 degrees Fahrenheit

17. The term *stress cracking* is often included in discussions about which of the following?
    a. polyurethane
    b. ABS
    c. polyethylene
    d. none of the above

18. In high-speed welding, the ______________.
    a. rod is preheated
    b. base material is preheated
    c. both a and b
    d. neither a nor b

19. In high-speed welding, the average welding speed is ______________.
    a. 12 inches per minute
    b. 24 inches per minute
    c. 40 inches per minute
    d. 70 inches per minute

20. Which of the following factors must be considered when choosing a welding rod for the airless welder?
    a. size
    b. compatibility
    c. both a and b
    d. neither a nor b

21. Ultrasonic weldability depends on the plastics' ______________.
    a. melting temperature
    b. elasticity
    c. impact resistance
    d. all of the above

22. Which of the following is a factor underlining the importance of learning plastic repair technology?
    a. increasing use of plastics in the automobile
    b. repairing plastics is quicker than replacing
    c. changing attitude of insurance companies
    d. all of the above

23. Because of its extremely light weight and formability, this plastic has excellent aerodynamic properties.
    a. ABS
    b. PVC
    c. polypropylene
    d. polyurethane

24. During a burn test, the plastic in question burns with a yellow-orange sputtering flame and gives off black smoke. Welder A says the plastic is thermosetting polyurethane; Welder B says it is thermoplastic polyurethane. Who is right?
    a. Welder A
    b. Welder B
    c. Both A and B
    d. Neither A nor B

25. As long as alcohol and gasoline are applied for only a short time and in small amounts they pose no threat to ____________.
    a. polyethylene
    b. polystyrene
    c. polypropylene
    d. polyoxymethylene

26. When operating a hot-air welder, Welder A uses an operating air pressure that is slightly more than the starting pressure. Welder B uses an operating air pressure that is slightly less than the starting pressure. Who is right?
    a. Welder A
    b. Welder B
    c. Both A and B
    d. Neither A nor B

CHAPTER NINE

PLASTIC REPAIR METHODS

Objectives

After reading this chapter, you should be able to
- List the key factors that must be considered when deciding whether to repair or replace a damaged part.
- Choose which repair method—adhesive bonding or welding—should be used under any given circumstances.
- Describe the various chemical adhesive bonding techniques.
- Describe the basics of hot-air and airless welding.
- Explain tack-welding, hand-welding, and speed-welding techniques.

Plastic repair, like any other kind of body repair work, begins with an estimate. That is, it must be determined if the part should be repaired or replaced. Often a comparison of costs between the two alternatives dictates the final decision. However, cost is not the only factor that must be taken into consideration. For instance, a small crack, tear, gouge, or hole in a nose fascia or large panel that is difficult to replace or that is not readily available probably indicates that a repair is in order. On the other hand, extensive damage to the same component or damage to a part that is easy to replace (for example, a fender extension or trim piece) would dictate replacement. Another factor that cannot be overlooked when making this decision is whether or not the repairer has the skill required to make this particular job a quality repair.

When all things are taken into consideration, figuring out the comparative costs of repair versus replacement is not as easy as it looks. For instance, suppose a late-model car has a cracked plastic fan shroud. A replacement shroud would cost 25 dollars. To weld it would take about 20 minutes, including the time it takes to prepare the surface. Repair versus replacement might seem like equally acceptable choices based on the facts related so far, but there are other factors to consider: one is the labor required to remove the shroud (this could be 20 minutes if the fan has to be pulled to get the shroud off); also, a replacement part might not be readily available. Welding the shroud in place would eliminate the labor required to remove and replace it, thus tipping the economic scale in favor of repair. Repairing the existing shroud would also save the time and trouble of back ordering a shroud through a dealer.

Another example is a damaged radiator. The radiator core is in good condition, but the radiator will not hold water because the nylon end tank has a crack in it. In addition, one of the plastic mounting tabs on the end tank has been broken off. The conventional approach for someone who does not make plastic repairs would be to pull the radiator and send it to the local radiator shop. The shop would either replace the damaged end tank or, like an increasing number of shops today, would simply weld up the crack in the existing tank, weld the broken tab back in place, and send a bill for the labor. However, this type of repair could have been done in the original shop if the repairer had learned how to work with plastics.

The factors that must be considered when deciding to replace or repair a part are as follows:

- Cost
- Extent of damage
- Ease of replacement
- Availability of replacement parts
- Skill of the repairer

If repair is the answer, it must be determined if the part has to be removed from the vehicle. The entire damaged area must be accessible for a quality repair job to be performed; if it is not, the part must be removed. Although parts do not always have to be removed, in many cases it is the only way to do the job right. Keep in mind that refinishing will have to be done as well.

ADHESIVE BONDING VERSUS WELDING

There are two methods of repairing plastics: chemical adhesives and welding. Table 9-1 indicates the best repair systems for the plastics most often used in the automotive industry.

CAUTION: Keep in mind that although the techniques described in this chapter are proven repair methods, no technique is 100 percent reliable when used to repair a fuel tank or in similar critical structural applications.

Although most types of plastics can be repaired with adhesive materials, welding is usually preferred for thermoplastics because of its speed and ease. A plastic welder takes only a couple of minutes to heat up to operating temperatures; once hot, it can weld at speeds ranging from 5 to 40 inches per minute, depending on the application. It is also possible to go directly from welding to sanding and painting without waiting for the repair to cure. At most, only a few minutes are needed for the welded plastic to cool.

Many adhesives require mixing and can take anywhere from 30 minutes to several hours to cure, but heat can be used to shorten the curing time. This is not necessarily a major disadvantage, since the repair technician can always find something else to do while the adhesive cures.

As for surface preparation, both welding and adhesives require some grinding, sanding, and/or trimming in order to achieve good adhesion. More surface preparation is generally needed with adhesives, however, because the edges of the damaged area must often be featheredged before the adhesive is applied.

Some adhesives require reinforcing or support patches behind the damaged area as well, making the repair a several step process. With plastic welding, however, there is no need to feather out the edges; V-grooving along the cracked or damaged area is generally all that is needed. Recommended V-grooves for one- and two-sided welds are shown in Figure 9-1.

Keep in mind, however, that hot-air plastic welding cannot be used on thermosetting plastic parts. Only thermoplastics soften when heated. This includes such plastics as polypropylene (PP), polyethylene (PE), soft thermoplastic polyurethane (TPUR), acrylonitrile butadian styrene (ABS), polyamide (PA or nylon), polycarbonate (PC or lexan), polyphenylene oxide (PPO or noryl), polyvinyl chloride (PVC), and acrylics such as plexiglass (PMMA). As a rule, all thermoplastics—with the exception of ethylene propylene diene (EPDM)—are weldable, although some are more easily welded than others.

For plastics in the thermosetting category, an alternative to hot-air plastic welding is needed. Thermosetting plastics undergo a chemical change when they cure that prevents them from softening when heated. This makes thermosetting plastics impossible to weld because the parent material will not melt and mix with the filler material. Trying to heat such a plastic will only result in charring the edges. Members of this category include polyurethane (PUR), polyester (UP or fiberglass), and fiberglass-reinforced ABS and PUR.

One alternative to hot-air plastic welding is the more versatile airless welding method. As shown in

TABLE 9-1: PLASTIC PARTS REPAIR SYSTEMS

KEY

AR	**Adhesive repair**	**S**	**Anerobic (instant) adhesive**
FGR	**Fiberglass repair**	**PC**	**Patching compound**
HAW	**Hot-air welding**	**AW**	**Airless welding**

ISO Code	Name	Repair System
ABS	Acrylonitrile-butadiene-styrene(hard)	HAW, S FGR, AW
ABS/PVC	ABS/Vinyl (soft)	PC, AW
EPI II or TPO	Ethylene propylene	AR, AW
PA	Nylon	S, FGR, AW
PC	Lexan	S, FGR
PE	Polyethylene	HAW, AW
PP	Polypropylene	HAW, AW
PPO	Noryl	FGR, AW
PS	Polystyrene	S
PUR, RIM, or RRIM	Thermoset polyurethane	AR, AW
PVC	Polyvinyl chloride	PC, AW
SAN	Styrene acrylonitrile	HAW, AW
SMC	Sheet Molded Compound (polyester)	FGR
TPR	Thermoplastic rubber	AR, AW
TPUR	Thermoplastic polyurethane	AR, AW
UP	Polyester (Fiberglass)	FGR

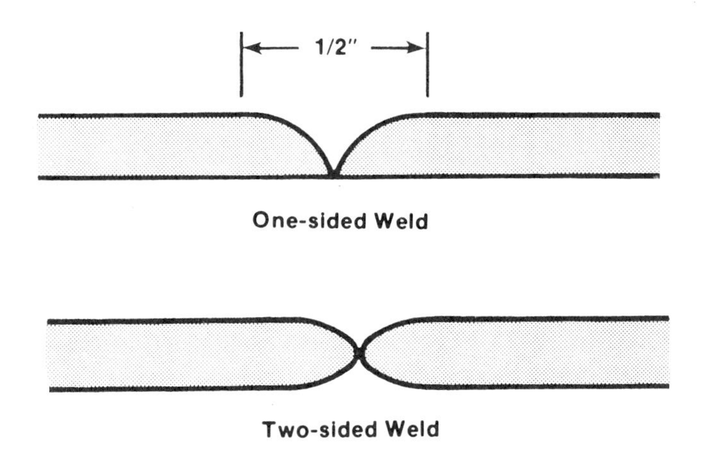

FIGURE 9-1 Proper V-grooves

Table 9-1, airless welding can be used to repair a number of thermosetting plastics, including nylon (PA), thermoset polyurethane (PUR), and thermoplastic rubber (TPR). As mentioned in the previous chapter, one difference between hot-air plastic welding and airless plastic welding is that the airless method more closely resembles a gluing process. As a result, the airless method is more versatile.

Of course, the other alternative to hot-air plastic welding is adhesive bonding. It is suitable for nearly every type of thermosetting plastic. However, a surface gouge on a rigid thermoset part where structural strength is not involved can be more economically repaired with polyester body filler. On the other hand, puncture damage that requires a backup or a structural repair that involves reinforcing the backside is best done, from a practical standpoint, with a combination of structural adhesive and polyester body filler. Since epoxy resins possess superior adhesive properties, all repair work done on the backside of the part should be done with fiberglass cloth and structural adhesive. The cosmetic repair on the front side of the part can then be completed with polyester body filler. Again, as stated earlier, information about which repair method is preferred for a particular repair can be found in shop manuals or special guides.

CHEMICAL ADHESIVE BONDING TECHNIQUES

As mentioned, the main advantage of adhesives is their versatility. Not all plastics can be welded, but adhesives can be used in most instances. In this section, various chemical adhesive bonding techniques are presented.

EPOXY AND URETHANE REPAIR MATERIALS

Both epoxy and urethane adhesive repair systems are "two component" systems that must be mixed in equal parts just prior to application. Some types of urethane are available in various degrees of flexibility that can be matched to the particular part being repaired.

Both epoxy and urethane applications are essentially the same. Whenever adhesives are used, keep the following points in mind:

- As with welding, surface preparation and cleanliness are extremely important. The part must be washed and cleaned with plastic cleaner before starting any repair.
- Both the damaged part and the repair material must be at room temperature (at least 60 degrees Fahrenheit) to achieve proper curing and adhesion.
- Both parts of the repair material must be equally and completely mixed before application.
- Whenever possible, particularly when repairing cuts or tears that go all the way through the paint, a fiberglass mat reinforcement should be used.

Two-Part Epoxy Repair of Polyolefin Component

Until recently, adhesive repairs were limited to urethane plastics only. With the introduction of adhesion promoters, polyolefins can now be repaired using virtually the same adhesive bonding techniques (Table 9-2). The only difference is the use of the adhesion promoter before each application of repair material; the rest of the procedure is the same. Automobile manufacturers are using polyolefins for large exterior parts such as fascias and bumper covers. Common polyolefins found in automotive applications include TPO, PP, and EP. To determine whether the base material is a polyolefin, grind the damaged area. If it grinds cleanly, an adhesion promoter is not needed. If the material melts or smears, it is a polyolefin and must have an adhesion promoter.

Use the following procedure to make a two-part epoxy repair of a polyolefin bumper cover.

1. Clean the entire cover with soap and water, wipe or blow-dry, then clean with a good plastic cleaner.

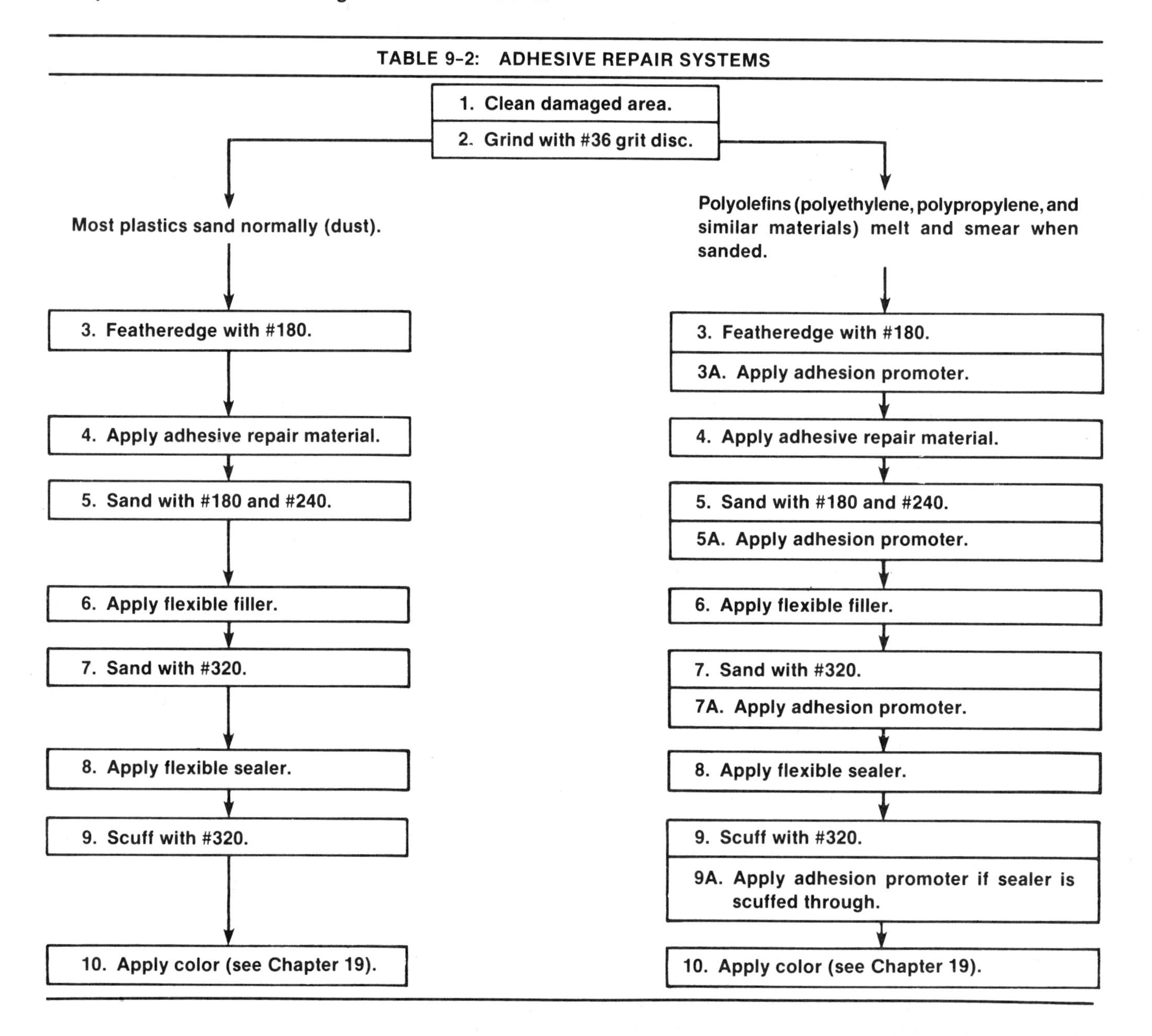

TABLE 9-2: ADHESIVE REPAIR SYSTEMS

1. Clean damaged area.
2. Grind with #36 grit disc.

Most plastics sand normally (dust).	Polyolefins (polyethylene, polypropylene, and similar materials) melt and smear when sanded.
3. Featheredge with #180.	3. Featheredge with #180. 3A. Apply adhesion promoter.
4. Apply adhesive repair material.	4. Apply adhesive repair material.
5. Sand with #180 and #240.	5. Sand with #180 and #240. 5A. Apply adhesion promoter.
6. Apply flexible filler.	6. Apply flexible filler.
7. Sand with #320.	7. Sand with #320. 7A. Apply adhesion promoter.
8. Apply flexible sealer.	8. Apply flexible sealer.
9. Scuff with #320.	9. Scuff with #320. 9A. Apply adhesion promoter if sealer is scuffed through.
10. Apply color (see Chapter 19).	10. Apply color (see Chapter 19).

2. V-groove and taper the damaged area using a slow speed grinder and 3-inch #36 grit disc. Grind about a 1-1/2-inch taper around the damage for good adhesion (Figure 9-2).
3. Use a sander with #180 grit paper to featheredge the paint around the damaged area; then blow dust-free. Depending on the extent of the damage, the backside might need reinforcement. To do this, follow steps 4 through 6:
4. Clean the backside with plastic cleaner (Figure 9-3), then apply a coat of adhesion promoter according to the manufacturer's recommendations (Figure 9-4).
5. Dispense equal amounts of both parts of the flexible epoxy adhesive and mix to a

FIGURE 9-2 Grinding around the damage for good adhesion

uniform color. Apply the material to a piece of fiberglass cloth using a plastic squeegee (Figure 9-5).

6. Attach the saturated cloth to the backside of the bumper cover and fill in the weave with additional adhesive material.
7. With the backside reinforcement in place as shown in Figure 9-6, apply a coat of adhesion promoter to the sanded repair area on the front side (Figure 9-7). Let the adhesion promoter dry completely.
8. Fill in the groove with adhesive material (Figure 9-8). Allow it to cure according to the manufacturer's recommendations.
9. Rough grind the repair area with #80 grit paper, then sand with #180 grit, followed by #240 grit. If additional adhesive material is needed to fill in a low spot or pinhole, be

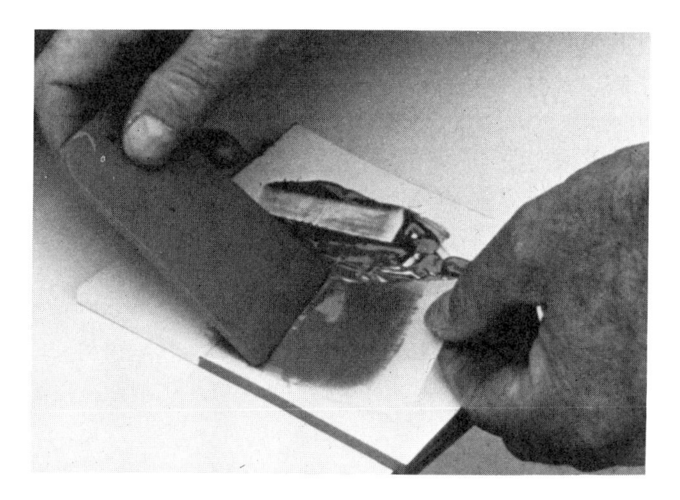

FIGURE 9-5 Applying flexible epoxy adhesive to fiberglass cloth

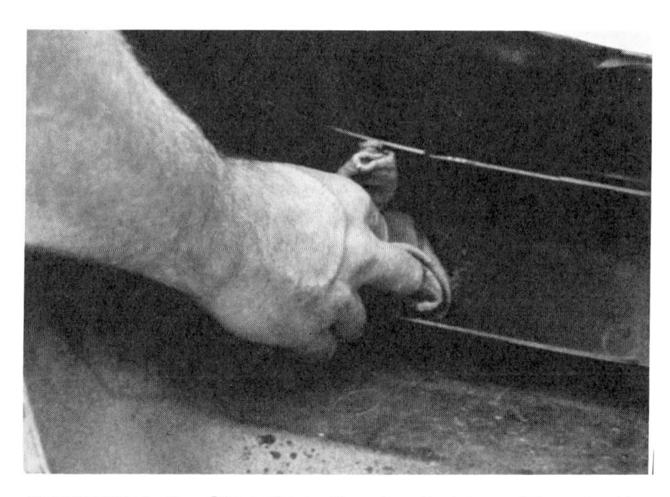

FIGURE 9-3 Cleaning the backside with plastic cleaner

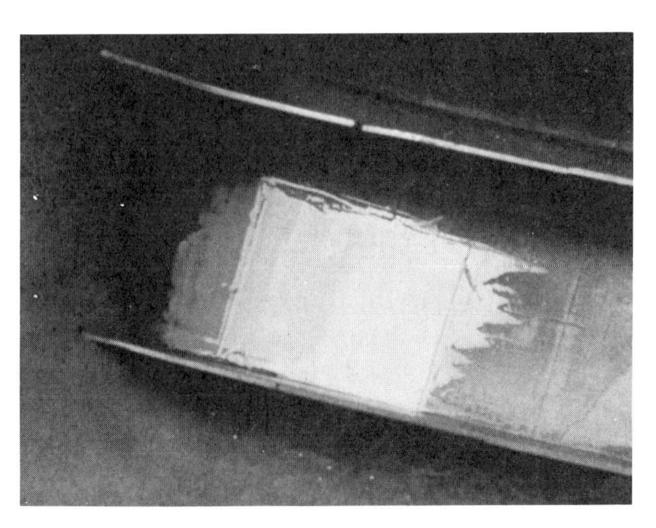

FIGURE 9-6 Finished backside reinforcement

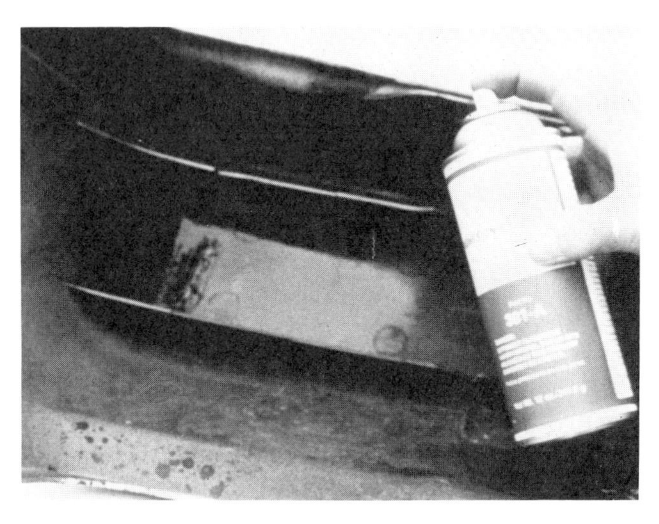

FIGURE 9-4 Applying an adhesive promoter to the backside

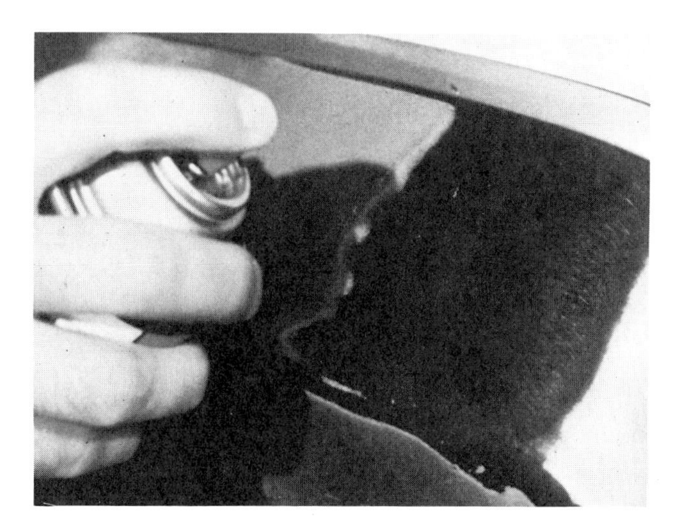

FIGURE 9-7 Applying adhesive promoter to the front side

FIGURE 9-8 Filling in the groove with adhesive material

sure to apply a coat of adhesion promoter first, followed by a skim coat of adhesive material.

SHOP TALK

Never attempt to use flexible putty on polyolefins—it will not work.

Adhesive Repair of Other Flexible Parts

When repairing other flexible parts with adhesive materials, the procedure to be followed is similar to the one presented above with these exceptions:

1. Omit the adhesion promoter.
2. Use either a skim coat of adhesive material or a flexible filler to fill in imperfections—not both.

Minor Cut and Crack Repairs. Adhesives are usually used for repairing minor cuts and cracks. First, the repair area must be wiped clean with water and a plastic cleaner. It is very important for the mating surfaces to be clean and free of wax, dust, grease, or any other substance. It is not necessary to use solvents (other than a plastic cleaner) for cleaning plastic parts. Allow the part(s) to warm to 70 degrees Fahrenheit before applying adhesives.

After cleaning, the next step is to prepare the damage using an adhesive kit. The kit should have two elements: an accelerator and an adhesive. Spray

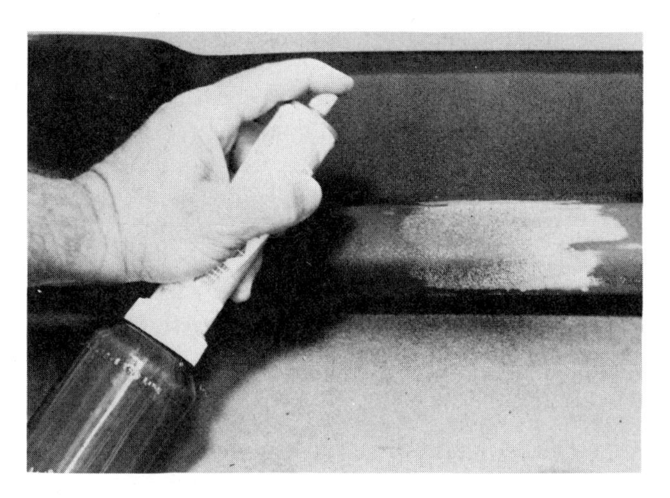

FIGURE 9-9 Spraying a crack with accelerator

one side of the crack with the accelerator, as shown in Figure 9-9, then apply the adhesive to the same side of the crack.

Carefully position the two sides of the cut or crack in their original position and press them together with firm pressure. Hold for a full minute to achieve good bond strength. Cure for 3 to 12 hours for maximum strength, following the instructions and precautions on the label.

If the original paint was not damaged and the repair was properly positioned, painting might not be required. Where painting is required, be sure to read the manufacturer's instructions that appear on the paint container.

Gouge, Tear, and Puncture Repair. The procedure for repairing gouges, tears, and punctures is a chemical bonding process; it requires no special tools or skills. As always, the first step is to clean around the damaged area thoroughly with a wax, grease, and silicone-removing solvent. Apply it with a water-dampened cloth, then wipe dry.

To prepare for the structural adhesive, bevel the edges of the hole back about 1/4 to 3/8 inch. The technician in Figure 9-10 is using a 3-inch, medium grit grinding disc. The beveling has produced a coarse surface for good adhesion; in any repair, the mating surfaces should be scuffed. Use a slow speed when grinding (2000 RPM or less). If the sanded area has a "greasy" appearance, it would be a good idea to apply a coat of adhesion promoter. Also, apply more adhesion promoter after each sanding step.

The next step is to featheredge the paint around the repair area with a finer grit disc (Figure 9-11). Remove the paint, but very little of the urethane plastic, and blend the paint edges into the plastic.

Continue removing paint until there is a paint-free band about 1 to 1-1/2 inches wide around the hole. The repair material must not overlap the painted surface. Carefully wipe off all paint and urethane dust in preparation for the next step. The repair area must be absolutely clean before proceeding.

Next, flame treat the beveled area of the hole. This is done to improve the adhesion of the structural adhesive. Use a torch with a controlled flame, and develop a 1-inch cone tip. Direct the flame onto the beveled area carefully and keep it moving until the area is slightly brown. Be extremely careful not to warp the urethane or burn the paint.

The next step is to apply auto backing tape to the repair area; an aluminum foil with a strong adhesive and a moisture-proof backing is best. Clean the inner surface of the repair area with silicone and wax remover, then install the tape. Cover the hole completely, with about a 1-inch adhesion surface around the edges.

Before applying the structural repair adhesive, the back of the opening should be thoroughly cleaned. Tape it with aluminum auto body repair tape to provide support for the repair. For best results, slightly dish the tape so that the adhesive overlaps the repair area on the backside. It might be possible to install the tape without loosening or removing any parts from the car. On some makes, partial disassembly is necessary.

Prepare the adhesive repair material as directed by the manufacturer. Most adhesive compounds come in two tubes. On a clean, flat, nonporous surface such as metal or glass, squeeze out equal amounts of each tube. Then, with an even paddling motion to reduce air bubbles, completely mix the two components until a uniform color and consistency is achieved.

Paddle the adhesive into the hole, using a squeegee or plastic spreader (Figure 9-12). This must be done quickly and carefully; the adhesive material will begin to set in about 2 to 3 minutes. Two applications are usually required. The first application simply fills the bottom of the hole, so it is not necessary to worry about contour at this time. However, an attempt should be made to fill most of the hole's volume. Cure for approximately 1 hour at room temperature or 20 minutes with a heat lamp or gun at approximately 200 degrees Fahrenheit.

Before the final application of adhesive, use a fine grit disc to grind down the high spots (Figure 9-13). Wipe the dust from the repair area. After the first application has been ground and wiped clean, mix the second application of the adhesive as before, and paddle it into an overfill contour of the repair area (Figure 9-14). A flexible squeegee or spatula is useful in approximating the panel contours.

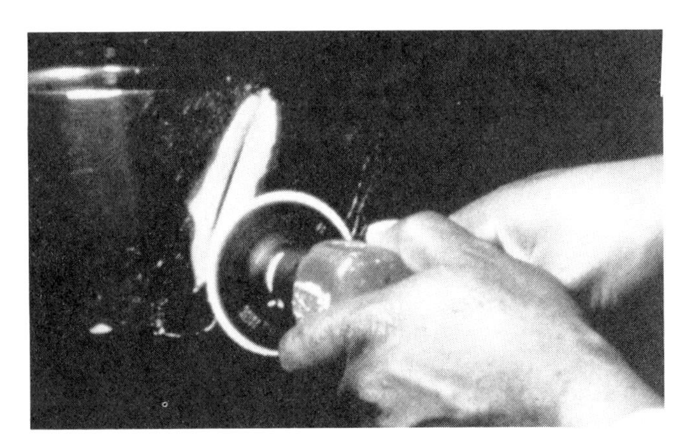

FIGURE 9-10 Beveling the damaged area with a 3-inch grinding disc

FIGURE 9-11 Using a finer grit disc to featheredge the paint

FIGURE 9-12 Applying structural adhesive with a squeegee

FIGURE 9-13 Grinding down high spots

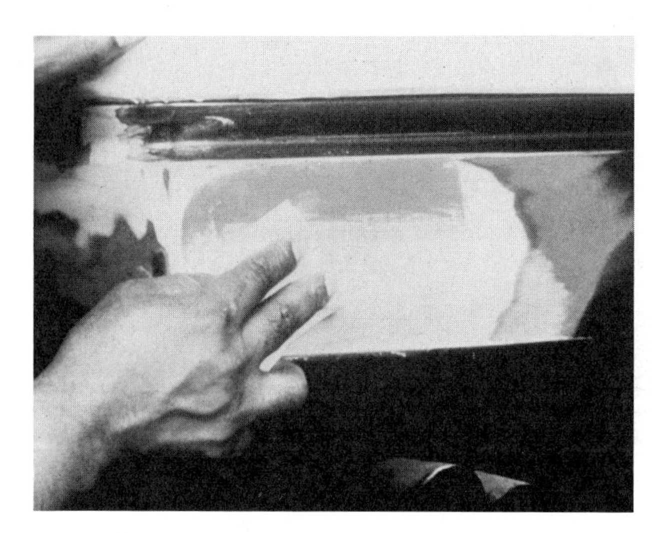

FIGURE 9-14 Applying a second coat of adhesive

When the adhesive material has dried, establish a rough contour to the surrounding area with a #80 grit abrasive on a sanding block. Then feather-sand with a disc sander and #180 sandpaper, followed by #240 sandpaper to achieve levelness with the surface of the part. Check the repair for any low areas, pits, or pinholes. Additional material can be spread over any defects.

Final feathering and finish sanding can be done with a disc sander and #320 grit disc. When final sanding is completed, remove all dust and loose material. The surface is now ready for priming and painting.

PLASTIC WELDING

The basic procedures for hot-air and airless welding are very similar. To make a good plastic weld with either procedure, keep the following factors in mind:

- *Welding Rod Material.* If it is not compatible with the base material, the weld will not hold.
- *Temperature.* Too much heat will char, melt, or distort the plastic.
- *Pressure.* Too much pressure stretches and distorts the weld.
- *Angle between Rod and Part.* If too shallow, a proper weld will not be achieved.
- *Speed.* If the torch movement is too fast, it will not permit a good weld; too slow a speed can char the plastic.

The basic repair sequence is generally the same for both procedures:

1. Prepare the damaged area.
2. Align the damaged area.
3. Make the weld.
4. Allow it to cool.
5. Sand. If pinholes, voids, and the like exist, bevel the edges of the defective area and add another bead of weld. Resand.
6. Paint or finish.

BASICS OF HOT-AIR WELDING

The typical hot-air plastic welding procedure is as follows:

1. Set the welder to the proper temperature (if a temperature adjustment is provided).
2. Wash and clean the part with plastic cleaner; do not use conventional prep solvents or dewaxers. To remove silicone-type materials, use a conventional cleaner first, making sure to completely remove all residue.
3. V-groove the damaged area.
4. Bevel the part 1/4 inch beyond the damaged area.
5. Tack weld or tape the break line with aluminum body tape.
6. Select the welding tip best suited to the type of damage, and the proper welding rod.
7. Make the weld. Allow it to cool and cure for about 30 minutes.
8. Grind, sand, or scrape the weld to the proper contour and shape.

BASICS OF AIRLESS PLASTIC WELDING

The typical airless plastic welding procedure is as follows:

1. Wash the damaged part with soap and water and wipe or blow-dry.
2. Clean the damaged part with plastic cleaner.
3. Align and tape the broken or split sections with aluminum body tape.
4. V-groove at least 50 percent of the way through the panel for a two-sided weld and 75 percent of the way through for a one-sided weld.
5. Use a slow speed grinder and a #60 or #80 grit disc to remove the paint from around the damaged area. Blow dust free.
6. After setting the welder to the proper temperature (Figure 9-15), slowly feed the rod into the melt tube.
7. Apply light pressure to the rod to slowly force it out into the grooved area.
8. As the rod melts, start to move the torch tip very slowly in the direction of the intended weld. Overlap the edges of the groove with melted plastic while progressing forward.
9. After completing the weld, use the flat "shoe" part of the torch tip to smooth it out.

GENERAL WELDING TECHNIQUES

Welding plastic is not difficult when done in a careful and thorough manner. Keep in mind the following general tips:

- The welding rod must be compatible with the base material in order for the strength, hardness, and flexibility of the repair to be the same as the part. To this end, the rods are color coded for easy identification.
- Always test a welding rod for compatibility with the base material. To do this, melt the rod onto a hidden side of the damaged part, let the rod cool, then try to pull it from the part. If the rod is compatible, it will adhere.
- Pay close attention to the temperature setting of the welder; it must be correct for the type of plastic being welded.
- Never use oxygen or other flammable gases with a plastic welder.
- Never use a plastic welder, heat gun, or similar tool in wet or damp areas. Remember: electric shock can kill.
- Become proficient at horizontal welds before attempting the more difficult vertical and overhead types.
- Make welds as large as they need to be. The greater the surface area of a weld, the stronger the bond.

FIGURE 9-15 Setting the temperature dial on an airless welder *(Courtesy of Urethane Supply Co.)*

- Before beginning an airless weld, run a small piece of the welding rod through the welder to clean out the torch tip.
- Consult a supplier for the brands of tools and materials that best fit the shop's needs. Always read and follow the manufacturer's instructions carefully.

When welding plastic, single- or double-V butt welds (Figure 9-16) produce the strongest joints; lap fillet welds are also good. When using a round- or V-shaped filler rod, the damaged area is prepared by slowly grinding, sanding, or shaving the adjoining surfaces with a sharp knife to produce a single or double V. For flat ribbon filler rods, V-grooving is not necessary. Wipe any dust or shavings from the joint with a clean, dry rag. The use of cleaning solvents is not generally recommended because they can soften the edges and cause poor welds.

TACK WELDING

The edges to be joined must be aligned. For long tears where backup is difficult, small tack welds can be made along the length of the tear to hold the two sides in place while doing the weld. For larger areas, a patch can be made from a piece of plastic and tacked in place. To tack weld, proceed as follows:

1. Hold the damaged area in its correct position with clamps or fixtures.
2. Using a tacking welding tip, fuse the two sides to form a thin hinge weld along the

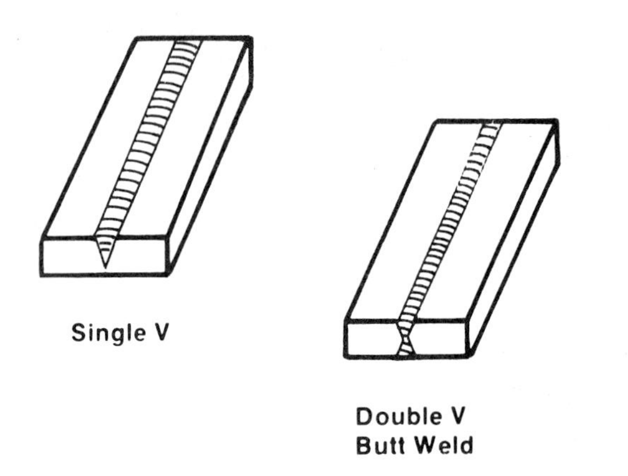

FIGURE 9-16 Single- and double-V butt welds produce strong joints.

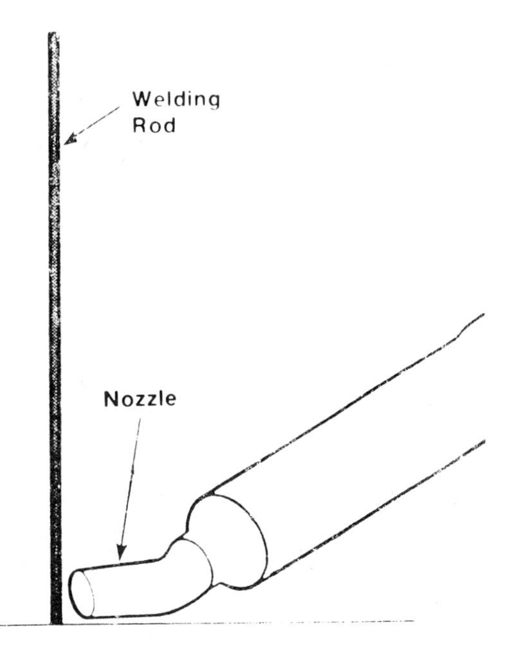

FIGURE 9-17 Keep the nozzle parallel to the base material and the rod at a right angle to the surface.

root of the crack. This is especially useful for long cracks because it allows easy adjustment and alignment of the edges.

3. Start the tacking by drawing the point of the welding tip along the joint. Press the tip in firmly, making sure to contact both sides of the crack; draw the tip smoothly and evenly along the line of the crack.
4. The point of the tip will fuse both sides in a thin line at the root of the crack. The fused parts will hold the sides in alignment, though they can be separated and retacked if adjustment is necessary. Fuse the entire length of the crack.

HAND WELDING

Several techniques must be mastered for quality hand welding to be performed.

Starting the Weld

Prepare the rod for welding by cutting the end at an angle of approximately 60 degrees. When starting a weld, the tip of the welder should be held about 1/4 to 1/2 inch above and parallel to the base material. The filler rod is held at a right angle to the work as shown in Figure 9-17, with the cut end of the rod positioned at the beginning of the weld.

Direct the hot air from the tip alternately at the rod and the base, but concentrating more on the rod. Always keep the filler rod in line with the V while pressing it into the seam. Light pressure (about 3 psi) is sufficient for achieving a good bond. Once the rod begins to stick to the plastic, start to move the torch and use the heat to control the flow. Be careful not to melt or char the base plastic or to overheat the rod. As the welding continues, a small bead should form ahead of the rod along the entire weld joint. A good start is essential, because this is where most weld failures begin. For this reason, starting points on multiple-bead welds should be staggered whenever possible.

Continuing the Weld

Once the weld has been started, the torch should continue to fan from rod to base material. Because the rod now has less bulk, a greater amount of heat must be directed at the base material. Experience will develop the proper technique.

Feeding the Rod

In the welding process the rod is gradually being used up, making it necessary for the welder to renew his or her grip on the rod. Unless this is performed carefully, the release of pressure might cause the rod to lift away from the weld and allow air to become trapped under the weld. As a result, the weld is weakened. To prevent this, the welder must develop the skill of continuously applying pressure on the rod while repositioning the fingers. This can be done by applying pressure with the third and fourth fingers while moving the thumb and first finger up the rod (Figure 9-18). Another way is to hold the rod down into the weld with the third and fourth finger while repositioning the thumb and first finger. The rod is cool enough to do this because only the bottom of it should be heated. However, care should

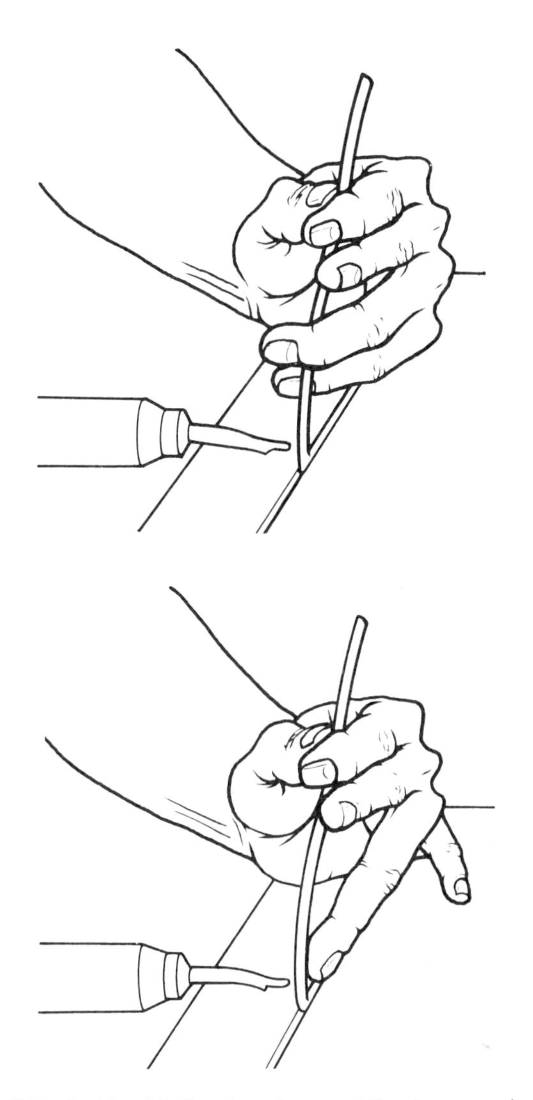

FIGURE 9-18 Methods of repositioning a grip on the rod

be observed in touching new welds or aiming the torch near the fingers.

Finishing the Weld

As the end of a weld is approached, maintain pressure on the rod as the heat is removed. Hold it still for a few seconds to make sure the rod has cooled enough so it will not pull loose, then carefully cut the rod with a sharp knife or clippers. Do not attempt to pull the rod from the joint. About 15 minutes cooling time is needed for rigid plastic and 30 minutes for thermoplastic polyurethane.

Rough Grinding the Weld

The welded area can be made smooth by grinding with coarse (36 grit) emory or sandpaper. A 9- or 10-inch disc on a low-speed electric grinder will remove large weld beads. Excess plastic can be removed with a sharp knife before grinding. Care must be taken not to overheat the weld area because it will soften. To speed up the work without damaging the weld, periodic cooling with water is necessary.

Checking the Weld

After rough grinding, the weld should be checked visually for defects. Any voids or cracks will make the weld unacceptable. Bending should not produce any cracks because a good weld is as strong as the part itself. Table 9-3 shows some typical welding defects, their causes, and corrections.

Finishing

The weld area can be finish sanded by using 220 grit sandpaper followed by a 320 grit. Either a belt or orbital sander may be used, plus hand sanding as required. If refinishing is required, follow the procedure designed specifically for plastics.

SPEED WELDING

On panel work, speed welding is very popular. Here are some techniques that are essential for quality speed welding.

Starting the Weld

With the high-speed torch held like a dagger and the hose on the outside of the wrist, bring the tip over the starting point about 3 inches from the material so the hot air will not affect the material (Figure 9-19). Cut the welding rod at a 60-degree angle, insert it into the preheating tube, and immediately place the pointed shoe of the tip on the material at the starting point. Hold the welder perpendicular to the material and push the rod through until it stops against the material at the starting point. If necessary, lift the torch slightly to allow the rod to pass under the shoe. Keeping a slight pressure on the rod with the left hand and only the weight of the torch on the shoe, pull the torch slowly toward you. The weld is now started.

CAUTION: Once the weld is started, do not stop. If the forward movement must stop for any reason, pull the tip off the rod immediately; if this is not done, the rod will melt into the tip's feed tube. Clean the feeder foot with a soft wire brush as soon as the welding is completed.

TABLE 9-3: PLASTIC WELDING TROUBLESHOOTING GUIDE

Problem	Cause	Remedy
Porous Weld	1. Porous weld rod 2. Balance of heat on rod 3. Welding too fast 4. Rod too large 5. Improper starts or stops 6. Improper crossing of beads 7. Stretching rod	1. Inspect rod 2. Use proper fanning motion 3. Check welding temperature 4. Weld beads in proper sequence 5. Cut rod at angle, but cool before releasing 6. Stagger starts and overlap splices 1/2″
Poor Penetration	1. Faulty preparation 2. Rod too large 3. Welding too fast 4. Not enough root gap	1. Use 60° bevel 2. Use small rod at root 3. Check for flow liners while welding 4. Use tacking tip or leave 1/32″ root gap and clamp pieces
Scorching	1. Temperature too high 2. Welding too slowly 3. Uneven heating 4. Material too cold	1. Increase airflow 2. Hold constant speed 3. Use correct fanning motion 4. Preheat material in cold weather
Distortion	1. Overheating at joint 2. Welding too slowly 3. Rod too small 4. Improper sequence	1. Allow each bead to cool 2. Weld at constant speed; use speed tip 3. Use larger sized or triangular shaped rod 4. Offset pieces before welding 5. Use double V or back-up weld 6. Back-up weld with metal
Warping	1. Shrinkage of material 2. Overheating 3. Faulty preparation 4. Faulty clamping of parts	1. Preheat material to relieve stress 2. Weld rapidly—use back-up weld 3. Too much root gap 4. Clamp parts properly; back up to cool 5. For multilayer welds, allow time for each bead to cool
Poor Appearance	1. Uneven pressure 2. Excessive stretching 3. Uneven heating	1. Practice starting, stopping and finger manipulation on rod. 2. Hold rod at proper angle 3. Use slow uniform fanning motion, heating both rod and material (for speed welding: use only moderate pressure, constant speed, keep shoe free of residue)
Stress Cracking	1. Improper welding temperature 2. Undue stress or weld 3. Chemical attack 4. Rod and base material not same composition 5. Oxidation or degradation of weld	1. Use recommended welding temperature 2. Allow for expansion and contraction 3. Stay within known chemical resistance and working temperatures of material 4. Use similar materials and inert gas for welding 5. Refer to recommended application

TABLE 9-3: PLASTIC WELDING TROUBLESHOOTING GUIDE (CONTINUED)

Problem	Cause	Remedy
Poor Fusion	1. Faulty preparation 2. Improper welding techniques 3. Wrong speed 4. Improper choice of rod 5. Wrong temperature	1. Clean materials before welding 2. Keep pressure and fanning motion constant 3. Take more time by welding at lower temperatures 4. Use small rod at root and large rods at top—practice proper sequence 5. Preheat materials when necessary 6. Clamp parts securely

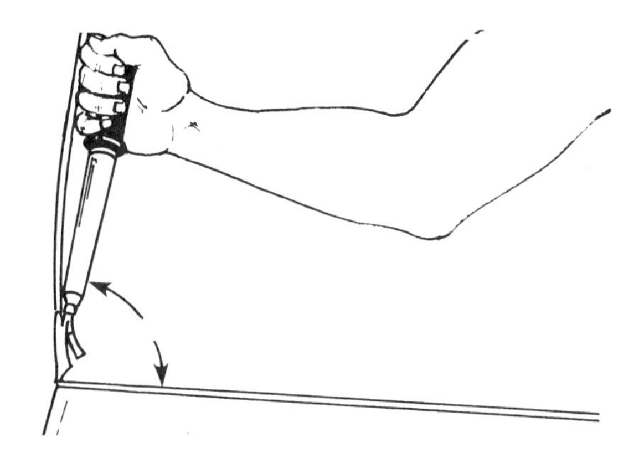

FIGURE 9-19 Starting a speed weld

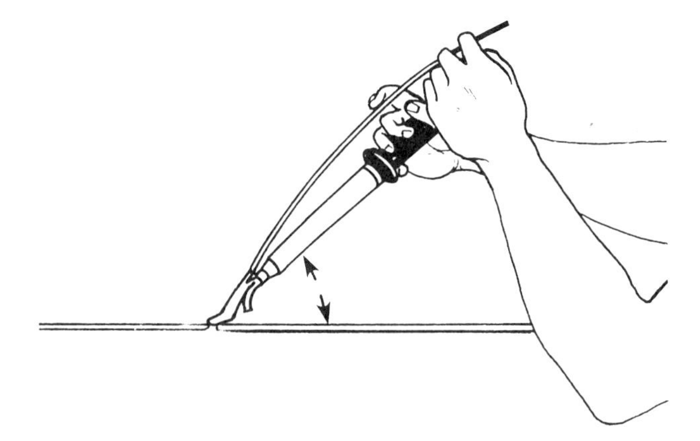

FIGURE 9-20 Continuing a speed weld

Continuing the Weld

In the first inch or two of travel, the rod should be helped along by pushing it into the tube with slight pressure. Once the weld has been properly started, the torch is brought to a 45-degree angle; the rod will now feed automatically without further help. As the torch moves along, visual inspection will indicate the quality of the weld being produced.

The angle between the welder and base material determines the welding rate. Since the preheater hole in the speed tip precedes the shoe, the angle of the welder to the material being welded determines how close the hole is to the base material and how much preheating is being done. It is for this reason the torch is held at a 90-degree angle when starting the weld and at 45 degrees thereafter (Figure 9-20). When a visual inspection of the weld indicates a welding rate that is too fast, the torch should be brought back to the 90-degree angle temporarily to slow down the welding rate, then gradually moved to the desired angle for proper welding speed. It is important that the welder be held in such a way that the preheater hole and the shoe are always in line with the direction of the weld, so that only the material in front of the shoe is preheated. A heat pattern on the base material will indicate the area being preheated. The rod should always be welded in the center of that pattern.

Finishing the Weld

It is important to remember that, once started, speed welding must be maintained at a fairly constant rate of speed (Figure 9-21). The torch cannot be held still. To stop welding before the rod is used up, bring the torch back past the 90-degree angle and cut off the rod with the end of the shoe. This can also be accomplished by pulling the speed tip off the remaining rod. When cutting the rod with the shoe, the remaining rod must be removed promptly from the preheater tube. Rod not removed promptly from the preheater tube will char or melt, clogging the tube and making it necessary for the tube to be cleaned out by inserting a new rod in the tube.

A good speed weld in a V-joint will have a slightly higher crown and more uniformity than the nor-

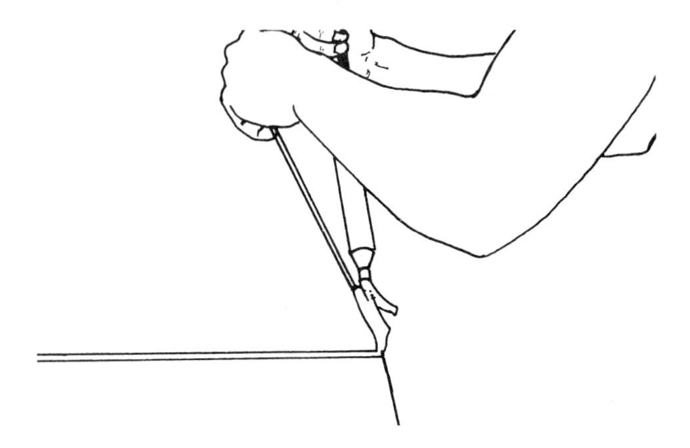

FIGURE 9-21 Finishing a speed weld

mal hand weld. It should appear smooth and shiny, with a slight bead on each side. For best results and faster welding speed, the shoe on the speed tip should be cleaned occasionally with a wire brush to remove any residue that might cling to it and create drag on the rod.

REVIEW QUESTIONS

1. Which of the following must be considered when deciding whether to replace or repair a damaged part?
   a. cost
   b. extent of damage
   c. availability of replacement parts
   d. both a and c
   e. all of the above

2. Welder A says that welding is usually preferred for thermoplastics. Welder B says that thermoplastics can only be repaired with adhesive bonding. Who is right?
   a. Welder A
   b. Welder B
   c. Both A and B
   d. Neither A nor B

3. What can be used to shorten the time adhesives take to cure?
   a. cooling
   b. heating
   c. both a and b
   d. neither a nor b

4. Welder A prepares the surface before using an adhesive. Welder B does not. Who is right?
   a. Welder A
   b. Welder B
   c. Both A and B
   d. Neither A nor B

5. Which of the following is a thermosetting plastic?
   a. polypropylene
   b. acrylonitrile butadian styrene
   c. polyvinyl chloride
   d. all of the above
   e. none of the above

6. Which of the following plastics can be repaired using airless welding?
   a. nylon
   b. thermoset polyurethane
   c. thermoplastic rubber
   d. all of the above
   e. none of the above

7. Which is more versatile?
   a. chemical adhesive bonding
   b. hot-air welding
   c. airless welding
   d. none of the above

8. When Welder A grinds the base material, it melts and smears, so he or she uses an adhesion promoter to make the repair. Welder B says that an adhesion promoter is needed only when repairing a polyolefin component. Who is right?
   a. Welder A
   b. Welder B
   c. Both A and B
   d. Neither A nor B

9. What should never be used on polyolefins?
   a. plastic cleaner
   b. flexible putty
   c. adhesion promoter
   d. slow-speed grinder

10. When repairing minor cuts and cracks with adhesives, Welder A uses solvent for cleaning plastic parts. Welder B uses only a plastic cleaner. Who is right?
    a. Welder A
    b. Welder B
    c. Both A and B
    d. Neither A nor B

11. Welder A says that the greater the surface area of a weld, the stronger the bond. Welder B says that there is no relationship between the surface area of a weld and the bond strength. Who is right?
    a. Welder A
    b. Welder B
    c. Both A and B
    d. Neither A nor B

12. Which of the following is not permanent?
    a. tack weld
    b. hand weld
    c. speed weld
    d. all of the above
    e. none of the above

13. At what angle should the welding rod be cut before welding begins?
    a. 45 degrees
    b. 50 degrees
    c. 60 degrees
    d. 90 degrees

14. How much cooling time is needed for a rigid plastic weld?
    a. 15 minutes
    b. 20 minutes
    c. 30 minutes
    d. 45 minutes

15. Which weld will have a higher crown?
    a. hand weld
    b. speed weld
    c. both will be equally high
    d. none of the above

16. Which of the following plastics cannot be repaired by hot-air welding?
    a. styrene acrylonitrile
    b. polystyrene
    c. polypropylene
    d. polyethylene

17. When doing an adhesive repair, when is it most important that a fiberglass mat reinforcement be used?
    a. when the damaged part is a polyolefin
    b. when the damaged part is a thermosetting plastic
    c. when repairing a cut or tear that goes all the way through the paint
    d. when both the damaged part and the repair material are at room temperature

18. After mixing both parts of a flexible epoxy adhesive, apply it to the fiberglass cloth using a ____________.
    a. paintbrush
    b. spray can
    c. plastic squeegee
    d. none of the above

19. Low spots are discovered after completing a two-part epoxy repair of a polyolefin. Welder A applies a fresh coat of adhesion promoter, followed by a skim coat of adhesive material. Welder B skips the adhesion promoter and applies only the adhesive material. Who is right?
    a. Welder A
    b. Welder B
    c. Both A and B
    d. Neither A nor B

20. Which of the following statements concerning minor cut and crack adhesive repairs is incorrect?
    a. Allow the parts to warm to 70 degrees Fahrenheit before applying the adhesive.
    b. The adhesive is applied after the accelerator.
    c. Both sides of the cut or crack must be sprayed with accelerator.
    d. For maximum strength, allow 3 to 12 hours of curing time.

21. After completing an airless plastic weld, what is used to smooth it out?
    a. Disc sander and #320 grit (or finer) disc
    b. Cheese grater file
    c. Flat "shoe" part of the torch tip
    d. Slow-speed grinder and #36 grit disc

22. Welder A starts a hand weld by holding the tip of the welder at a right angle to the work; Welder B holds it parallel to the work. Who is right?
    a. Welder A
    b. Welder B
    c. Both A and B
    d. Neither A nor B

23. Which of the following is a possible cause of warping of a plastic weld?
    a. insufficient root gap
    b. welding too slowly
    c. faulty clamping of parts
    d. both a and b

24. Which of the following is a possible cause of a porous plastic weld?
    a. improper starts or stops
    b. welding too fast
    c. rod too large
    d. all of the above

CHAPTER TEN

SPECIAL PLASTIC AND FIBERGLASS REPAIR TECHNIQUES

Objectives

After reading this chapter, you should be able to

- Perform a one-sided weld on ABS.
- Perform a two-sided weld on polypropylene.
- Restore a distorted bumper cover by reshaping it and a stretched bumper cover by shrinking it.
- Repair a urethane bumper that has sustained damage in the mounting area.
- Repair a cut or tear in a urethane bumper cover.
- Repair vinyl-clad urethane foam that has sustained surface dents, cuts, tears, and cracks.
- List the safety points unique to working with fiberglass.
- Perform fiberglass repairs, including fixing holes in fiberglass panels, fixing shattered fiberglass panels, and replacing fiberglass panels.

As stated earlier, the use of plastic in cars and light trucks is on the rise with each new model year. The actual number of automotive plastics is also expanding, as new hybrids are constantly refined for use in the vehicle manufacturing process. And not all such plastics can be repaired using the conventional methods outlined in the previous chapter; some require special repair materials and procedures to restore them to their original condition. This chapter will examine the following techniques: one- and two-sided welds; reshaping, shrinking, and repairing bumper covers; repairing high-stress areas of a bumper; repairing and refinishing vinyl-clad urethane foam; and repairing and replacing fiberglass. In the case of the one- and two-sided welds, although the airless method is used here because of its previously mentioned advantages, the hot-air method can of course be used instead.

In all cases, follow the procedures exactly as they are presented. Do not skip any of the steps, and do not perform them out of order. And remember: always be sure to read and follow the instructions carefully for the repair products used.

ONE-SIDED ABS WELD

ABS is one of the most common hard plastics found in today's vehicles. A cosmetic filler is not required when working with hard plastics; the welding rod serves as the filler. Although two-sided welds are preferred when the damaged part can be removed from the vehicle, the one-sided weld is perfectly adequate if done properly. Use the following step-by-step procedure to do a one-sided ABS weld using an airless welder:

1. Set the temperature dial on the welder for ABS. Let the welder warm up while the part is being prepared.

CAUTION: Never set up a welder in wet or damp areas. Electrical shock can kill.

2. As always, proper weld preparation is very important. Wash the part with soap and water, then wipe or blow-dry.

FIGURE 10-1 V-grooving in preparation for the weld

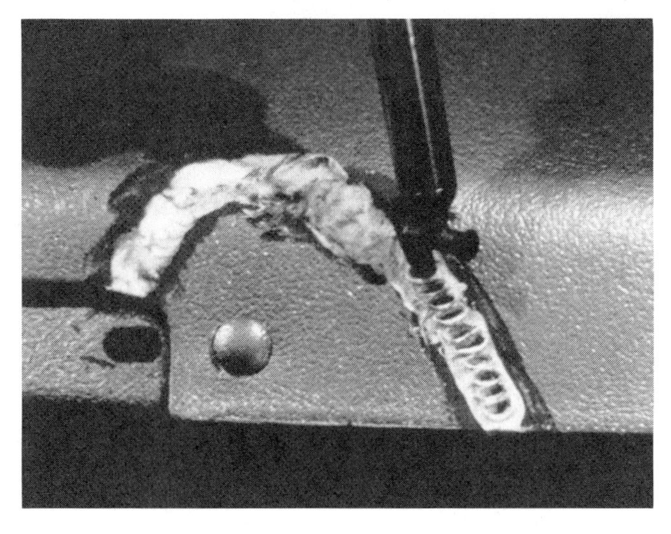

FIGURE 10-3 The stitch-tamp procedure is unique to ABS plastic.

3. Wash the part with plastic cleaner to remove grease, wax, and other contaminants that can cause problems with the repair. Several applications of plastic cleaner might be necessary.
4. Align the break with aluminum body tape; this will hold the plastic securely for welding. Press the tape in place with a plastic squeegee or paint paddle.
5. Because the welding will be done from one side only, it will be necessary to V-groove the damaged area 75 percent of the way through the base material. Use a die grinder, 1/4-inch drill with a cutter bit or grinding pad, small grinder, or rotary file (Figure 10-1). The part is now ready to be welded.
6. Begin the weld by inserting the rod into the melt tube. Place the flat "shoe" part of the tip over the V-groove and very gently feed the rod through the tube as the melting begins (Figure 10-2). Move the tip along the groove slowly for good melt-in and heat penetration.
7. For optimum strength, use the stitch-tamp procedure; this technique is used solely on ABS plastic. To do this, move the pointed end of the tip slowly into the weld area. This will force the rod into the base material and bond the two together for a strong weld (Figure 10-3). Stitch tamp the entire length of the weld. The rod and base material should be well mixed and become a uniform color.

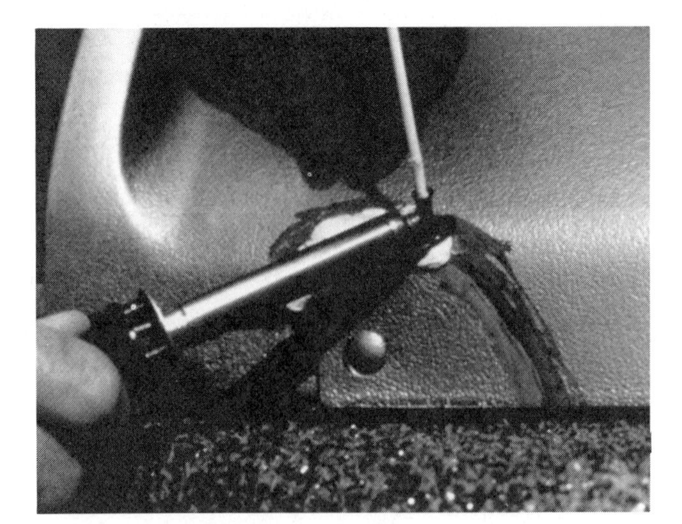

FIGURE 10-2 Gently feed the welding rod through the melt tube.

FIGURE 10-4 Use a single-edge razor blade to smooth the excess rod buildup.

8. Resmooth the weld area using the flat "shoe" part of the tip, again working slowly for good heat penetration.
9. Shape the excess rod buildup to a smooth contour using a slow-speed grinder and #80 grit disc. This can also be accomplished by scraping with a single-edged razor blade (Figure 10-4).
10. When the weld is sufficiently smooth, scuff sand the area with #320 grit paper; this will provide good adhesion for refinishing. Blow dust free.

TWO-SIDED POLYPROPYLENE WELD

Polypropylene is a hard plastic used in interior trim panels, inner fender liners, bumper covers, fan shrouds, and heater housings. The two-sided weld is recommended whenever the damaged part can be removed from the vehicle because it restores the total strength of the part. It is especially important that high-stress parts such as mounting tabs and edge reinforcements get two-sided welds. Naturally, one-sided welds are fine for solely cosmetic repairs. V-groove 50 percent of the way through the base material on each side of the cut or tear so that the welds touch each other.

Following is the step-by-step procedure for doing a two-sided polypropylene weld using an airless welder:

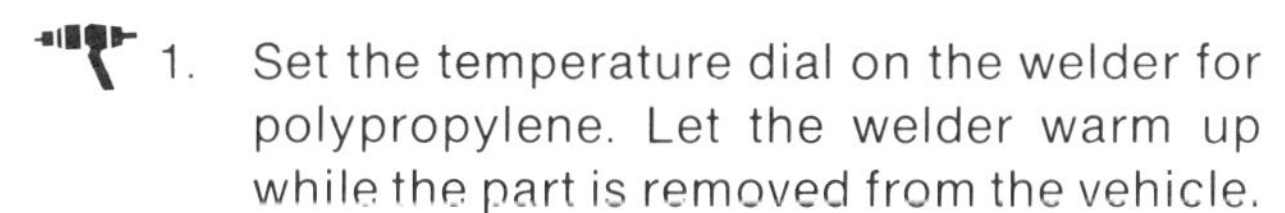

1. Set the temperature dial on the welder for polypropylene. Let the welder warm up while the part is removed from the vehicle.

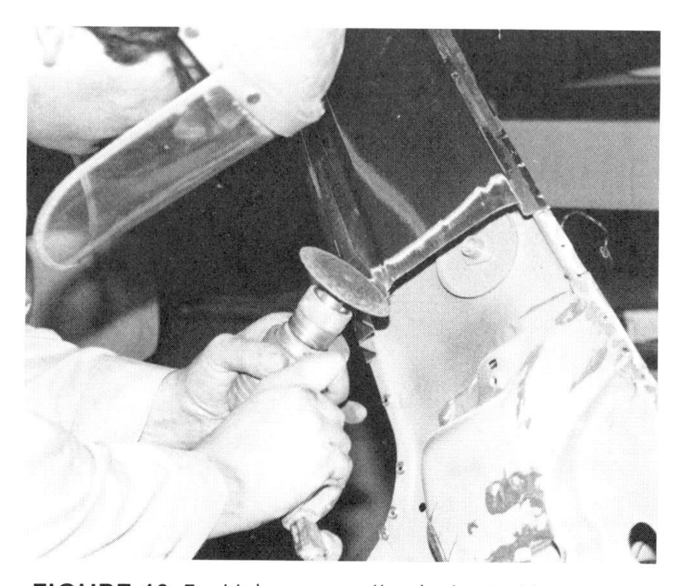

FIGURE 10-5 Using a small grinder to V-groove

FIGURE 10-6 The melt-flow procedure

2. Wash the part with soap and water. Wipe or blow-dry.
3. Wash the part with a good plastic cleaner to remove grease, wax, and other contaminants.
4. Align the break with aluminum body tape; proper alignment of the joint is critical. Press the tape into place with a plastic squeegee or paint paddle.
5. V-groove 50 percent of the way through the backside of the panel, using a die grinder, 1/4-inch drill with a cutter bit or grinding pad, small grinder (Figure 10-5), or rotary file.
6. The backside weld should be done using the melt-flow method. With the rod inserted in the melt tube, place the flat "shoe" part of the tip directly over the V-groove and hold it in place until the melted rod begins to flow out around the shoe (Figure 10-6).
7. Let the rod melt out on its own; do not force it. You should feel the rod begin to collapse as it melts.

FIGURE 10-7 Use a drum sander attachment on a die grinder to reshape the weld contour.

FIGURE 10-8 The finished repair after color coating

8. Move the shoe very slowly and crisscross the groove as it fills with melted plastic. Do not move too fast or the welder will not have sufficient time to properly heat the base material and melt the rod.
9. Quick-cool the weld with a damp sponge or cloth. Remove the tape in preparation for the front side repair.
10. V-groove 50 percent of the way through the front of the panel and into the backside weld. This will enable the two welds to be tied together for a strong repair.
11. Use the same welding and smoothing procedures done on the ABS repair for the front side weld. A slow-speed grinder, a single-edge razor blade, or a drum sander attachment on a die grinder (Figure 10-7) can be used to reshape the contour.
12. Scuff sand the area to be refinished with #320 grit paper and blow dust free. Prime and color coat to complete the repair (Figure 10-8).

RESHAPING A DISTORTED BUMPER COVER

A concentrated heat source, such as a torch lamp or high-temperature heat gun, is used to reshape a distorted bumper cover. It is important that the heat totally penetrates the plastic. Use the following procedure:

1. Thoroughly wash the cover with soap and water, then clean with plastic cleaner to remove all road tar, oil, grease, and undercoating. Wipe or blow-dry.
2. Apply the heat directly to the distorted area. Check for adequate heat penetration by feeling the surface on the opposite side of the cover. When it becomes too uncomfortable to touch (about 140 degrees Fahrenheit), the heat has sufficiently penetrated the plastic.
3. Heat alone is rarely enough to reshape a bumper cover. A paint paddle or squeegee is ideal for helping to shape the cover (Figure 10-9).
4. When the reshaping is complete, quick-cool the cover with a damp sponge or cloth; an ice cube also works well.

SHRINKING A STRETCHED BUMPER COVER

Flexible plastics do not always tear when impacted. Sometimes they have a tendency to stretch, much like metal. Use the following procedure to shrink a stretched bumper cover:

1. Clean the cover with soap and water, followed by plastic cleaner. Wipe or blow-dry.
2. Use a torch lamp or high-temperature heat gun to "hot spot" the stretched area (Figure 10-10).
3. Quick-cool the repair area immediately with a damp sponge or cloth; an ice cube works even better (Figure 10-11). Whatever cooling method is used, the point is to cool it as quickly as possible.

REPAIRING HIGH-STRESS AREAS OF A URETHANE BUMPER

When a bumper cover is ripped away from a vehicle, the likely result will be broken mounting tabs. Any damage to the mounting area of a bumper

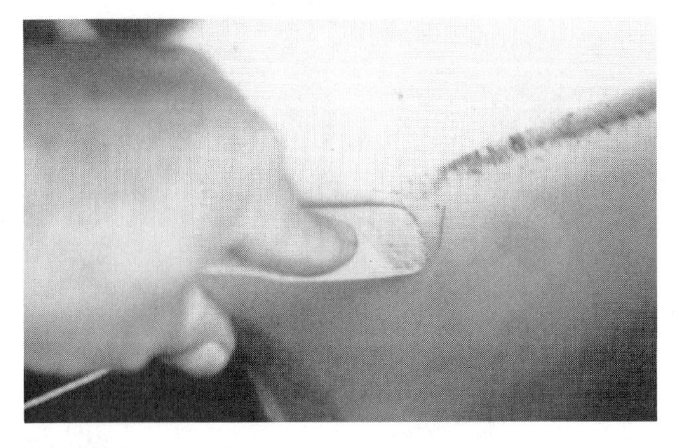

FIGURE 10-9 Shaping a bumper cover with a paint paddle

FIGURE 10-10 Using a heat gun to "hot spot" a stretched bumper cover

FIGURE 10-11 An ice cube works best for quick cooling.

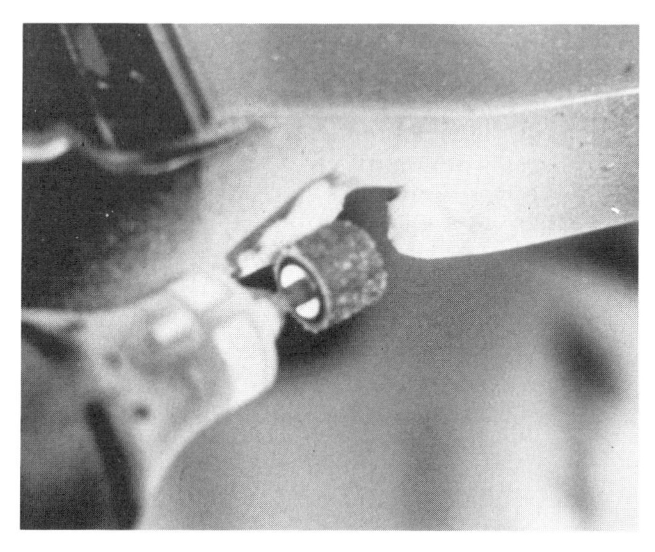

FIGURE 10-12 Beveling back the torn edges of the mounting tabs

FIGURE 10-13 Build a form in the shape of the missing tab.

must have a quality repair or it will break loose when the bumper is remounted. If an airless welder is being used to make the repair, there is no need to determine whether the urethane is a thermoset or a thermoplastic, because both are repairable with this type of welder. The urethane must be a thermoplastic in order to use a hot-air system.

Use the following step-by-step procedure to make a two-sided repair:

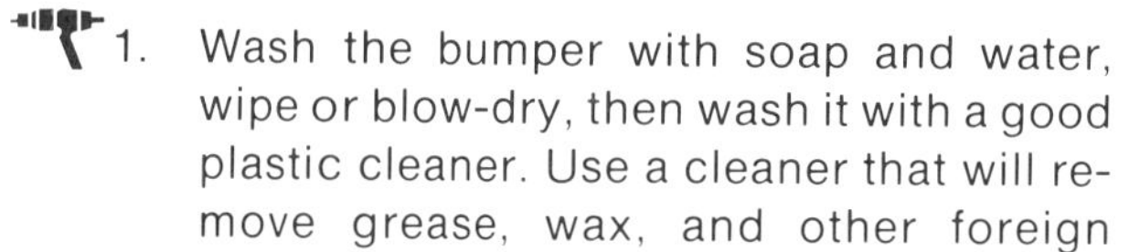

1. Wash the bumper with soap and water, wipe or blow-dry, then wash it with a good plastic cleaner. Use a cleaner that will remove grease, wax, and other foreign matter.
2. Bevel back the torn edges of the mounting tabs at least 1/4 inch using a drum sander (Figure 10-12).

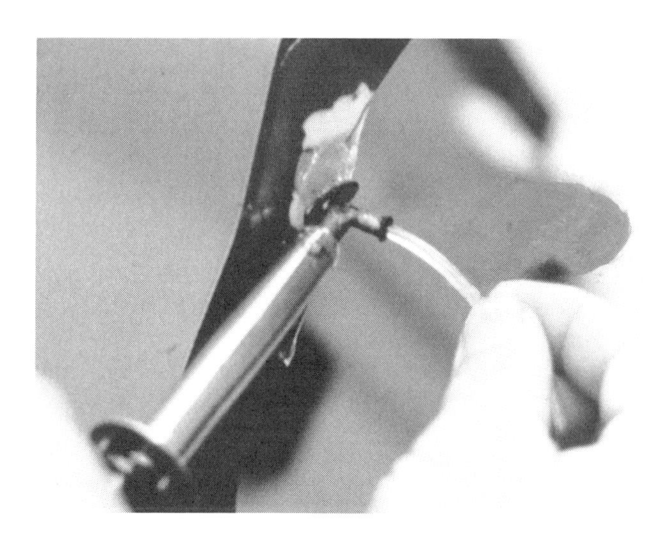

FIGURE 10-14 Overfill the form slightly with melted plastic.

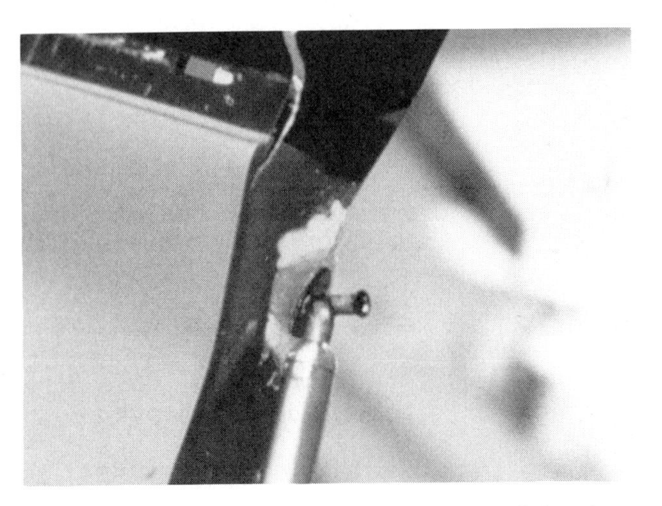

FIGURE 10-15 Using the "shoe" part of the tip to smooth the weld

FIGURE 10-16 Filling the groove with melted plastic

3. With a small grinder, rough up the plastic as far as the weld will extend. Blow the area dust free.
4. Use aluminum body tape to build a form in the shape of the missing tab (Figure 10-13). Turn up the edges of the tape to form the thickness of the weld.
5. Set the temperature dial on the welder to accommodate a clear urethane rod and allow it time to warm up.
6. Push the rod slowly through the melt tube, slightly overfilling the form with melted plastic. Feather the plastic into the beveled area. Make sure the mold cavity is full and the melted plastic is feathered out over the beveled edges of the broken mounting tabs (Figure 10-14).
7. Use the flat "shoe" part of the tip to smooth and shape the weld (Figure 10-15). Quick-cool with a damp sponge or cloth, then remove the tape.
8. Use a die grinder and small cutter bit to make a V-groove, then bevel the backside of the torn edges. This will allow the edges to be welded on both sides for optimum strength.
9. Fill in the groove with melted plastic (Figure 10-16). Use the flat "shoe" part of the tip to smooth out the plastic, then quick-cool with a damp sponge or cloth.
10. Finish the weld to the desired contour using a slow-speed grinder and #60 or #80 grit disc. The bumper can be reinstalled immediately, but it will not regain its total strength until it has cooled completely (Figure 10-17).

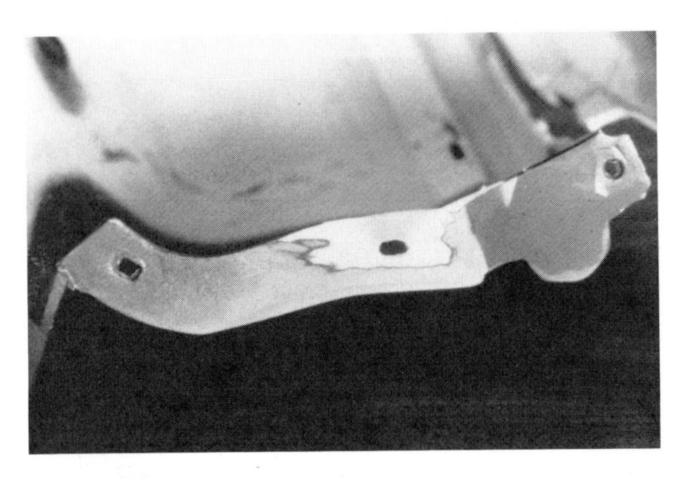

FIGURE 10-17 The finished bumper repair

REPAIRING A CUT OR TEAR IN A URETHANE BUMPER COVER

In most cases, a one-sided weld is sufficient to repair a cut or tear in a bumper cover. However, if the damage is in a high-stress area, a two-sided weld is required for greater strength, followed by application of a flexible filler. The following step-by-step procedure outlines a typical two-sided weld and filler application:

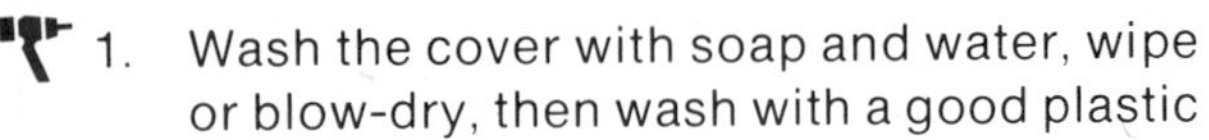

1. Wash the cover with soap and water, wipe or blow-dry, then wash with a good plastic cleaner.
2. Align the cut or tear by applying aluminum body tape to the painted side of the cover (Figure 10-18). The tape must be applied to the painted side because it provides better adhesion.

3. Use a die grinder and a small cutter bit to V-groove the cut or tear on the backside (Figure 10-19). Be sure to penetrate at least 50 percent of the way through the bumper.
4. Use a drum sander or similar tool to bevel back the edges of the V-groove at least 1/4 inch on each side.
5. Set the temperature dial on the welder to accommodate a clear urethane rod. Give it sufficient time to warm up.
6. Start the weld by slowly feeding the rod into the melt tube, applying only enough pressure to force the melted rod out into the grooved area. As the rod begins to melt, very slowly move the tip in the direction of the intended weld, and overlap the edges of the groove with melted plastic.

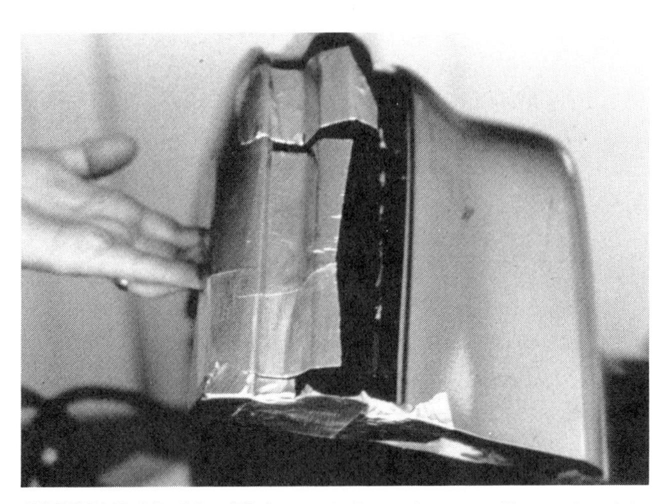

FIGURE 10-18 This cut has been aligned with aluminum body tape.

FIGURE 10-19 V-grooving with a die grinder and small cutter bit

FIGURE 10-20 Air hoses are useful for cleaning grinding dust from a vehicle's finish.

7. After completing the full length of the weld, remove the rod from the melt tube. Use the flat "shoe" part of the tip to smooth out the weld area. Proceed slowly to give the plastic sufficient time to melt under the shoe.
8. Quick-cool the weld with a damp sponge or cloth, then remove the tape.
9. Wash the front side with soap and water and plastic cleaner. Wipe or blow-dry.
10. V-groove the front side, making sure the groove is deep enough to penetrate the backside weld so the two welds can be blended together for added strength.
11. Using a drum sander or similar tool, bevel the sides of the groove. This will permit the weld to feather out over the edges of the base material.
12. Use a slow-speed grinder and #60 or #80 grit disc to remove the paint from around the weld area. Do not overgrind; 2 or 3 inches is plenty. Then blow the area dust free with an air hose (Figure 10-20).
13. Use the same welding technique that was used on the backside, making sure to get a good melt-in. Quick-cool the weld.
14. Smooth the weld using a slow-speed grinder and #60 or #80 grit disc. Because heat buildup from the grinder can cause the edges of the weld to peel up, grind for only a few seconds at a time, then allow the plastic to cool. Plastic reacts to heat from a grinder very quickly and can experience warpage.
15. Mix and apply the flexible filler as per the manufacturer's instructions. One thin application is sufficient; overfilling will reduce flexibility.
16. When the filler has cured according to the package instructions, use a slow-speed grinder and #60 or #80 grit disc to grind the filled area to a smooth contour. Follow

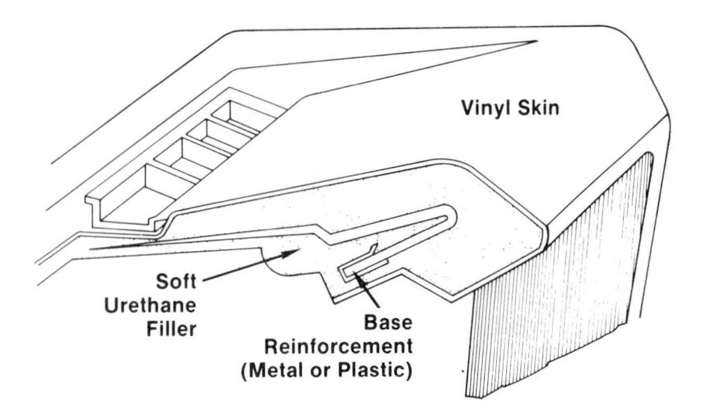

FIGURE 10-21 Cutaway of a typical crash pad

with #180 grit paper to featheredge the paint and put the final touches on the filler.

17. If any imperfections remain that must be filled, first apply a skim coat of flexible putty. Allow it to dry, then sand it smooth with #220 grit paper. Do not use body putty or primer to fill imperfections.

REPAIRING VINYL-CLAD URETHANE FOAM

The use of vinyl-clad urethane foam in vehicle interiors is commonplace these days. Most padded instrument panels, or crash pads, are sections of urethane foam covered with a thin vinyl overlay. The foam is molded to a plastic or metal reinforcement base plate and can be repaired easily using an airless welder. Although the basic welding technique is fine for repairing cuts, tears, and cracks, it is also necessary to be able to repair dents and similar deformities.

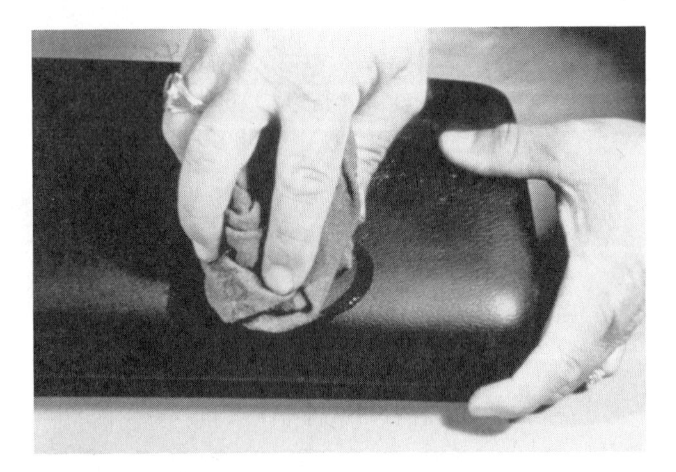

FIGURE 10-22 Applying a damp cloth to a dent

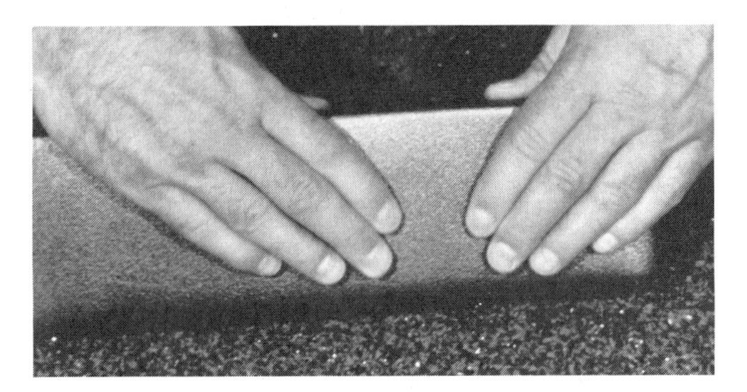

FIGURE 10-23 Massaging to remove a dent

REMOVING SURFACE DENTS

Besides being unsightly, a surface dent in effect stretches the plastic—similar to a hail dent in metal. When done carefully and precisely, the following technique is very effective in removing dents in crash pads, armrests, and other padded interior panels. Whenever possible, remove the part from the vehicle before beginning. If the part remains in the vehicle, be sure to cover the seats for protection. Figure 10-21 shows a cutaway view of a typical crash pad.

1. Apply a damp sponge or cloth to the dent; soak it for about 30 seconds (Figure 10-22). The vinyl is porous and will absorb some of the water. Do not wipe it dry; let the area remain moist.
2. Use a high-temperature heat gun or torch lamp to heat the dented area. Hold the heat source 10 to 12 inches from the surface, and keep it moving in a circular motion at all times. Gradually move the heat inward into the center of the dent, but be careful not to overheat the vinyl or it will blister.
3. Continue heating until the dented area is uncomfortable to touch (about 140 degrees Fahrenheit), then turn off the heat source and set it outside the vehicle to prevent accidental damage.

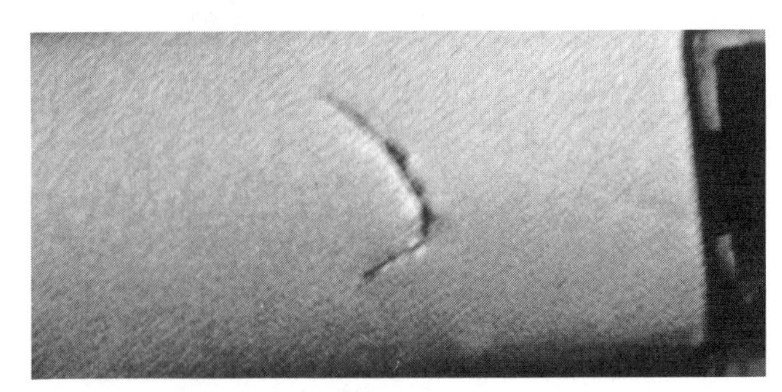

FIGURE 10-24 An unsightly crash pad crack

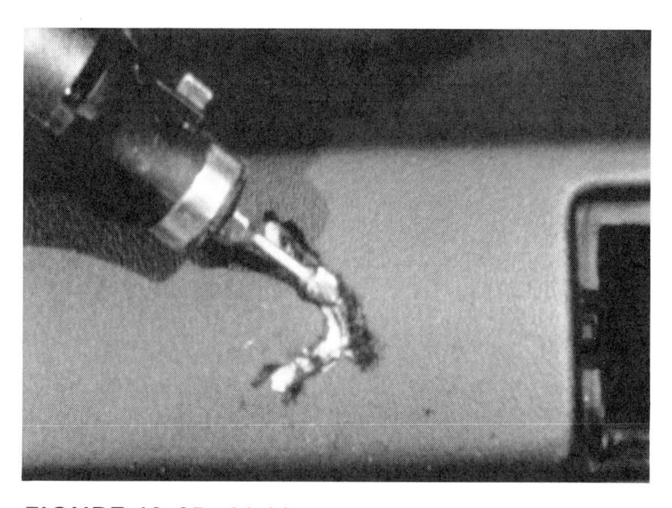

FIGURE 10-25 Making a V-groove in the foam padding

4. If necessary, massage the pad with a forcing motion toward the center of the dent (Figure 10-23). Some dents will not require massaging because the heat is enough to remove them. It might be necessary to heat and massage more than once.
5. After reshaping, quick-cool the area with a damp sponge or cloth.

REPAIRING AND REFINISHING CUTS, TEARS, AND CRACKS

A cut, tear, or crack in a crash pad detracts greatly from a car's appearance, as evidenced in Figure 10-24. Many such defects are the direct result of abuse; others are caused by weather and age. The basic repair procedure is as follows:

1. Set the temperature dial on the welder to accommodate a clear urethane rod. Let the welder warm up.
2. Use the same cleaning procedure as for the ABS and polypropylene welds. Wash the entire pad with soap and water, blow it dry, then clean it with plastic cleaner.
3. Examine the damaged area:
   - If it is brittle, warm it with a heat gun to soften the vinyl before doing the repair.
   - If the edges are curled or jagged, cut them away with a small cutter bit before making the V-groove.
4. V-groove at least 1/4 inch deep in the foam padding (Figure 10-25). This will enable the melted rod to flow into the inner pad and provide extra strength.
5. The vinyl is very thin; despite this, bevel the edges as much as possible. Then rough them up at least 1/4 inch on both sides for good weld penetration.
6. Start the weld at the bottom of the groove and rotate the welder so the flat "shoe" part of the tip is up and the melt tube is down (Figure 10-26). Slowly feed the rod through the melt tube until the groove is filled with melted plastic flush with the surface.
7. Rotate the welder so the shoe is turned toward the weld; smooth out the excess rod buildup (Figure 10-27). Feather it out over the beveled edges at least 1/4 inch on each side of the groove.
8. Use a die grinder with a drum sander attachment to remove any remaining excess rod buildup. Rough up the vinyl about 2 inches beyond the weld on each side for good filler adhesion. Be careful not to grind away the entire weld.

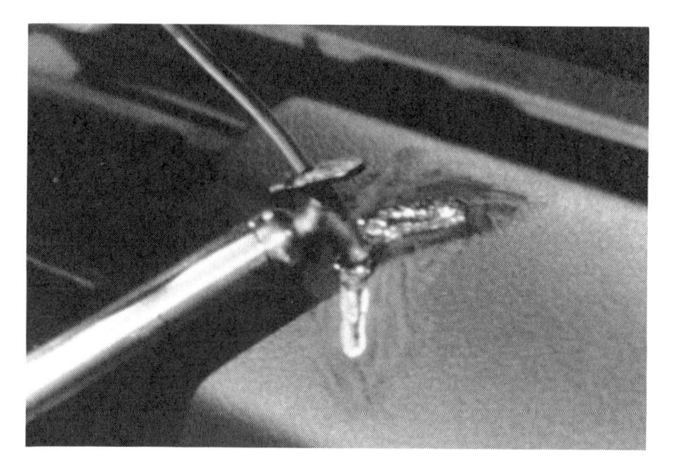

FIGURE 10-26 Start the weld at the bottom of the groove.

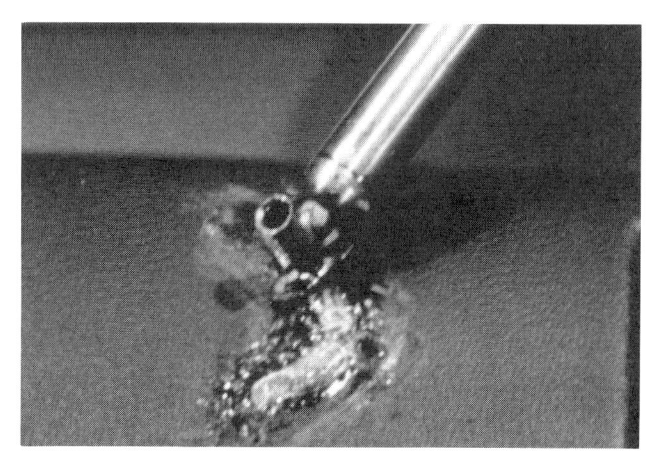

FIGURE 10-27 Smoothing the excess rod buildup

FIGURE 10-28 Blending in the color coat

9. The filler used must be flexible and designed specifically for vinyl. Follow the manufacturer's mixing and application instructions. With a plastic squeegee, apply the filler to the smooth area only. Spread it to the desired contour.
10. Allow the filler to cure, then begin contour sanding with #40 grit paper to remove the tacky glaze and rough up the contour. Follow up with #80 grit and finish with #180 grit to remove the deeper sanding marks. If any of the filler is accidentally sanded through, apply a skim coat and resand.
11. To retexture, follow the same steps used in the ABS and polypropylene repairs. Remember to apply the texture material only to the repair area at first. Then blend it out to a natural break line or retexture the entire panel (Figure 10-28).
12. Nib sand the newly textured area with #220 grit paper and blow dust free. Follow the manufacturer's recommendations when applying the color coat. Remember that weather cracks are caused by prolonged exposure to the sun, which causes the plasticizer to evaporate from the flexible PVC material. However, they can be repaired if the panel is not too badly deteriorated. To test for this, press on the panel to flex the vinyl covering; if more cracks appear, the panel is probably beyond repair.

FIBERGLASS REPAIRS

The fiberglass plastics used on car and truck bodies are made from molten glass drawn out into

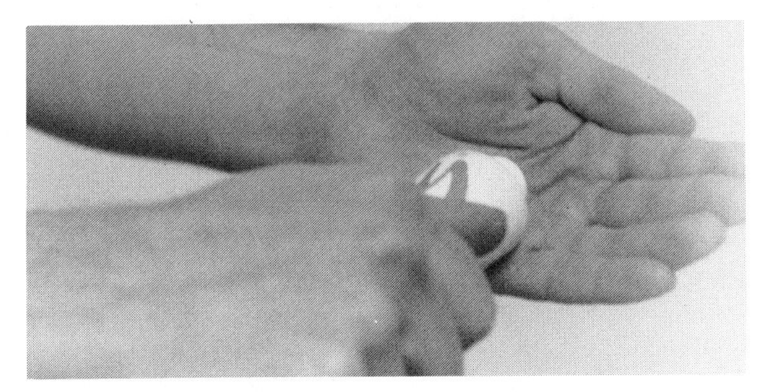

FIGURE 10-29 Use protective skin cream on any parts of the body exposed to fiberglass.

tiny, threadlike fibers. These fibers are reinforced with plastic resin and catalyst, then gel coated to a shiny finish. The combination of fiberglass and resin forms an extremely tough, durable, and corrosion-resistant material that is used in both OEM construction and repair work. However, in the case of the latter, it should be approached with caution; the techniques required are quite different from those in metalwork and even from those plastic repairs discussed previously.

WORKING WITH FIBERGLASS

Working with fiberglass requires thinking safety at all times. The resin and related ingredients can irritate the skin and stomach lining. The curing agent or hardener is generally a methyl ethyl ketone peroxide, which produces harmful vapors. Read and understand these safety points before using any fiberglass products:

- Read all label instructions and warnings carefully.

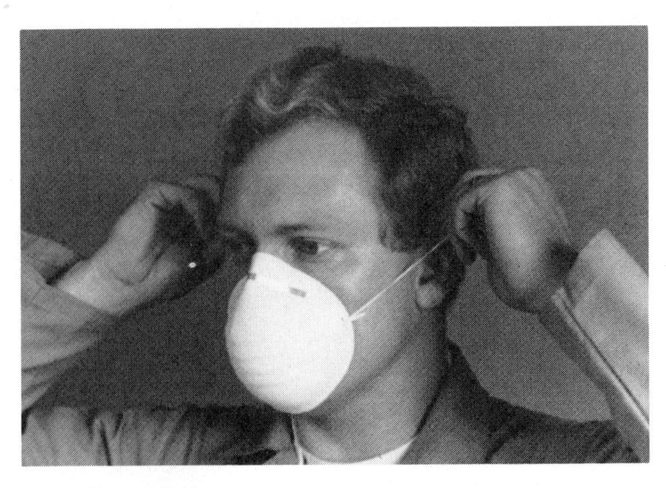

FIGURE 10-30 Respirators are a must when working with fiberglass.

- Wear rubber gloves when working with fiberglass and resin or hardener. Long sleeve shirts with buttoned collar and cuffs are helpful in preventing sanding dust from getting on the skin.
- A protective skin cream should be used on any exposed areas of the body (Figure 10-29).
- If the resin or hardener comes in contact with the skin, wash with borax soap and water or denatured alcohol.
- Always work in a well-ventilated area of the shop.
- Wear a respirator to avoid inhaling sanding dust and resin vapors (Figure 10-30).
- When making fiberglass repairs, mask the surrounding areas to avoid spilling resin on them.
- Clean all tools and equipment with lacquer thinner immediately after use. Dispose of the leftover mixed material in a safe container.

REPAIRING HOLES IN FIBERGLASS PANELS

Repairing holes in fiberglass is a simpler operation than most people think; it is actually easier than filling holes in metal body panels. If the underside of the damaged panel is easily accessible, use the following repair procedure:

1. Clean the surface surrounding the damage with a good commercial grease and wax remover. Use a #36 grinding disc to remove all paint and primer at least 3 inches beyond the repair area.

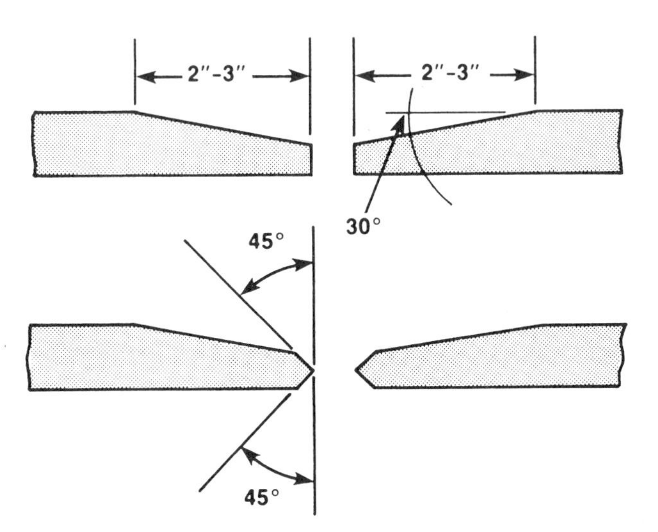

FIGURE 10-31 Beveling the inside and outside of the repair area permits better adhesion.

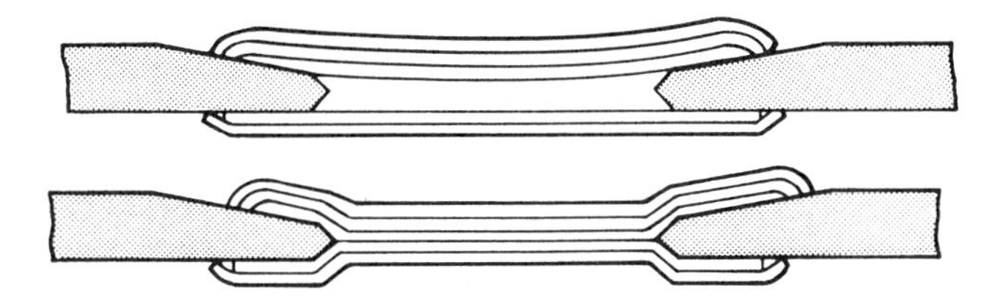

FIGURE 10-32 The layers of saturated glass cloth should contact the outside repair area.

2. Grind, file, or use a hacksaw to remove all cracked or splintered material away from the hole on both the inside and outside of the repair area.
3. Remove any dirt, sound deadener, and the like from the inner surface of the repair area. Clean with reducer, lacquer thinner, or a similar solvent.
4. Scuff around the hole with #80 grit paper to provide a good bonding surface.
5. Bevel the inside and outside edge of the repair area about 30 degrees to permit better patch adhesion (Figure 10-31).
6. Clean the repair surface thoroughly with reducer or thinner.
7. Cut several pieces of fiberglass cloth or mat large enough to cover the hole and the scuffed area. The exact number will vary depending on the thickness of the original panel, but five is usually a good number to start with.
8. Prepare a mixture of resin and hardener following the label recommendations. Using a small paintbrush, saturate at least two layers of the fiberglass cloth with the activated resin mix and apply it to the inside or back surface of the repair area. Make sure the cloth fully contacts the scuffed area surrounding the hole.
9. Saturate three more layers of cloth with the mix. Apply it to the outside surface, making certain that these layers fully contact the inner layers and the scuffed outside repair area (Figure 10-32).
10. With all of the layers of cloth in place, form a saucer-like depression in them. This is necessary to increase the depth of the repair material. Use a rubber squeegee to work out any air bubbles.
11. Clean all tools with lacquer thinner immediately after use.
12. Let the saturated cloth become tacky. An infrared heat lamp can be used to speed up the process; if one is used, keep it 12 to 15 inches away from the surface. Do not heat

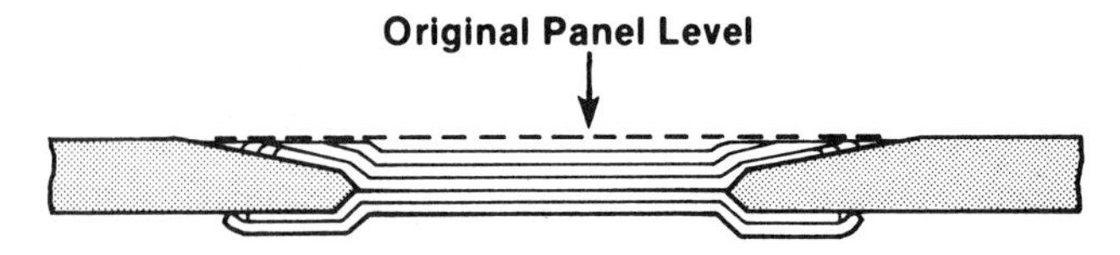

FIGURE 10-33 Sand the patch slightly below the contour of the panel.

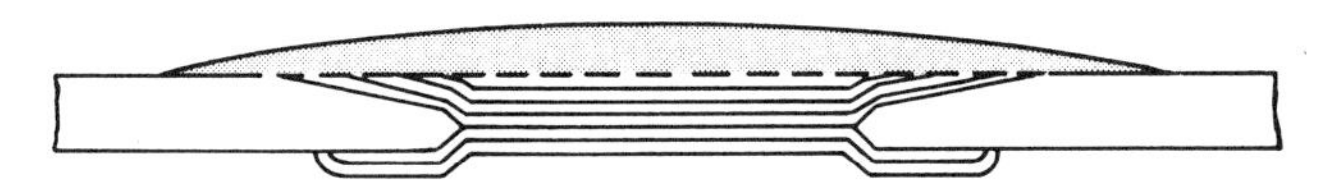

FIGURE 10-34 Hardened patching material

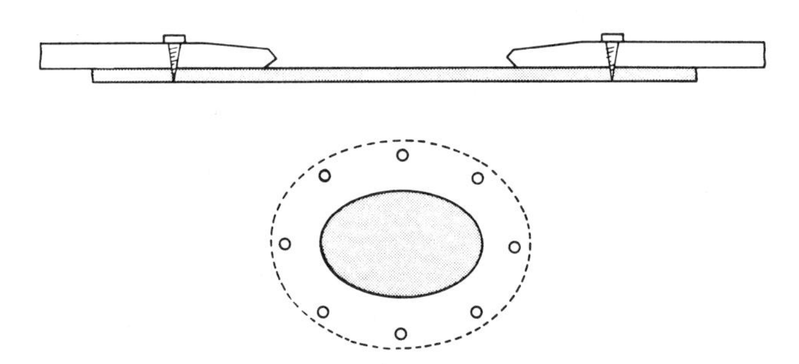

FIGURE 10-35 Sheet metal attached to the backside of a damaged panel

the repair area above 200 degrees Fahrenheit, because too much heat will distort the material.

13. With #50 grit paper, disc sand the patch slightly below the contour of the panel (Figure 10-33).
14. Prepare more fiberglass resin mix. Use a plastic spreader to fill the depression in the repair area, leaving a sufficient mound of material to grind down smooth and flush.
15. Allow the patch to harden (Figure 10-34). Again, a heat lamp can be used to speed the curing process.
16. When the patch is fully hardened, sand the excess material down to the basic contour, using #80 grit paper and a sanding block or pad. Finish sand with #120 or finer grit paper.

This same repair can be made by attaching sheet metal to the backside of the panel with sheet metal screws (Figure 10-35). Sand the metal and both the inner and outer sides of the hole to provide good adherence for the repair material. Before fastening the sheet metal, apply resin mix to both sides of the rim of the hole. Follow the above procedure for the remainder of the repair.

When the inner side of the hole is not accessible, apply a fiberglass patch to the outer side only. After the usual cleaning and sanding operations, apply several layers of fiberglass cloth to the outer side of the hole. Before it dries, make a saucer-like depression in the cloth to provide greater depth for the repair material.

REPLACING FIBERGLASS PANELS

The fiberglass bodies used on Corvettes and some large truck noses are made of individual panels. These panels are joined together in such a way that the seams and joints are invisible. Bonding strips are often used at the joints to provide a sound bonding surface, as seen in Figure 10-36. Most kit cars are also made of fiberglass panels with invisible seams. It is important that any replacement be performed to exacting standards so as to preserve the visual integrity of the vehicle body.

If struck hard enough, an impacted fiberglass panel could crack or break away, depending on the force of the blow. A judgment must be made regarding repair versus replacement; the extent of the damage determines whether or not the panel can be salvaged. The following procedure details specifically the replacement of an outer rear quarter panel on a Corvette, but it is a good example of the general fiberglass panel replacement technique. It can also be applied to replacement of the fiberglass parts used on truck cab bodies. Keep in mind that the car cannot be jacked up or resting on safety stands for the fitting of the new panels; it must be in its normal resting position until you are satisfied with the alignment and fit.

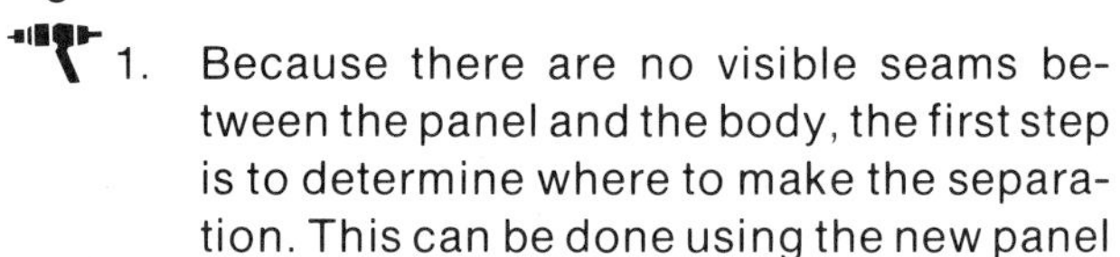

1. Because there are no visible seams between the panel and the body, the first step is to determine where to make the separation. This can be done using the new panel

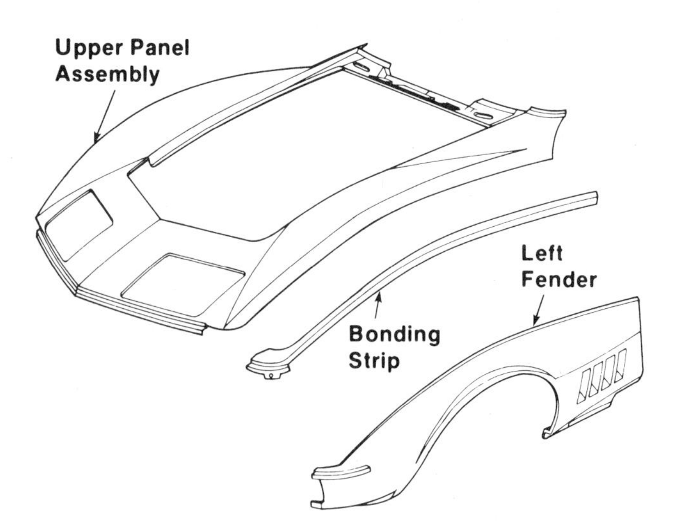

FIGURE 10-36 Bonding strips are useful at fiberglass panel joints.

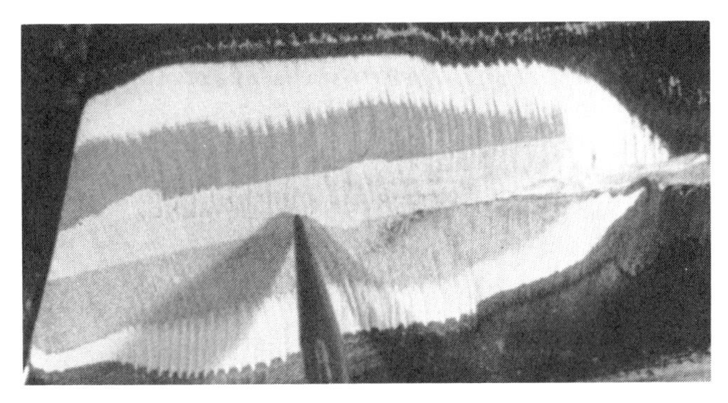

FIGURE 10-37 Exposing the seam reveals where the separation is to be made.

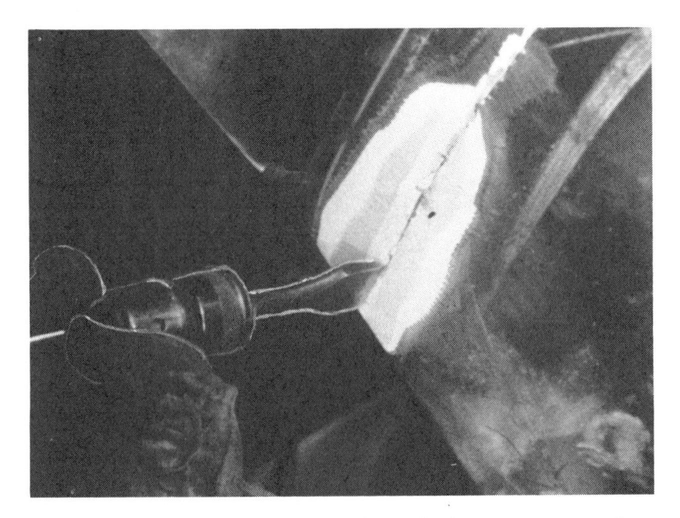

FIGURE 10-38 Using an air chisel to separate the panel from the body

as a guide or by following along the crack created when the panel was struck.

2. Use a disc sander and a coarse grit disc to remove the finish along the seam. This will expose the seam and leave no question as to where the separation is to be made (Figure 10-37).

WARNING: For personal safety, always wear safety goggles, a respirator, long sleeved shirt, and gloves when sanding fiberglass.

---

3. To separate the panel from the body, a hammer and wide-blade chisel can be used, though an air chisel fitted with a wide blade makes the job easier (Figure 10-38).
4. Use light, intermittent finger pressure on the air chisel's trigger, just enough to crack the seam. Too much force on the chisel can break the bonding strip, which provides the underneath surface for joining the two panels.

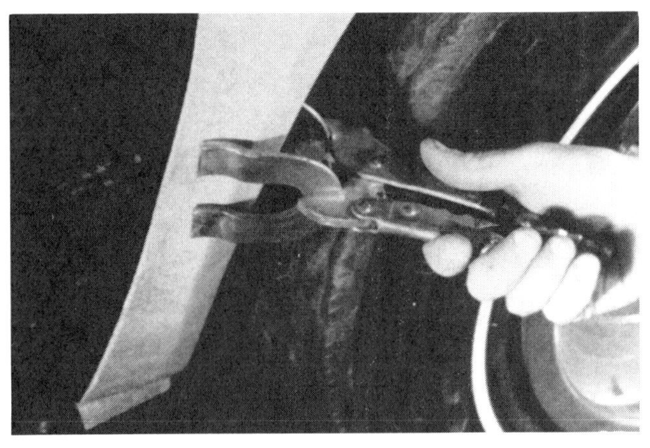

FIGURE 10-39 Clamps are used to align the panels.

5. Continue separating the panels at the seams. If possible, work from the underside of the panel and force the chisel between the seams to separate and remove it.
6. Use a disc sander with a coarse-grit disc along all seam areas; sand 3 or 4 inches beyond the seam. Remove all the finish and reduce the thickness of the panel where it meets the lower bonding strip. This will allow space for the adhesive and filler that will be applied later.
7. Continue sanding all the surfaces to which the new panel will be bonded, in order to remove all the remaining adhesive and fiberglass particles.
8. Hold the new panel in place to check that all the bonding areas have been sanded and properly prepared for the new panel.
9. While still holding the new panel in place, carefully check the fit at the door opening and the areas where it will be joined. Proper fit is necessary for good appearance, as well as to provide sufficient allowance for the adhesive filler.
10. To be absolutely sure that the panel will be in correct alignment, use clamps and sheet metal screws to temporarily hold it in position (Figure 10-39). Set a clamp to hold the panel and align the fit at the door.
11. Drill a hole near the top edge of the panel and install a sheet metal screw to hold it in place (Figure 10-40). Check the fit once again and install screws wherever a separation between the panel and the bonding strip appears. If desired, clamps or other types of quick release clips can be used in place of screws.

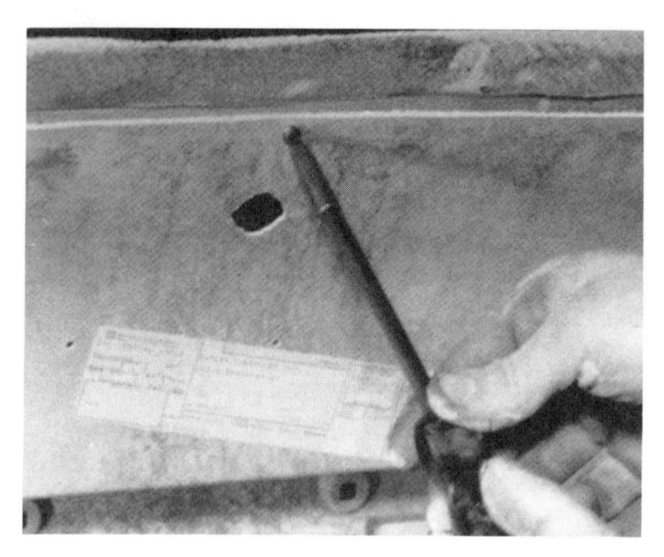

FIGURE 10-40 Installing sheet metal screws to hold the panel in place

12. Double-check all the joints to be sure there is clearance for the adhesive and filler. Also, make sure that all the places where there might be a gap between the panel and bonding strip have been clamped.
13. With the panels held in place with screws or clamps, closely check the alignment and fit of the panels. Step back and view the assembly from a distance; all panels must be straight and level, because once they are cemented in place they cannot be readjusted.
14. Cut strips of fiberglass matting approximately 3 inches wide. The lengths should be short and easy to handle. Prepare enough so their length equals the length of all the areas to which the new panels will be joined. When mixed with the resin, the fiberglass matting forms the adhesive that joins the panels together.
15. Apply several layers of masking tape to the adjacent areas to protect them from adhesive.
16. Raise the vehicle and place it on safety stands. Remove the wheel, sheet metal screws, clamps, and panel.
17. Pour a small amount of fiberglass resin into a container, add the hardener, and stir thoroughly. The amount of hardener used affects the hardening rate of the resin. Because of the time needed for fitting, adjusting, and aligning the panels, prepare the mix so that it will not harden too fast.
18. On a clean, disposable work surface, apply resin to the mat strips (Figure 10-41). Thoroughly saturate the strips. Coat the bonding areas with resin, then apply the mat to the bonding strips and brush out all air bubbles (Figure 10-42).
19. Brush resin onto the bonding areas of the panel. With the help of an assistant, position the panel onto the bonding area and reinstall the first of the sheet metal screws into its original hole. Do not tighten the screw yet.
20. Check the alignment of the panel with the door, then reinstall the screws and clamps. After rechecking the alignment, tighten the screws just enough for the resin to ooze from the seams.
21. Working from the underside of the panel, brush the excess fiberglass mat so that it contacts the panel; this will add strength to the joint. Install the rear panel and brush out the resin and matting in the same way.

FIGURE 10-41 Applying resin to the mat strips

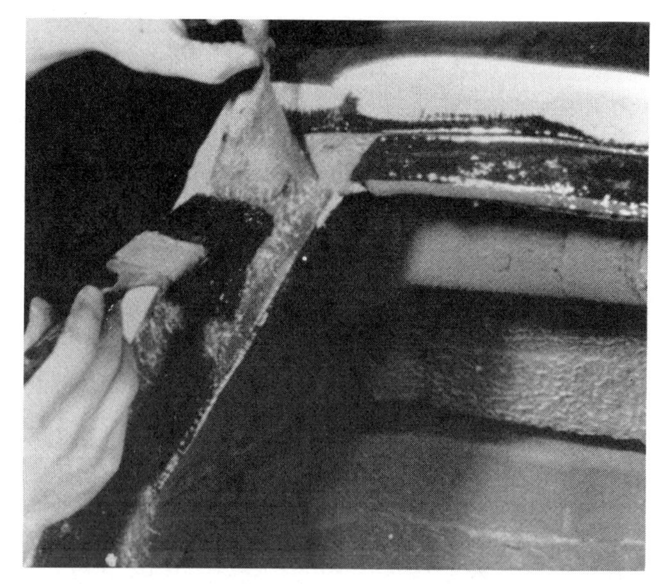

FIGURE 10-42 Brushing out the air bubbles

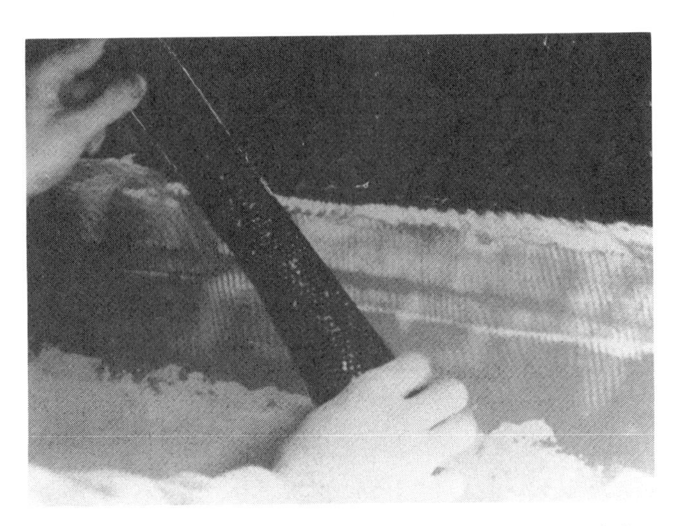

FIGURE 10-43 Using a cheese grater to level the filler

22. Fill any remaining holes or broken areas with resin and matting, and allow the resin to harden completely. Depending on the temperature and the amount of hardener used, the length of the curing process varies.

CAUTION: It is best not to rush the hardening process. If the panel is moved before the adhesive has set fully, the bond will be broken and an early failure will result.

---

23. After the adhesive has cured completely, remove the clamps and sheet metal screws. Disc sand all the seam areas.
24. The strength of the joint relies on the bond between the bonding strip and each panel. All spaces and screw holes must be filled with a very strong fiberglass-reinforced plastic filler. Mix the hardener and filler thoroughly.
25. Wipe the panel with a clean cloth. Apply the filler to all seams and joints, and let it harden slightly.
26. Use a cheese grater to level the filler so that it conforms to the panel's shape (Figure 10-43). A long-line air sander can be used to smooth the filled areas (Figure 10-44). Difficult-to-reach corners and joints can be hand sanded.
27. Wipe the surface clean and apply a level coat of common plastic filler to fill any minor imperfections or irregularities. Use the long-line air sander to smooth the filler. Finally, check the filled areas and sand out any imperfections.

With the seams filled and smoothed, the repair area must be thoroughly cleaned and prepared for refinishing. The restored vehicle now shows no signs of the severe collision damage it sustained (Figure 10-45).

REPAIRING SHATTERED FIBERGLASS PANELS

Many times, an impact will cause part of a fiberglass panel to shatter or break into several pieces. Rather than replace the whole panel, it might be possible to reassemble the pieces, much like putting a puzzle together. This procedure is not only less expensive than replacing the panel, it also saves the

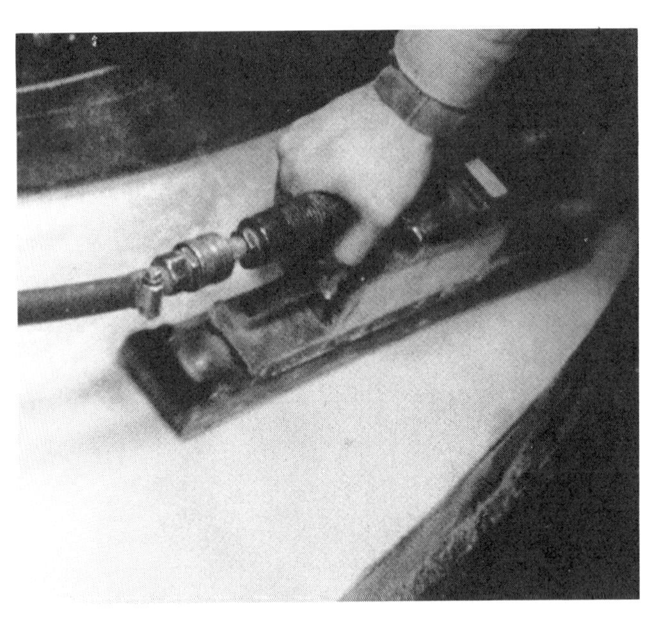

FIGURE 10-44 Using a long line air sander to smooth the filled areas

FIGURE 10-45 The restored vehicle

time usually spent waiting for new replacement parts. The following procedure can be used to repair most shattered fiberglass panels:

1. Clean the repair area with a wax and grease remover. Use a #36 grit sanding disc to remove the finish 3 inches around the damage.
2. Grind 3 inches around the backside of the repair area to remove any undercoating, dirt, and other foreign material.
3. Prepare a mixture of fiberglass resin and hardener. Follow the manufacturer's instructions for mixing and applying. Add fiberglass fibers to give the mixture greater bridging strength and mix only enough filler that can be applied before it begins to set up and harden.
4. When assembling a large number of pieces, some cure time might be necessary periodically before subsequent pieces can be added.
5. If most of the pieces are available, reassemble the panel one piece at a time. Begin by smearing a thin coat of filler on the mating edges. Use a C-clamp or vise grip to hold the piece in its original position on the panel.
6. Smear a coating of filler over the panel and the piece; the aim is to build a "bridge" over the joint. Repeat this process with each of the pieces until they are all clamped and glued together (Figure 10-46).
7. If the shattered area is large or if some of the pieces are missing, fiberglass bonding strips or a piece of sheet metal can be fastened to the backside of the panel to provide support for the repair area. Bonding strips can be obtained from other wrecked panels or from a salvage yard.
8. To apply the bonding strip, hold it in position and drill holes through both the panel and the strip. Roughen the face of the strip with a #36 sanding disc and coat the panel-to-strip mating surfaces with filler.
9. Use sheet metal screws to screw the outside of the panel into the bonding strip (Figure 10-47). After the repair is finished, these screws will be removed. If a piece of sheet metal is being used as backing, wrap it with plastic before screwing it to the panel; this will allow it to be easily removed after the repair is finished.
10. After the backing is in place, coat the panel edges and the edges of the pieces with the resin/fiber mixture. Assemble the pieces, then coat the repair area with the filler. Be sure to completely fill all voids.

FIGURE 10-46 The broken fragments have been clamped and glue together.

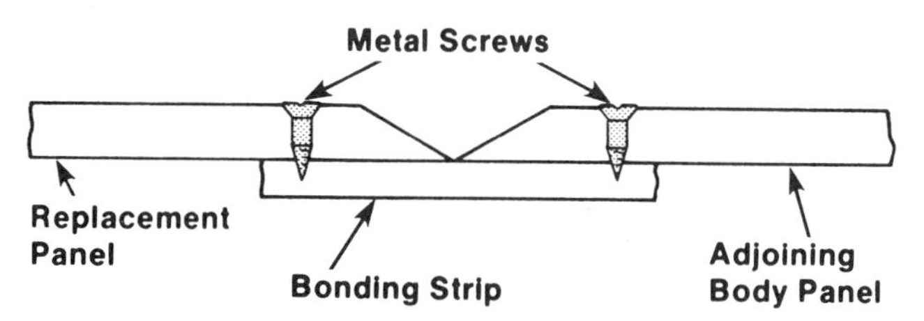

FIGURE 10-47 The outside of the panel screwed into the bonding strip

11. If fiberglass bonding strips are not used to support the pieces, the rigidity of the repair area can be increased with fiberglass cloth. Cut two or three pieces of cloth or mat approximately 3 inches larger than the repair area.
12. Saturate the cloths with fiberglass filler and place them over the repair area. Use a plastic spreader to press the cloth into any voids or low spots and to work out air bubbles.
13. After the panel has been reassembled and the filler has hardened, remove any sheet metal screws and grind the repair area with a #36 sanding disc to remove all high spots.
14. Cut three pieces of fiberglass cloth or mat 3 inches larger than the repair area. Thoroughly saturate the cloth pieces with filler and place them over the repair (Figure 10-48). Carefully smooth them flat and work out any air bubbles.
15. After the filler has cured, sand to the correct contour and prime.

USING MOLDED CORES

Naturally, holes are much more difficult to repair in a curved portion of a fiberglass panel than those on a flat surface. Basically, the only solution in such a case (short of purchasing a new panel section) is to use the mold core method of replacement. This is often the quickest and cheapest way to repair a curved fiberglass surface; the entire process takes almost an hour. While the procedure illustrated in Figure 10-49 and described here relates to a rear fender section, the principle can be applied to any type of curved fiberglass panel damage. The mold core is made as follows:

1. First off, locate an undamaged panel on another vehicle that matches the damaged one; this will be used as a model. A new or used car can be employed, since the model vehicle will not be harmed if care is taken.
2. On the model vehicle, mask off an area slightly larger than the damaged area. Apply additional masking paper and tape to the surrounding area, especially on the low side of the panel. This will prevent any resin from getting on the finish.
3. Coat the area being used as a mold with paste floor wax. Leave a wet coat of wax all over the surface. A piece of waxed paper can be substituted for the coat of wax; be sure that it is taped firmly in place (Figure 10-49A).
4. Cut several pieces of special fiberglass mold veil (thin fiberglass mat material) in sizes ranging from 2 by 4 inches to 4 by 6 inches. Standard fiberglass mats can be used if the panel does not have reverse curves.
5. Mix the fiberglass resin and hardener following the label instructions on both products.
6. Starting from one corner of the mold area, place pieces of veil on the waxed area so each edge overlaps the next one; use just one layer of veil (Figure 10-49B).
7. Apply the resin/hardener to the veil material with a paintbrush (Figure 10-49C). Force the mixture into the curved surfaces and around corners with the tips of the bristles.
8. Use the smaller pieces of veil along the edges and on difficult curves. Additional resin can be applied if needed, brushing in one direction only to force the material into the indentations. In all cases, use only one layer of veil.

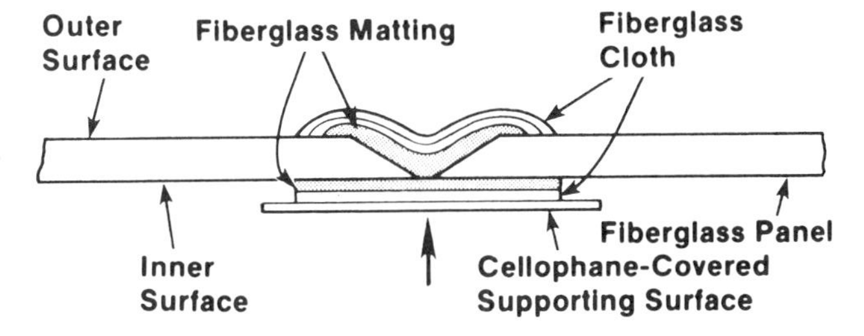

FIGURE 10-48 The repair area covered with fiberglass cloth

CAUTION: Be sure that the resin does not get on any part of the model vehicle that is not coated with wax.

---

9. After the veil pieces have been applied to the entire waxed area, allow the mold core to cure a minimum of 1 hour.
10. Once the mold core has hardened, gently work the piece loose from the model vehicle (Figure 10-49D). The core should be an exact reproduction of this section of the panel.
11. Remove the floor wax protecting the model vehicle's paint finish using a wax and grease remover. Then polish this section of the panel.
12. Since the mold core is generally a little larger than the original panel, place it under the damaged panel and align. If necessary, trim down the edges, the core, and the damaged panel slightly where needed for better alignment. The edges of the damaged panel and core must also be cleaned.
13. Using fiberglass adhesive, cement the mold core in place (Figure 10-49E). Allow the core and panel to cure.
14. Grind back the original damaged edges to a taper or bevel, maintaining the desired contour.
15. Lay the fiberglass mat, soaked in resin/hardener, on the taper or bevel and over the entire inner core. Once the mat has hardened, level it with a coat of fiberglass filler. Then prepare it for painting (Figure 10-49F).

In some instances, it might not be possible to place the core on the inside of the damaged panel. In

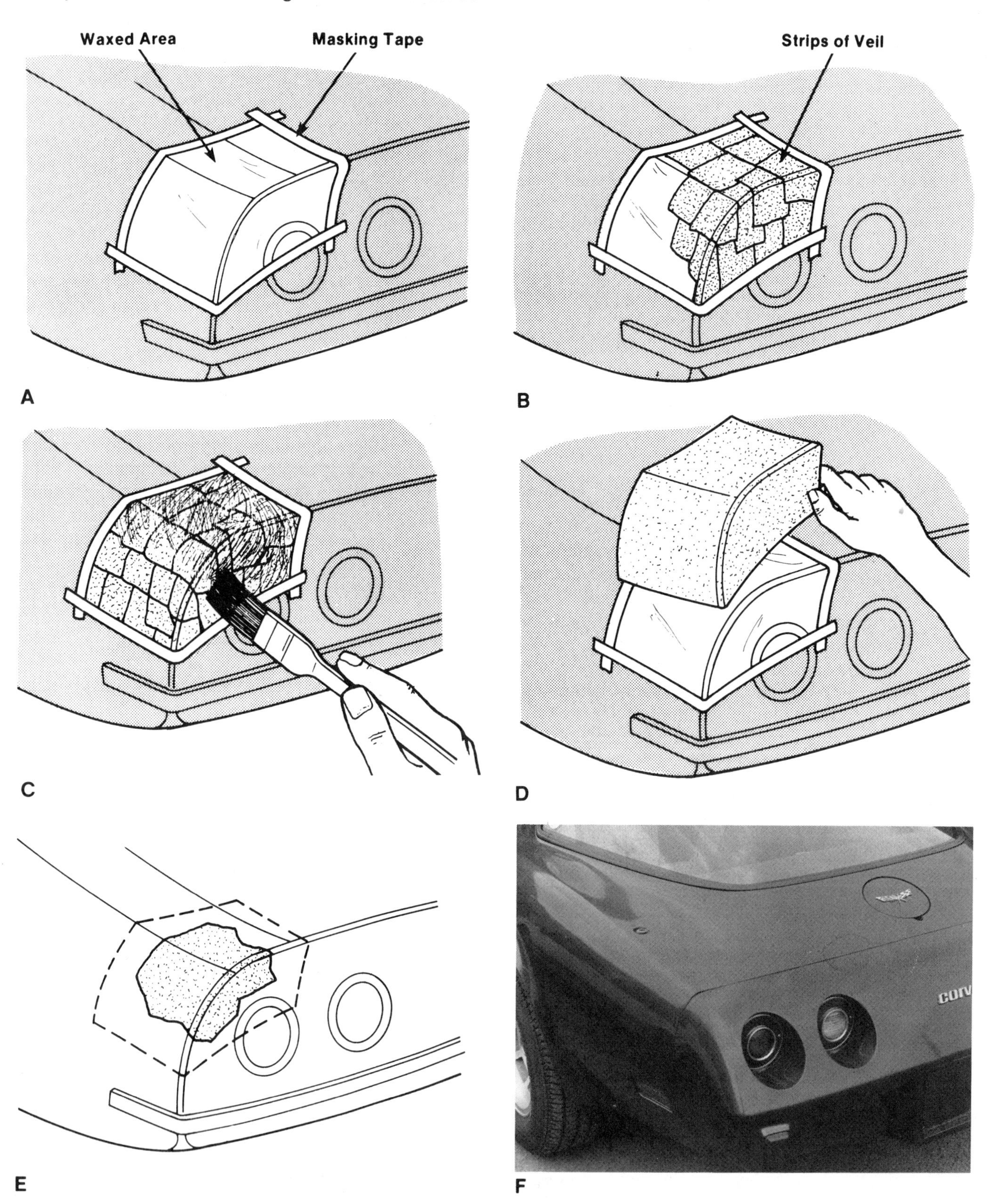

FIGURE 10-49 Steps in making a fiberglass core: (A) coat the area being used as a mold model with paste floor wax on a piece of waxed paper; (B) place pieces of fiberglass veil over the waxed or waxed paper surface; (C) apply resin/hardener to the veil material; (D) remove the mold core from the model; (E) cement the core piece in place; (F) the completed job. *(Courtesy of Unican Corp.)*

this case, the damaged portion must be cut out to the exact size of the core. After the panel has been trimmed and its edges beveled, tabs must be installed to support the core from the inside. These tabs can be made from pieces of the salvaged fiberglass panel or from fiberglass cloth strips saturated in resin/hardener.

After cleaning and sanding the inside sections, attach the tabs to the inside edge of the panel and bond with fiberglass adhesive. Vise grips can be used to hold the tabs in place while bonding. Taper the edge of the opening and place the core on the tabs. Fasten the core to the tabs with fiberglass adhesive. Grind down any high spots so that layers of fiberglass mat can be added.

Place the saturated mats over the core, extending about 1-1/2 to 2 inches beyond the hole in all directions. Work each layer with a spatula or squeegee to remove all air pockets. Additional resin can be added with a paintbrush to secure the layers. Allow the resin/hardener time to cure sufficiently, then sand the surface level. For a smooth surface, use fiberglass filler to finish the job.

REVIEW QUESTIONS

1. What is not a requirement when welding hard plastics?
   a. cosmetic filler
   b. V-grooving
   c. plastic cleaner
   d. shaping the excess rod buildup to a smooth contour

2. In preparation for a one-sided weld, Welder A V-grooves 50 percent of the way through the base material. Welder B V-grooves 25 percent of the way through the base material. Who is correct?
   a. Welder A
   b. Welder B
   c. Both A and B
   d. Neither A nor B

3. Which of the following welding techniques is used solely on ABS plastic?
   a. stitch tamp
   b. trial and error
   c. melt flow
   d. all of the above

4. Which of the following is used to V-groove automotive plastic?
   a. rotary file
   b. 1/4-inch drill
   c. grinder
   d. all of the above

5. After shrinking a stretched bumper cover, what works best for quick-cooling?
   a. damp sponge
   b. fan
   c. ice cube
   d. damp cloth

6. When repairing a torn urethane bumper cover, when is a two-sided weld required?
   a. never
   b. at all times
   c. when the damage is in a high-stress area
   d. when a flexible filler is not part of the repair process

7. When removing a surface dent from vinyl-clad urethane foam, Welder A holds the heat source steady, approximately 2 to 4 inches from the surface. Welder B holds the heat source 10 to 12 inches from the surface and keeps it moving in a circular motion. Who is correct?
   a. Welder A
   b. Welder B
   c. Both A and B
   d. Neither A nor B

8. To repair a cut, tear, or crack in vinyl-clad urethane foam, start the weld ____________.
   a. at the bottom of the groove
   b. in the middle of the groove
   c. at the top of the groove
   d. just outside the groove

9. Heat alone is usually sufficient to reshape ____________.
   a. crash pads
   b. bumper covers
   c. both a and b
   d. neither a nor b

10. Which of the following statements concerning repairing holes in fiberglass panels is incorrect, assuming the underside of the panel is easily accessible?
    a. Bevel the inside and outside edge of the repair area about 30 degrees to permit better patch adhesion.
    b. Prepare a mixture of resin and hardener following the label recommendations.

c. After the fiberglass patch has become tacky, use #50 grit paper to disc sand it slightly below the contour of the panel.
d. When the fiberglass patch has fully hardened, no further sanding is required.

11. When sanding fiberglass, always wear ______________.
a. gloves
b. safety goggles
c. respirator
d. all of the above

12. After making an ABS weld, Welder A uses a single-edged razor blade to shape the excess rod buildup. Welder B uses a slow-speed grinder and #80 grit disc. Who is correct?
a. Welder A
b. Welder B
c. Both A and B
d. Neither A nor B

13. When repairing a high-stress area of a urethane bumper cover, what is used to build a form in the shape of the missing tab?
a. masking tape
b. aluminum body tape
c. flexible filler
d. bonding strip

14. When replacing a shattered fiberglass panel, what can be added to the resin and hardener mix to provide greater bridging strength?
a. fiberglass fibers
b. fiberglass filler
c. common plastic filler
d. none of the above

15. In the fiberglass mold core replacement procedure, what can be substituted for the coat of paste floor wax?
a. waxed paper
b. masking tape
c. standard fiberglass mats
d. fiberglass adhesive

APPENDIX A

GLOSSARY

Abrasive Material such as sand, crushed steel grit, aluminum oxide, silicon carbide, or crushed slag used for cleaning or surface roughening.
ABS (Acrylonitrile/Butadiene/Styrene) A common thermoplastic.
Acceptable Weld A weld that meets all the requirements prescribed by the welding specifications.
Acetone Flammable, volatile liquid used in acetylene cylinders to dissolve and stabilize acetylene under high pressure.
Acetylene Highly combustible gas composed of carbon and hydrogen. Used as a fuel gas in the oxyacetylene welding process.
Adhesion Promoter A spray material that, when applied to a polyolefin, enables it to be repaired using the adhesive bonding method.
Adhesive Bonding Materials joining process in which an adhesive, placed between the mating surfaces, solidifies to produce an adhesive bond.
Airless Plastic Welding A relatively new method of auto body repair in which the temperature setting of the welder is adjusted to suit the specific plastic.
Alloy Mixture of two or more metals.
Alternating Current (AC) Electricity that reverses its direction of electron flow.
Ampere A measure of electric current used to refer to the input of electrical energy.
Annealing Comprehensive term used to describe the heating and cooling cycle of steel in the solid state, usually implying relatively slow cooling. In annealing, the temperature of the operation, the rate of heating and cooling, and the time the metal is held at heat depend upon the composition, shape, and size of the steel being treated and the purpose of the treatment.
Arc The flow of electricity through an air gap. The arc flowing through the air produces high temperatures.
Arc Brazing A brazing process in which the heat required is obtained from an electric arc.
Arc Burn A temporary but painful eye condition experienced when the eyes are exposed to ultraviolet light for a short period of time; this is also called arc flash.
Arc Cutting A group of cutting processes wherein the severing or removing of metals is brought about by melting with the heat of an arc between an electrode and the base metal.
Arc Gouging An application of arc cutting wherein a bevel or groove is formed.
Arc Voltage The voltage across the welding arc.
Arc Welding A group of welding processes in which coalescence is produced by heating with an arc or arcs, with or without the application of pressure, and with or without the use of filler metal.
Argon A chemically inert gas that will not combine with the products of the weld zone. It is an excellent shielding gas for the gas metal arc process.
Automatic Welding Welding with equipment that performs the entire welding operation without constant observation and adjustment of the controls by an operator. The equipment may or may not perform the loading and unloading of the work.

Back Gouging The forming of a bevel or groove on the other side of a partially welded joint to assure complete penetration upon subsequent welding from that side.
Backing Material (metal, weld metal, asbestos, carbon, flux, and so on) backing up the joint during welding.
Back Weld A weld deposited at the back of a single groove weld.
Balling Up The formation of globules of molten brazing filler metal or flux due to lack of wetting of the base material.
Base Metal The metal to be welded, soldered, or cut.
Bead The weld; used to describe the neat ripples formed by semiliquid metal.
Bead Weld A type of weld composed of one or more string or weave beads deposited on an unbroken surface.
Bevel An angular type of edge preparation.
Bevel Angle The angle formed between the prepared edge of a member and a plane perpendicular to the surface of the member.
Blind Joint A joint, no portion of which is visible.

Bond The junction of the welding metal and the base metal.

Bonding Strip A piece of fiberglass. Several strips are used as backside support when repairing a large shattered fiberglass panel or when some of the broken pieces are missing.

Braze Welding A method of welding using a filler metal. Unlike brazing, the filler metal is not distributed in the joint by capillary attraction.

Brazing A group of welding processes that produces coalescence of mateials by heating them to a suitable temperature and by using a filler metal. The filler metal is distributed between the closely fitted surfaces of the joint by capillary attraction.

Brazing Filler Metal The metal that fills the capillary gap and liquefies above 840 degrees Fahrenheit.

Bridging A welding defect caused by poor penetration. A void at the root of the weld is spanned by weld metal.

Buckling Distortion caused by the heat of a welding process.

Buildup Sequence The order in which the weld beads of a multipass weld are deposited with respect to the cross section of a joint.

Burn Test A means of identifying plastics in which a small piece of the plastic is ignited and its burn characteristics are evaluated.

Butt Joint A joint between two members aligned approximately in the same plane.

Capillary Attraction The condition in which adhesion between the molten filler metal and the base metals, together with surface tension of the molten filler metal, causes distribution of the filler metal between the properly fitted surfaces of the joint to be brazed.

Carburizing Flame An oxyacetylene flame in which there is an excess of acetylene. Also known as excess acetylene or reducing flame.

Coalescence The growing together or growth into one body of the materials being welded.

Coating A relatively thin layer of material applied by surfacing for the purpose of corrosion prevention, resistance to high-temperature scaling, wear resistance, or lubrication.

Cold Soldered Joint A joint with incomplete coalescence caused by insufficient application of heat to the base metal during soldering.

Complete Joint Penetration Joint penetration that extends completely through the joint.

Cone The conical part of an oxyfuel gas flame next to the orifice of the tip.

Continuous Weld A weld that extends continuously from one end of a joint to the other.

Corrosive Flux A flux with a residue that chemically attacks the base metal.

Crack A fracture-type discontinuity characterized by a sharp tip and high ratio of length and width to opening displacement.

Crater In arc welding, a depression at the termination of a weld bead or in the weld pool beneath the electrode.

Crown The surface of the finished bead.

Crush Zones Buckling points designed into certain structural components for absorbing the energy impact in a collision.

Cutting Tip That part of an oxygen cutting torch from which the gases are emitted.

Cutting Torch A device used in oxygen cutting for controlling and directing the gases used for preheating and the oxygen used for cutting the metal.

Cylinder A portable container used for transportation and storage of a compressed gas.

Cylinder Cart A portable cart used for moving cylinders.

Defective Weld A weld containing one or more defects.

Deposited Metal Filler metal that has been added during a welding operation.

Direct Current (DC) Electricity that flows in only one direction.

Direct Current Reverse Polarity The arrangement of direct current arc welding leads wherein the work is the negative pole and the electrode is the positive pole of the welding arc.

Direct Current Straight Polarity Electrical current flowing from the electrode to a base metal.

Discontinuity An interruption of the typical structure of a weld. A discontinuity is not necessarily a defect.

Duty Cycle The number of minutes out of ten that a welding machine can run without overheating.

Edge Joint A joint between the edges of two or more parallel or nearly parallel members.

Edge Preparation The contour prepared on the edge of a member for welding.

Electrode A component of the arc welding circuit through which current is conducted between the electrode holder and the arc.

Electrode Holder A device used for mechanically holding the electrode and conducting current to it.

Electrode Spitting The condition that occurs when electrode particles are ejected across the arc.

Face Shield A protective device to be worn on the head for shielding the face and neck.

Fiberglass Plastic A very durable and corrosion-resistant material being used increasingly on car and truck bodies.
Filler Metal The metal to be added in making a welded, brazed, or soldered joint.
Filler Rod Metal wire that is melted into the puddle of the weld.
Fillet Weld. A weld joining two surfaces approximately at right angles to each other in a lap joint, tee joint, or corner joint.
Filter Lens A filter, usually colored glass, used in goggles, helmets, and handshields to exclude harmful light rays.
Flexible Filler A filler that must be applied over top of the weld when repairing a high-stress area of a urethane part.
Flux Material used to prevent, dissolve, or remove oxides and other undesirable substances.
Fuel Gases Gases usually used in addition to oxygen for heating, including acetylene, natural gas, hydrogen, propane, and other synthetic fuels and hydrocarbons.
Fusion The melting together of filler metal and base metal, or of base metal only, which results in coalescence.

Gas Metal Arc Cutting An arc cutting process used to sever metals by melting them with the heat of an arc between a continuous metal (consumable) electrode and the work. Shielding is obtained entirely from an externally supplied gas or gas mixture.
Gas Metal Arc Welding An arc welding process that produces coalescence of metals by heating them with an arc between a continuous filler metal (consumable) electrode and the work. Shielding is obtained entirely from an externally supplied gas or gas mixture.
Globular Metal Transfer A form of metal transfer in MIG welding that occurs in large, irregularly shaped drops.
Goggles A device with colored lenses that protects the eyes from harmful radiation during welding and cutting operations.
Groove The opening provided between two members to be joined by a groove weld.
Ground Connection An electrical connection of the welding machine frame to the earth for safety.
Gun In semiautomatic, machine, and automatic welding, a manipulating device to transfer current and guide the electrode into the arc. It may include provisions for shielding and arc initiation.

Hardening Any process of increasing the hardness of metal by suitable treatment, usually involving heating and cooling.
Heat-Affected Zone The portion of the base metal that has not been melted, but whose mechanical properties or structure have been altered by the heat of welding, brazing, soldering, or cutting.
Helmet A protective device, used in arc welding, for shielding the face and neck. A helmet is equipped with a suitable filter lens and is designed to be worn on the head.
Hold Time The time that pressure is maintained at the electrodes after the welding current has stopped.
Hot Air Plastic Welding A method of welding automotive plastic in which a ceramic or stainless steel electric heating element produces hot air. The air, in turn, blows through a nozzle and onto the plastic.
HSLA Steel High-strength, low-alloy steel used in the structural components of many domestic vehicles.
HSS Steel High tensile strength steel whose strength is derived from heat treatment. This steel will tear or fracture if the collision stresses exceed the tensile strength.

Impact Test A test in which one or more blows are suddenly applied to a specimen. The results are usually expressed in terms of energy absorbed or number of blows of a given intensity required to break the specimen.
Inadequate Joint Penetration Joint penetration that is less than what is specified.
Inclusion Nonmetallic material(s) that becomes trapped in the weld metal, between weld beads, or between the weld and the base metal.
Incomplete Fusion Fusion that is less than complete.
Induction Brazing A process in which bonding is produced by the heat obtained from the resistance of the work to the flow of induced electric current and by using a nonferrous filler metal having a melting point above 800 degrees Fahrenheit, but below that of the base metals. The filler metal is distributed in the joint by capillary attraction.
Induction Welding A process in which fusion is produced by heat obtained from resistance of the work to the flow of induced electric current, with or without the application of pressure.
Inert Gas A gas that does not normally combine chemically with the base metal or filler metal.
Infrared Rays Dangerous rays produced by the light of arc welding that are injurious to the eyes and skin.
Intermittent Weld A weld whose continuity is broken by recurring unwelded spaces.

Joint The location where two or more members are to be joined.

Joint Penetration The minimum depth a groove or flange weld extends from its face into a joint, exclusive of reinforcement.

Lamellar Tear Terrace-like separations in the base metal of a weld, usually caused by shrinkage stresses.
Leg of a Fillet Weld The distance from the root of the joint to the toe of the fillet weld
Local Preheating Preheating a specific portion of a structure.

Manual Welding Welding in which the entire operation is performed and controlled by hand.
MIG (Metallic Inert Gas) Gas metallic arc welding.
Mixing Chamber That part of a gas welding or oxygen cutting torch in which the gases are mixed.
Molded Core A core made from fiberglass mold veil, resin, and hardener that is used to repair a curved portion of a fiberglass panel.
Molten Weld Pool The liquid state of a weld prior to solidification as weld metal.

Neutral Flame An oxyfuel gas flame in which the portion used is neither oxidizing nor reducing.
Nondestructive Testing A method of checking for weld surface defects; can be visual, penetrant, or ultrasonic.
Nonferrous Metals that contain no iron. Aluminum, brass, bronze, copper, lead, nickel, and titanium are nonferrous.
Nozzle Spray Used to clean the nozzle of a MIG welding machine; preferred over paste because the spray does not leave a heavy residue inside the nozzle.

One-Sided Weld In plastic welding, a repair method in which only one side of the damaged part is welded. This technique is used most often for solely cosmetic repairs.
Overhead Position The position in which the welding is performed from the underside of the joint.
Overlap Protrusion of weld metal beyond the toe or root of the weld.
Oxidizing Flame An oxyfuel gas flame having an oxidizing effect (excess oxygen).
Oxyacetylene Cutting An oxygen cutting process in which the necessary cutting temperature is maintained by flames obtained from the combustion of acetylene with oxygen.
Oxyacetylene Welding An oxyfuel gas welding process that produces coalescence of metals by heating them with a gas flame or flames obtained from the combustion of acetylene with oxygen. The process may be used with or without the application of pressure and with or without the use of filler metal.

Panel Spotting The means by which both lap and flange joints can be made with a spliced or full panel installation.
Pass A single longitudinal progression of a welding operation along a joint or weld deposit. The result of a pass is a weld bead.
Penetration The depth of fusion into the metal being welded.
Phosgene A potentially dangerous compound formed when fumes from chlorinated solvents decompose in the welding arc.
Pitch Center to center spacing of welds.
Plasma A gas that has been heated to an at least partially ionized condition, enabling it to conduct an electric current.
Plasma Arc Cutting An arc cutting process in which the metal is severed by melting a localized area with a constricted arc and removing the molten material with a high velocity jet of hot, ionized gas issuing from the orifice.
Plug Weld A circult weld made through a hole in one member of a lap or tee joint joining that member to the other.
Polyethylene A common thermoplastic whose maximum weld strength is achieved ten hours after the weld is completed.
Polyolefin A type of automotive plastic used most often for large exterior parts.
Polypropylene A thermoplastic that is very similar to polyethylene.
Polyurethane An extremely lightweight and formable plastic available as both a thermoplastic and thermosetting plastic.
Polyvinyl Chloride (PVC) A common thermoplastic.
Porosity Gas pockets or voids in metal.
Positions of Welding All welding is accomplished in one of four positions: flat, horizontal, overhead, and vertical. The limiting angles of the various positions depend somewhat on whether the weld is a fillet or groove.
Preheat Temperature The specified temperature that the base metal must attain in the welding, brazing, soldering, or cutting area immediately before these operations are performed.
Pressure Welding A welding process in which the pieces of metal are heated to a softened state by electrodes. After the pieces are soft, pressure is used to complete the weld.

Preweld Interval In spot welding, the time between the end of squeeze time and the start of weld time during which the material is preheated.
Puddle The molten part of the weld where the arc is supplied.

Regulators A device for controlling the delivery of gas at some substantially constant pressure regardless of variation in the higher pressure at the source.
Residual Stress Stress remaining in a structure or member as a result of thermal or mechanical treatment or both.
Resistance Welding A group of welding processes that produces coalescence of metals with the heat obtained from resistance of the work to electric current in a circuit of which the work is a part, and by the application of pressure.
Reverse Polarity The arrangement of direct current arc welding leads with the work as the negative pole and the electrode as the positive pole of the welding arc.
Root Opening The separation between the members to be joined at the root of the joint.
Root Penetration The depth a weld extends into the root of a joint measured on the centerline of the root cross section.

Seam Weld A continuous weld made between or upon overlapping members in which coalescence may start and occur on the mating surfaces or may have proceeded from the surface of one member. The continuous weld may consist of a single weld bead or a series of overlapping spot welds.
Sectioning A means of replacing parts at factory seams.
Sequence The order in which the beads (passes) are welded on the joint.
Shielded Welding An arc welding process in which protection from the atmosphere is obtained through use of a flux, decomposition of the electrode covering, or an inert gas.
Short-Circuiting Metal Transfer A form of metal transfer in MIG welding that does not occur across the arc. It occurs instead when the electrode wire contacts the base metals.
Solder A filler metal that liquefies prior to 840 degrees Fahrenheit.
Spatter In arc and gas welding, the metal particles expelled during welding that do not form a part of the weld.
Speed Welding In plastic welding, a popular method of doing panel work in which a fairly constant rate of speed must be maintained.
Spool Gun A self-contained system consisting of a small drive and a wire supply that allows a welder to move freely around a job.
Spot Weld A weld made between or upon overlapping members in which coalescence may start and occur on the mating surfaces or may proceed from the surface of one member. The weld cross section is approximately circular.
Spray Arc Gas metal arc welding that uses an arc voltage high enough to transfer the electrode metal across the arc in small globules.
Spray Transfer A mode of metal transfer in gas metal arc welding in which the consumable electrode is propelled across the arc in small droplets.
Stitch Welding The use of intermittent welds to join two or more parts.
Straight Polarity The arrangement of direct current arc welding leads in which the work is the positive pole and the electrode is the negative pole of the welding arc.
Stress Cracking Cracking of a weld or base metal containing residual stresses.
Stud Welding A general term for the joining of a metal stud or similar part to a workpiece. Welding can be accomplished by arc, resistance, friction, or other suitable process with or without external gas shielding.
Surface Preparation The operations necessary to produce a desired or specified surface condition.

Tack Weld A weld made to hold parts in proper alignment until the final welds are made.
Thermoplastics Plastics that are capable of being repeatedly softened and hardened by heating or cooling, with no change in their appearance or chemical makeup. They are weldable with a plastic welder.
Thermosetting Plastics Plastics that undergo a chemical change by the action of heating, a catalyst, or ultraviolet light leading to an infusible state. Thermosets are not weldable, although they can be "glued" using an airless welder.
TIG (Tungsten Inert Gas) Gas tungsten arc welding
Tip-to-Base Metal Distance An important factor in obtaining good welding results, usually 1/4 to 5/8 inch.
Toe of Weld The junction between the face of a weld and the base metal.
Torch Brazing A brazing process in which bonding is produced by heating with a gas flame and by using a nonferrous filler metal having a melting point above 800 degrees Fahrenheit, but below that of the

base metal. The filler metal is distributed in the joint by capillary attraction.

Two-Sided Weld In plastic welding, a repair method in which both sides of the damaged part are welded. This is the preferred method whenever the part can be removed from the vehicle.

Ultrasonic Plastic Welding A method of repairing auto plastics in which the welding time is controlled by the power supply. This method is best suited to rigid plastics.

Ultrasonic Stud Welding A variation of the shear joint used to join plastic parts in which the weld is made along the circumference of the stud.

Ultraviolet Rays Harmful energy waves given off by the arc that are dangerous to the eyes and skin.

Undercut A groove melted into the base metal adjacent to the toe or root of a weld and left unfilled by weld metal.

Undercutting An undesirable crater at the edge of the weld caused by poor weaving techniques or excessive welding speed.

Vinyl-Clad Urethane Foam The material used in most padded instrument panels. This foam is repairable using an airless welder.

Voltage Regulator An automatic electrical control device for maintaining a constant voltage supply to the primary of a welding transformer.

Weld A localized coalescence of metal in which coalescence is produced either by heating to suitable temperatures, with or without the application of pressure, or by the application of pressure alone, and with or without the use of filler metal. The filler metal either has a melting point approximately the same as the base metals or has a melting point below that of the base metals but above 800 degrees Fahrenheit.

Weldability The capacity of a metal to be welded under the fabrication conditions imposed into a specific, suitably designed structure and to perform satisfactorily in the intended service.

Weld Bead A weld deposit resulting from a pass.

Welding Current The current in the welding circuit during the making of a weld.

Welding Machine Equipment used to perform the welding operation. For example, spot welding machine, arc welding machine, seam welding machine, and so on.

Welding Rod A form of filler metal used for welding or brazing that does not conduct the electrical current.

Welding Sequence The order of making the welds in a weldment.

Welding Tip A welding torch tip designed for welding.

Welding Torch A device used in oxyfuel gas welding or torch brazing for mixing and controlling the flow of gases.

Weldment An assembly whose component parts are joined by welding.

Weld Metal The portion of a weld that has been melted during welding.

Wetting The bonding or spreading of a liquid filler metal or flux on a solid base metal.

Wire Feed Speed The rate of speed at which a filler metal is consumed in arc welding.

Work Angle The angle that the electrode makes with the surface of the base metal in a plane perpendicular to the axis of the weld.

Work Lead The electric conductor between the source of arc welding current and the work.

APPENDIX B

REFERENCE TABLES

COMMON TYPES OF WELD DISCONTINUITIES

Discontinuity	Location	Remarks
Porosity (can be uniformly scattered, cluster, or linear type)	Weld	Weld only as described in text.
Nonmetallic slag	Weld	
Metallic tungsten		
Incomplete fusion (also called lack of fusion)	Weld	At joint boundaries or between passes
Inadequate joint penetration (also called lack-of-joint preparation)	Weld	Root of weld penetration
Undercut	Base metal	Junction of weld and base metal at surface
Underfill	Weld	Outer surface of joint preparation
Overlap	Weld	Junction of weld and base metal at surface
Laminations	Base metal	Base metal, generally near mid-thickness of section
Delamination	Base metal	Base metal, generally near mid-thickness of section
Seams and laps	Base metal	Base metal surface almost always longitudinal
Lamellar tears	Base metal	Base metal, near weld, heat-affected zone
Cracks (includes hot cracks and cold cracks described in text)		
Longitudinal	Weld, heat-affected zone	Weld or base metal adjacent to weld fusion boundary
Transverse	Weld, base metal, heat-affected zone	Weld (can spread into heat-affected zone and base metal)
Crater	Weld	Weld, at point where arc is terminated
Throat	Weld	Weld axis
Toe	Heat-affected zone	Junction between face of weld and base metal
Root	Weld	Weld metal, at root
Underbead and heat-affected zone	Heat-affected zone	Base metal in heat-affected zone
Fissures	Weld	Weld metal

ADVANTAGES OF SHIELDING GASES			
Metal	**Welding Type**	**Shielding Gas**	**Advantages**
Aluminum and Magnesium	Manual Welding	Argon	Better arc starting, cleaning action, and weld quality; lower gas consumption
		Argon-helium	High welding speeds possible
	Machine Welding	Argon-helium	Better weld quality; lower gas flow than required with straight helium
		Helium (DCSP)	Deeper penetration and higher weld speeds than can be obtained with argon-helium
Carbon steel	Spot Welding	Argon	Generally preferred for longer electrode life; better weld nugget contour; ease of starting; lower gas flows than helium
	Manual Welding	Argon	Better pool control, especially for position welding
	Machine Welding	Helium	Higher speeds obtained than with argon
Stainless steel	Manual Welding	Argon	Permits controlled penetration on thin-gauge material (up to 14 gauge)
	Machine Welding	Argon	Excellent control of penetration on light-gauge materials
		Argon-helium	Higher heat input; higher welding speeds possible on heavier gauges
		Argon-hydrogen (up to 35% H_2)	Prevents undercutting; produces desirable weld contour at low current levels; requires lower gas flow
		Argon-hydrogen helium	Excellent for high-speed tube mill operation
		Helium	Provides highest heat input and deepest penetration
Copper, nickel, and Cu-Ni alloys		Argon	Ease of obtaining pool control, penetration, and bead contour on thin-gauge metal
		Argon-helium	Higher heat input to offset high heat conductivity of heavier gauges
		Helium	Highest heat input for welding speed on heavy metal sections
Silicon-bronze		Argon	Reduces cracking of this "hot short" metal
Aluminum-bronze		Argon	Less penetration of base metal

APPENDIX C

TROUBLESHOOTING GUIDES

TROUBLESHOOTING GUIDE FOR A TYPICAL ARC WELDING MACHINE

Problem	Cause	Remedy
Welder will not start	Line switch not turned on	Place line switch in *on* position.
	Supply line fuse blown	Replace (check reason for blown fuse first).
	Power circuit dead	Check input voltage.
	Overload relay tripped	Let cool; remove cause of overloading.
	Loose or broken power, electrode, or ground lead	Replace lead or tighten and repair connection.
	Wrong voltage	Check input voltage against instructions.
	Polarity switch not centered (AC-DC units only)	Center switch handle on +, –, or AC.
	Open circuit to starter button	Repair.
Welder starts but blows fuse after welding begins	Short circuit in motor or other connection	Check connections and lead insulation.
	Fuse too small	Check instruction manual for correct size.
Welder welds but soon stops	Proper ventilation hindered	Make sure all case openings are free for proper circulation of air.
	Overloading—welding in excess of rating	Operate welder at rated load and duty cycle.
	Fan inoperative	Check leads and connections.
	Motor generator sets—	
	Wrong direction of rotation	Check connection diagram.
	Brushes worn or missing	Check brushes for pressure on commutator.
	Wrong driving speed	Check nameplate for correct motor speed.
	Excessive dust accumulation in welder	Clean thermostat, coils, and other components.
Variable or sluggish welding arc	Current too low	Check recommended currents for rod type and size being used.
	Low line voltage	Check with power company.
	Welding leads too small	Check instruction manual for recommended cable sizes.
	Poor ground, electrode, or control-circuit connections	Check all connections. Clean, repair, or replace as required.
	Motor generator sets—Brushes improper; weak springs; not properly fitted	Check and repair.
	Rough or dirty commutator	Turn down or clean commutator.
Welding arc is loud and spatters excessively.	Current setting is too high	Check setting and output with ammeter.
	Polarity wrong	Check polarity; try reversing or an electrode of opposite polarity.

TROUBLESHOOTING GUIDE FOR A TYPICAL ARC WELDING MACHINE (CONTINUED)

Problem	Cause	Remedy
Polarity switch will not turn.	Contacts rough and pitted from improper turning under load.	Replace switch.
Welder will not shut off.	Line switch has failed mechanically	Replace switch.
Arcing at ground clamp	Loose connection or weak spring	Tighten connection or replace clamp.
Electrode holder becomes hot.	Loose connection, loose jaw, inadequate duty cycle	Tighten connection or replace holder.
Touching welder gives shock	Frame not grounded	See instruction manual for proper grounding procedure.

GUIDE TO EVALUATING POROSITY AND CRACKING IN TIG WELDS

Cause	Contributing Factors	Corrective Measures
Hydrogen	Dirt containing oils or other hydrocarbons; moisture in atmosphere or on metal, or a hydrated oxide film on metal; spatter; moisture in gas or gas lines. Base metal may be source of entrapped hydrogen (the thicker the metal, the greater the possibility for hydrogen).	Degrease and mechanically or chemically remove oxide from weld area. Avoid humidity; use dry metal or wipe dry. Reduce moisture content of gas. Check gas and water lines for leaks. Increase gas flow to compensate for increased hydrogen in thicker sections. To minimize spatter, adjust welding conditions.
Impurities	Cleaning or other compounds, especially those containing calcium	Use recommended cleaning compounds; keep work free of contaminants.
Incomplete root penetration	Incomplete penetration in heavy sections increases porosity in the weld.	Preheat; use higher welding current, or redesign joint geometry.
Temperature	Running too cool tends to increase porosity due to premature solidification of molten metal.	Maintain proper current, arc-length, and torch-travel speed relationship.
Welding speed	Too great a welding speed may increase porosity.	Decrease welding speed and establish and maintain proper arc-length and current relationship.
Solidification time	Quick freezing of weld pool entraps any gases present, causing porosity.	Establish correct welding current and speed. If work is appreciably below room temperature, use supplemental heating.
Chemical composition of weld metal	Pure aluminum weld metal is more susceptible to porosity than an aluminum alloy.	If porosity is excessive, use an alloy filler material.
Cracking	Temperature, welding time, and solidification may be contributing causes of cracking. Other causes include discontinuous welds, welds that intersect, repair welds, cold-working either before or after welding, and weld-metal composition. In general, crack-sensitive alloys include those containing 0.4 to 0.6 Si, 1.5 to 3.0% Mg, or 1.0% Cu.	Lower current and faster speeds often prevent cracking. A change to a filter alloy that brings weld metal composition out of cracking range is recommended where possible.

GUIDE TO GENERAL WELD QUALITY CONTROL TECHNIQUES

Technique	Equipment	Defects Detected	Advantages	Disadvantages	Other Considerations
Visual inspection	Pocket magnifier, welding viewer, flashlight, weld gauge, scale, etc.	Weld preparation, fit-up cleanliness, roughness, spatter, undercuts, overlaps, weld contour, and size; welding procedures	Easy to use; fast, inexpensive, usable at all stages of production.	For surface conditions only; dependent on subjective opinion of inspector.	Most universally used inspection method
Dry penetrant or fluorescent inspection	Fluorescent or visible penetrating liquids and developers; ultraviolet light for the fluorescent type.	Defects open to the surface only; good for leak detection.	Detects very small, tight surface imperfections; easy to apply and to interpret; inexpensive; use on magnetic or nonmagnetic materials.	Time consuming in the various steps of the processes; normally no permanent record.	Often used on root pass of highly critical pipe welds. If material is improperly cleaned, some indications can be misleading.
Magnetic particle inspection	Iron particles, wet, dry, or fluorescent; special power source; ultraviolet light for the fluorescent type.	Surface and near-surface discontinuities, cracks, etc.; porosity, slag, etc.	Indicates discontinuities not visible to the naked eye; useful in checking edges prior to welding and also for repairs; no size restriction.	Used on magnetic materials only; surface roughness may distort magnetic field; normally no permanent record.	Testing should be from two perpendicular directions to catch discontinuities that might be parallel to one set of magnetic lines of force.
Radiographic inspection	X-ray or gamma ray; source; film processing equipment; film viewing equipment; penetrameters.	Most internal discontinuities and flaws, limited by direction of discontinuity.	Provides permanent record; indicates both surface and internal flaws; applicable on all materials.	Usually not suitable for fillet weld inspection, film exposure and processing are critical; slow and expensive.	Most popular technique for subsurface inspection; required by some codes and specifications.

APPENDIX D

DECIMAL AND METRIC EQUIVALENTS

DECIMAL AND METRIC EQUIVALENTS					
Fractions	**Decimal (in.)**	**Metric (mm)**	**Fractions**	**Decimal (in.)**	**Metric (mm)**
1/64	.015625	.397	33/64	.515625	13.097
1/32	.03125	.794	17/32	.53125	13.494
3/64	.046875	1.191	35/64	.546875	13.891
1/16	.0625	1.588	9/16	.5625	14.288
5/64	.078125	1.984	36/64	.578125	14.684
3/32	.09375	2.381	19/32	.59375	15.081
7/64	.109375	2.778	39/64	.609375	15.478
1/8	.125	3.175	5/8	.625	15.875
9/64	.140625	3.572	41/64	.640625	16.272
5/32	.15625	3.969	21/32	.65625	16.669
11/64	.171875	4.366	43/64	.671875	17.066
3/16	.1875	4.763	11/16	.6875	17.463
13/64	.203125	5.159	45/64	.703125	17.859
7/32	.21875	5.556	23/32	.71875	18.256
15/64	.234275	5.953	47/64	.734375	18.653
1/4	.250	6.35	3/4	.750	19.05
17/64	.265625	6.747	49/64	.765625	19.447
9/32	.28125	7.144	25/32	.78125	19.844
19/64	.296875	7.54	51/64	.796875	20.241
5/16	.3125	7.938	13/16	.8125	20.638
21/64	.328125	8.334	53/64	.828125	21.034
11/32	.34375	8.731	27/32	.84375	21.431
23/64	.359375	9.128	55/64	.859375	21.828
3/8	.375	9.525	7/8	.875	22.225
25/64	.390625	9.922	57/64	.890625	22.622
13/32	.40625	10.319	29/32	.90625	23.019
27/64	.421875	10.716	59/64	.921875	23.416
7/16	.4375	11.113	15/16	.9375	23.813
29/64	.453125	11.509	61/64	.953125	24.209
15/32	.46875	11.906	31/32	.96875	24.606
31/64	.484375	12.303	63/64	.984375	25.003
1/2	.500	12.7	1	1.00	25.4

INDEX